AF824597

THE R·G·S STORY

RIO GRANDE SOUTHERN

1482

Dell A. McCoy
Russ Collman
William A. Graves

ANDY PAYNE COLLECTION – COURTESY OF BILL PLUNKETT

RIO GRANDE SOUTHERN Class 60 2-8-0 No. 21 was in Rico, Colorado, during 1892, when this beautiful, sharp photograph was taken.

THE R·G·S STORY

RIO GRANDE SOUTHERN

VOLUME V

RICO and the MINES

DELL A. McCOY • RUSS COLLMAN
AND
WILLIAM A. GRAVES

SUNDANCE Books

THE R • G • S STORY

RIO GRANDE SOUTHERN

— VOLUME V —

Rico and the Mines

DELL A. MCCOY • RUSS COLLMAN
AND
WILLIAM A. GRAVES

SUNDANCE PUBLICATIONS *Limited*
250 BROADWAY • DENVER, COLORADO 80203

Mile By Mile...
Station By Station...

THE R • G • S STORY allows you—the reader—to cover the entire route of the legendary Rio Grande Southern Railroad, beginning at Ridgway, Colorado, the lonely northern terminal of this little narrow-gauge line. Over time and space, the volumes of this detailed chronicle will lead you over every mile and past every station of the RGS—over the high mountains of the San Juan Region and across the spindly timber trestles to Durango, Colorado, the bustling southern terminal of this enchanting railroad.

Published by...
Sundance Publications, Limited
Denver, Colorado

Production Manager — Dell A. McCoy

Photographic Director — Tom Martin

Editorial Consultants —
Ed J. Haley, Robert W. Richardson, John W. Maxwell, Robert A. LeMassena, Stan Schwedler, Al Chione, Robert F. Shank Sr., Gabe Bradford, Herb Kelsey, Jim L. Key, William A. Graves and Mark S. Kramer

Copyright © 1996 by Sundance Publications, Ltd., Denver, Colorado 80203. Printed in U.S.A. All rights reserved. This book, or parts thereof, may not be reproduced in any form without written permission of the publisher.

ISBN 0-913582-61-1

JOHN W. MAXWELL PHOTO

FRONT COVER —
Galloping Goose No. 4 at Rico...

John Maxwell was riding aboard Galloping Goose No. 4 when the Rocky Mountain Railroad Club had chartered the railbus for an excursion over the RGS. This was on July 13, 1946, when the excursion had reached Rico, after clattering down the grade from Lizard Head Pass. The railfans were en route to Dolores on this warm summer day in Colorado's beautiful San Juan Region.

BACK COVER —
From Expectation Mountain...

Autumn colors were late in arriving on the hillsides surrounding the mining town of Rico during October of 1995. This photographic angle was taken from Beamis Flats on Expectation Mountain, showing residential and business buildings near Mantz Avenue (running top to bottom) and Glasgow Avenue (running horizontally).

DELL A. McCOY PHOTO

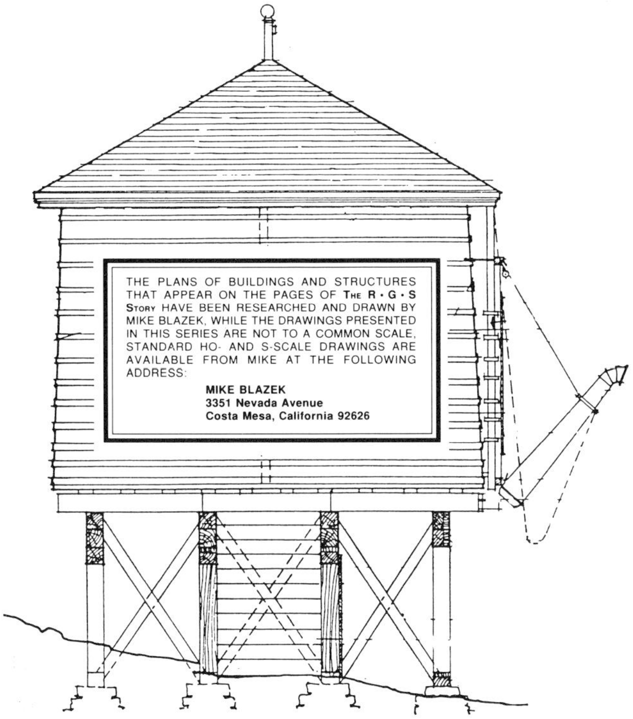

THE PLANS OF BUILDINGS AND STRUCTURES THAT APPEAR ON THE PAGES OF THE R·G·S STORY HAVE BEEN RESEARCHED AND DRAWN BY MIKE BLAZEK. WHILE THE DRAWINGS PRESENTED IN THIS SERIES ARE NOT TO A COMMON SCALE, STANDARD HO- AND S-SCALE DRAWINGS ARE AVAILABLE FROM MIKE AT THE FOLLOWING ADDRESS:

MIKE BLAZEK
3351 Nevada Avenue
Costa Mesa, California 92626

BRIDGE-66-B

Scale 40 feet = 1 inch Horizontal.
Scale 20 feet = 1 inch Vertical.

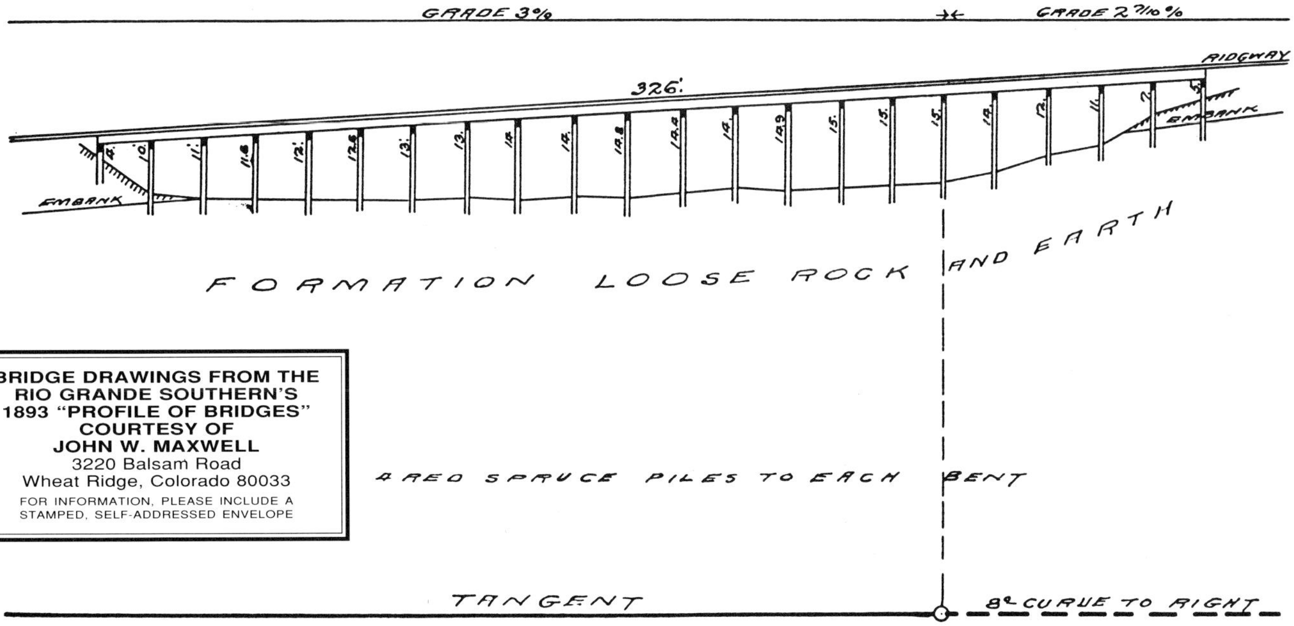

BRIDGE DRAWINGS FROM THE RIO GRANDE SOUTHERN'S 1893 "PROFILE OF BRIDGES" COURTESY OF JOHN W. MAXWELL
3220 Balsam Road
Wheat Ridge, Colorado 80033
FOR INFORMATION, PLEASE INCLUDE A STAMPED, SELF-ADDRESSED ENVELOPE

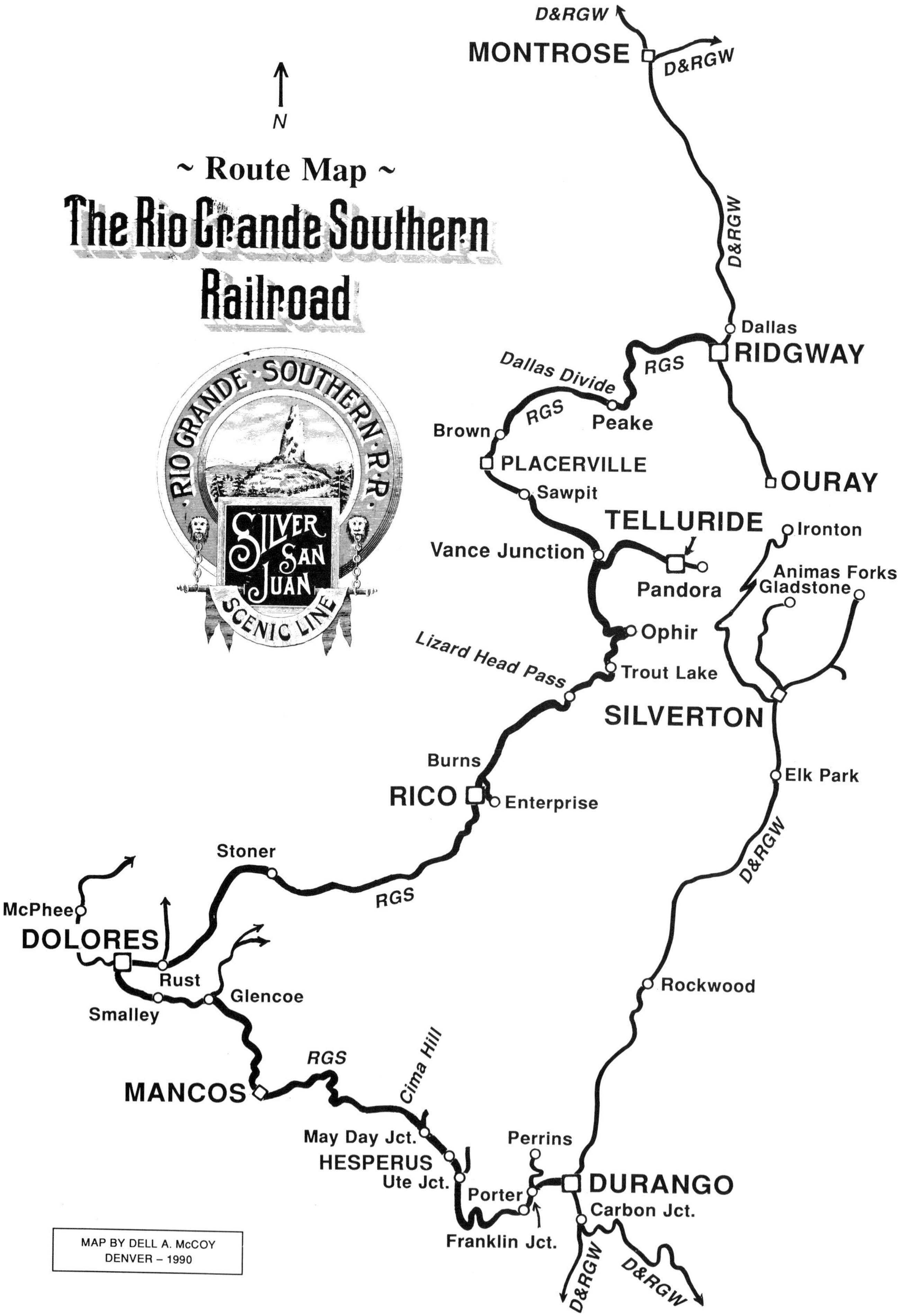
N
~ Route Map ~
The Rio Grande Southern Railroad
RIO GRANDE SOUTHERN R.R.
SILVER SAN JUAN
SCENIC LINE
D&RGW
MONTROSE
D&RGW
D&RGW
Dallas
RIDGWAY
RGS
Dallas Divide
RGS
Peake
Brown
PLACERVILLE
OURAY
Sawpit
TELLURIDE
Ironton
Vance Junction
Pandora
Animas Forks
Gladstone
Ophir
Lizard Head Pass
Trout Lake
SILVERTON
Burns
Elk Park
RICO
Enterprise
D&RGW
Stoner
RGS
McPhee
DOLORES
Rust
Glencoe
Smalley
Rockwood
RGS
Cima Hill
MANCOS
May Day Jct.
Perrins
HESPERUS
Ute Jct.
Porter
DURANGO
Carbon Jct.
Franklin Jct.
D&RGW
D&RGW
MAP BY DELL A. McCOY
DENVER – 1990

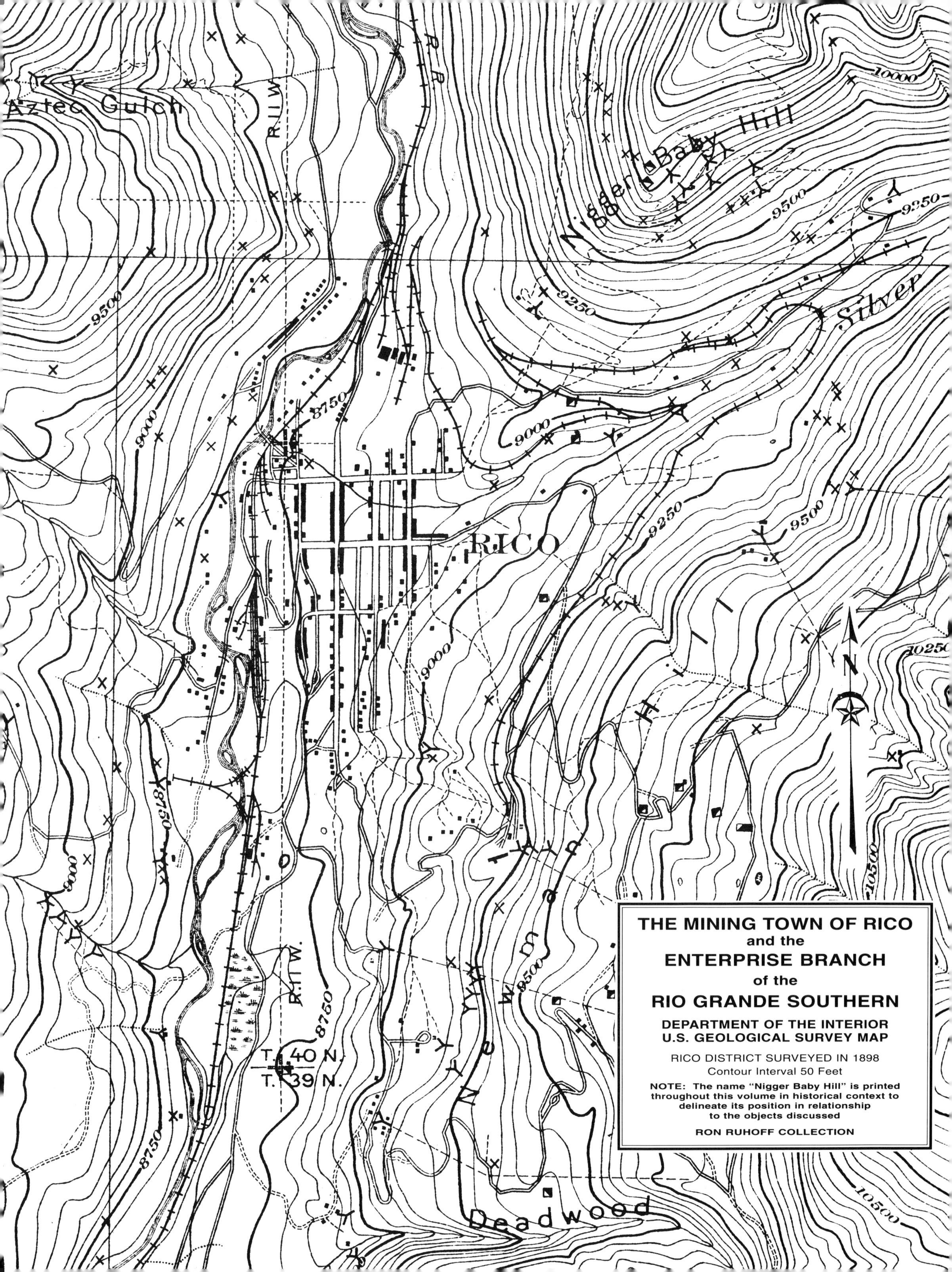

THE MINING TOWN OF RICO
and the
ENTERPRISE BRANCH
of the
RIO GRANDE SOUTHERN

DEPARTMENT OF THE INTERIOR
U.S. GEOLOGICAL SURVEY MAP

RICO DISTRICT SURVEYED IN 1898
Contour Interval 50 Feet

NOTE: The name "Nigger Baby Hill" is printed throughout this volume in historical context to delineate its position in relationship to the objects discussed

RON RUHOFF COLLECTION

~ ACKNOWLEDGEMENTS ~

THIS DETAILED ACCOUNT about the Rio Grande Southern Railroad in southwestern Colorado is based on a multitude of historic materials made available to the authors by many generous individuals and by helpful staff members of numerous organizations. These friendly people helped in many ways, much too numerous to list in detail here.

To a large extent, photographic collectors and railfan photographers are responsible for the inspiration to produce a series of volumes on this subject. In fact, so many photographs have become available, we have had to expand the number of volumes in ***THE R • G • S STORY***. Those responsible for photographs have their names printed in a credit line with each illustration in this volume.

A special note of appreciation goes to Bill Plunkett, who originally had in mind doing an illustrated book about the RGS, but decided—in his generosity—to hand over his accumulated photographs for use in ***THE R • G • S STORY***.

Mike Blazek has contributed dozens of his fine structure drawings for this series of books, and Jim Key, along with Joe Sacks, have spent many hours drawing plans of RGS trestles, as they appeared during the later years of the railroad, as well as detailed RGS station maps, researched and drawn by Jim Key. Mike and Jim have also provided additional information about details along the RGS right-of-way.

Special recognition also goes to Mallory Hope Ferrell and the late Josie Moore Crum. Mallory H. Ferrell's book, ***The Silver San Juan,*** was a primary reference source regarding facts about the RGS, as well as providing a source for answering questions that arose regarding previously unpublished material and photographs. Josie M. Crum in her book, ***The Rio Grande Southern Railroad,*** provided comparative historical information for material the authors collected for ***THE R • G • S STORY***.

We also want to thank those individuals involved in proofreading this detailed volume—most notably Ed J. Haley, who also provided many detailed bits of information about the last years of the Rio Grande Southern. Robert A. LeMassena, Stan Schwedler, John W. Maxwell, William A. Graves, Al Chione, Robert F. Shank Sr., Gabe Bradford and Herb Kelsey all proofread copies of the text and captions, and they came up with additional information, which has been included in this volume.

Special mention should be made of Mark Kramer and his friends, who spent countless hours examining RGS Dispatcher's Train Sheets at the Colorado Railroad Museum in Golden for facts relating to specific photographs of train movements. Walter Collins also came up with useful data for photographs from the RGS train sheets he owns.

RGS trestle and bridge notes have been provided by the respected and well-informed John W. Maxwell.

"Right-of-Way and Track Maps, The Rio Grande Southern Railroad, June 30, 1919," was provided courtesy of Pete and Kathy Porter.

Very useful personal observations were provided by Robert W. Richardson, the first curator of the Colorado Railroad Museum, and Ed J. Haley, a past president of the Rocky Mountain Railroad Club. Both of these men traveled aboard RGS trains on several occasions and made "train-chasing" trips along the line to take photographs.

One of the foremost contributors of both photographs and text information was Larry Pleasant. In years past, he owned and operated Old West Antiques and the Pleasant Press in Dolores, Colorado. He collaborated with the late Charles M. Engel of Rico in gathering newspapers published at Rico, in 1891, which contained local mining and railroad information.

One of the greatest sources of historic photographs of Rico was the late Freda Engel, the widow of Charles Miller Engel. She allowed the authors to use his fabulous collection of photographs, many of which he had personally taken. Rachel A. Hardwick, niece of the late Freda Engel, graciously offered the authors the use of additional photographs from Charles M. Engel's collection. From various sources he added many other rare photographs, taken before he was born. This collection now resides in her safekeeping, and the pictures we have used from her are listed as being from the collection of Charles Miller Engel.

Orval L. Jahnke, one of the last mine managers in Rico, provided many important facts and photographs. Ken Hazen of Rico was instrumental in providing access to collections of photographs from local residents in Rico. Ann and her husband, James Ferando, a former miner living in Rico, opened their albums to provide domestic insights into the life and times of the Rico area. David and Faye Wolford, relatives of the former RGS engineer, Henry Wolford, allowed us to use their photographs for ***THE R • G • S STORY***. Former Rico miner, Myron Jones, provided many forgotten facts about mining and the town of Rico. Glen Bair, a Rico resident, furnished additional particulars of history. Former Rico resident, Margaret Matzick, provided rare items from her family archives.

The late Richard A. Ronzio had previously allowed the authors to use his superior collection of rare photographs, which we have included in this volume. David C. Butler and his sister, Norma L. Butler, former residents of Rico—related to Charles J. Lee, a Rico photographer—allowed us to use their valuable photographic albums and information for ***THE R • G • S STORY***.

William M. Moedinger, a railfan-photographer from Pennsylvania, personally and graciously presented us with photographs for use in this series. In addition, Donald Duke, publisher of Golden West Books of California, was extremely generous in allowing the authors to use color views he photographed along the RGS.

Gabe Bradford and Herb Kelsey, editors of ***The Galloping Goose*** newsletter, published in California, were extremely generous in allowing the authors to use research material dealing with Rico and the Rio Grande Southern.

— The authors, Denver, Colorado, 1996

INTRODUCTION

THE AUTHORS OF THIS VOLUME of ***The R • G • S Story***—aided by research materials from Gabe Bradford, Mike Blazek and others—undertook the difficult task of piecing together an almost non-existent story of the little mining town of Rico, the Rio Grande Southern Railroad's operations at Rico and the mines, mills and smelters of Rico's Pioneer Mining District. Dell A. McCoy, publisher, made a number of new friends as he gathered photographs from residents of Rico, as well as from former residents who have moved to locations at lower elevations. These individuals had many pictures and experiences to share with us. Dell has assembled the historic photographs and written the captions for this volume. William A. Graves has produced the Chronology of Rico's mining district, as well as other stories, while Russ Collman has put together the history of the Rio Grande Southern at the railroad's division point, Rico.

VOLUME V of ***The R • G • S Story*** continues this series with the origin of Rico and the mining district. Pioneer prospectors returned to their home towns with stories of mineral wealth beyond imagination. This resulted in the establishment of many mining claims in what became the Pioneer Mining District, while the mining camp of Rico gradually became an organized town. This and the development of mines near Telluride sparked the interest of Otto Mears, and he seized the opportunity to build a railroad to transport the mineral wealth to distant smelters for refining. After completing the Rio Grande Southern's line into Telluride in 1890, Otto built his line over Lizard Head Pass and into Rico during September of 1891.

THE RICO DEMOCRAT described the arrival of the RGS as follows: "At last, the railroad is here, and the romantic days of the stage[coach] driver are past. After long years of waiting and watching, the people of Rico are connected with the outside world by bands of steel. The event was consummated at 10:10 p.m., Wednesday, September 30, when rails were laid across the north line of the town site, and a few minutes later, the construction train pushed into the city limits."

Rico was waiting with open arms for the arrival of the RGS, so the railroad could haul out the immense volume of ore from the mines and bring in supplies at reduced costs. Finally, Rico had an efficient means of transportation. The people were overjoyed. They could now travel in the comfort of a passenger car to distant points. Before the RGS arrived, they had to travel by stagecoach or on horseback. And laboring oxen or draft horses previously had slowly hauled in the essential staples for the people and moved ore out to the smelters in far-distant towns.

Immediately, Otto Mears extended a branch back of town to reach the mines above Rico, to the east. Boxcars then could be loaded from the Black Hawk and Enterprise mines. The account of the Enterprise Branch is included in this volume.

A detailed discourse about Rico—with its business buildings and houses—is aptly depicted in words and pictures, showing the close tie-in with the railroad. Of special note is the story of the Hotel Enterprise, the largest building in Rico.

Although the railroad yard of the RGS no longer exists, this volume visually tells the history of this once-busy place—in more detail than thought possible. This includes the purpose of the yard, and the part it played as a helper station for locomotives assisting freight trains up the steep grade to Lizard Head Pass.

A descriptive photographic account and story are presented on these pages about one of Rico's foremost residents during the heyday of the town—Charles M. Engel. It tells about his involvement in keeping Rico alive. Mr. Engel was personally responsible for collecting and saving many photographs presenting Rico's colorful past—in this little town's beautiful mountain setting, located in the San Juan Region of southwestern Colorado.

RIGHT-OF-WAY AND TRACK MAPS

Pete and Kathy Porter Collection

LIST OF BRIDGE PROFILE DRAWINGS

RGS STATION MAPS OF RICO

~ MAP LIST ~

PROFILE OF THE RIO GRANDE SOUTHERN RAILROAD, DATED JANUARY 1, 1913

COLLECTION OF CLARK R. BAILEY

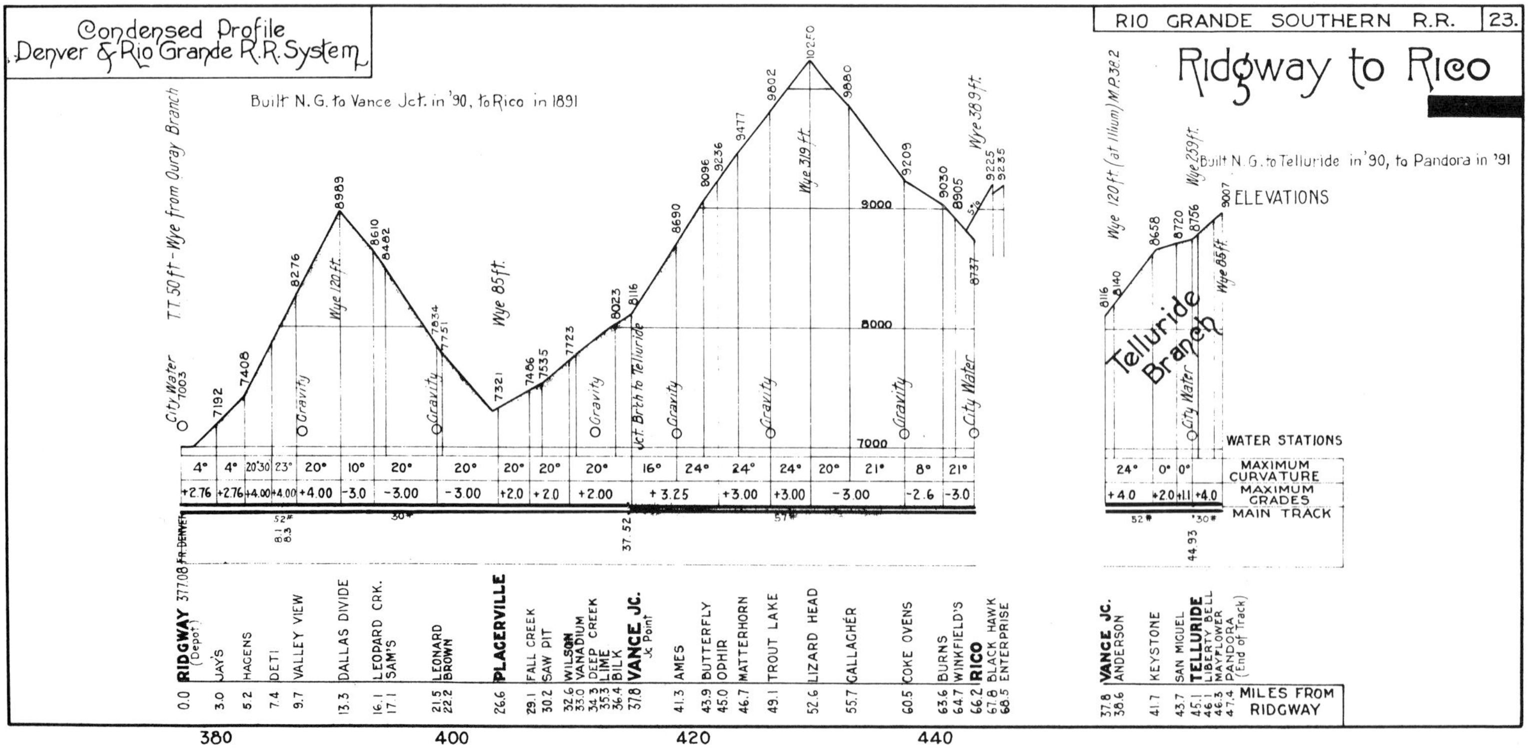

~ Table of Contents ~

Page

LAD A. AREND PHOTO – WILLIAM A. GRAVES COLLECTION

THE RICO DEPOT OF THE RIO GRANDE SOUTHERN

LARRY PLEASANT COLLECTION

CHARLES MILLER ENGEL (above) and Larry Pleasant of Dolores became close friends when the two men joined forces to enlighten the world with their historical accounts of the mining town of Rico. Larry was publishing the Dolores newspaper, which carried articles written by Charles. Larry presented this HO-scale model (3.5 mm. to one foot) of the Rico depot to Charles as a remembrance of their friendship and working relationship during the late 1960's. Charles unknowingly had become a railfan! This picture was taken in Larry Pleasant's business establishment, the Pleasant Press (and Old West Antiques).

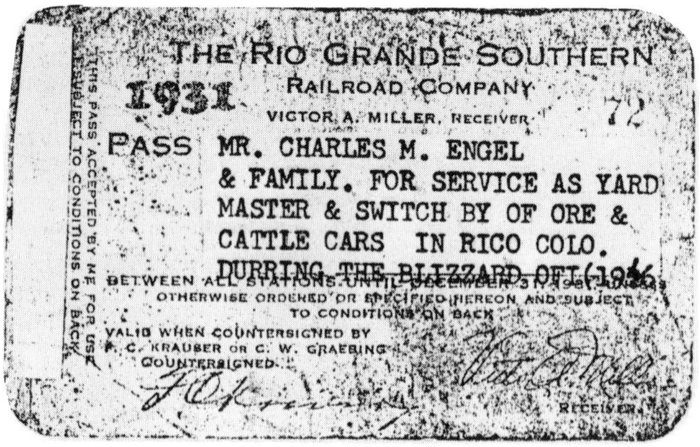

CHARLES MILLER ENGEL COLLECTION –
COURTESY RACHEL A. HARDWICK

CHARLES M. ENGEL had a humorous side to his personality, evidenced in this 1931 pass from the RGS. He altered it to give himself and his family a free pass for services rendered during the blizzard of 1946.

CHARLES MILLER ENGEL COLLECTION –
COURTESY RACHEL A. HARDWICK

FREDA RICE LUTENER and Charles Miller Engel celebrated their 50th wedding anniversary during September of 1973. A reception was held at the Community Church in Rico, and this cake was presented to the couple at the festive gathering. Freda was a most generous person, and she shared many photographs from Charles' collection—printed in this series of books. She enjoyed living in the family home in Rico, a true spirit of the Old West. She was born on May 5, 1902, and passed away on February 2, 1993.

CHARLES MILLER ENGEL

Rico Resident and Businessman

Courtesy Larry Pleasant

— From an Obituary —

No. 1

CHARLES MILLER ENGEL, a widely known native of Rico, Colorado, was born in this little mining town on December 12, 1902, and passed away on Thursday, January 28, 1982, at Southwest Memorial Hospital in Cortez following several week's illness. Mr. Engel was the son of Martha A. and Charles Engel Sr., who came to Rico in 1897 from Pittsburgh, Pennsylvania. The senior Mr. Engel was a pipefitter, and in 1909, he established the Engel Mercantile store in Rico—a major general-merchandise store in Rico for nearly half a century. His son worked in this store until the late 1930's.

During his 79 years, Charles M. Engel also became involved in many other endeavors. This resulted not only in acquiring a host of friends and acquaintances, but it later brought him recognition for his vast knowledge of the pioneers and history of the area—particularly in Rico's Pioneer Mining District.

Charles attended public school in Rico, and then traveled to Greeley, Colorado, where he went to the Industrial High School at Colorado State Teachers College, in 1921 and 1922. While living in Greeley, he worked at several trades, and later, he put the first Yellow Cab taxis on the streets of that city. This job resulted in much use of the telephone, during which time he became acquainted with a telephone operator named Freda Rice Lutener. The two young people later met, and after courting her for some time, Charles and Freda eloped to Laramie, Wyoming, where they were married on September 15, 1923. During the following year, they returned to his home in Rico, where he again became involved with his father in the general-merchandise store. And among other activities, he was the operator of the Rico Theater. He also played in the town's dance orchestra, which his father organized. During these years, he also worked for a time for the Rio Grande Southern Railroad as an extra gang boss over a Navajo section crew.

He spent much of his spare time prospecting and in geological exploration, while acquiring an impressive mineral-specimen collection. Charles also developed the Engel mining claim on Expectation Mountain, just west of Rico. In this endeavor, he was assisted at times by his brother-in-law, Oral Lutener. Charles was the Dolores County Coroner and an undertaker assistant for several years, and he also worked on highway projects in the area during the days of road improvements for automobiles.

(Continued on Page 18)

CHARLES ENGEL, SR., established his mercantile company during 1909, and he ran the establishment for approximately 50 years, with his son's help. The storefront appeared to be in new condition in this photograph, with the late afternoon sunlight warming it on this winter day. Apparently, Charles Engel, Sr., standing in the doorway at the left, was posing with his son, Charles Miller Engel, and Martha A. Engel, the mother of Charles. It is believed that the date for this picture was 1914, making Charles, Jr., 12 years old at the time. THE RICO DEMOCRAT reported on April 3, 1891, that: "Dr. G. D. Yokom, after a three-week visit with his family at Montrose, has returned and can be found at his post in the Rico Mercantile [the same building]." Notice the raised wooden sidewalk along Glasgow Avenue, the town's main street. A barber shop and millinery store shared the small building to the south during 1893. Notice that the Rico Post Office was in the drugstore next to the mercantile store.

CHARLES MILLER ENGEL COLLECTION
– COURTESY RACHEL A. HARDWICK

CHARLES MILLER ENGEL COLLECTION – COURTESY RACHEL A. HARDWICK

THIS PORTRAIT SHOWS: Charles Engel Sr. (right) 1870–1938, Martha A. Engel (left) 1872–1935, M. Anna Engel (top), Charles Miller Engel (bottom) 1902–1982.

CHARLES MILLER ENGEL COLLECTION –
COURTESY RACHEL A. HARDWICK

CHARLES MILLER ENGEL was photographed by his father on May 18, 1911, when the boy acted in a school play as a bachelor. This was taken in front of the Engel home, which was located on the east edge of Rico, on Dolores Mountain.

CHARLES MILLER ENGEL COLLECTION –
COURTESY RACHEL A. HARDWICK

CHARLES ENGEL, SR. was the type of man who loved horses, horseback-riding and all the equipment that went with it. He was at the Engel home, oiling his saddles to keep the leather from drying out.

CHARLES MILLER ENGEL COLLECTION –
COURTESY RACHEL A. HARDWICK

M. ANNA ENGEL and Charles Miller Engel were outdoors, "enjoying" the deep snow in front of their home, about 1915.

CHARLES MILLER ENGEL COLLECTION –
COURTESY RACHEL A. HARDWICK

MARTHA A. ENGEL, Charles Miller Engel's mother, was out panning for gold near the town of Rico—something many of the residents did as a part-time activity.

CHARLES MILLER ENGEL COLLECTION –
COURTESY RACHEL A. HARDWICK

THE LOVELY, ORNATE HOME of Charles and Martha Engel, sat higher on the east side of Rico than other homes. This is how the house looked about 1910, with the bay window facing west.

THE CHARLES ENGEL FAMILY had one of the best views in town. From their dining-room window they could look down over Rico from this lofty location. A coal stove provided the needed warmth for those long chilly days and evenings at this high altitude—and some nights required plenty of coal for the stove!

CHARLES MILLER ENGEL COLLECTION
– COURTESY RACHEL A. HARDWICK

CHARLES MILLER ENGEL COLLECTION –
COURTESY RACHEL A. HARDWICK

YOUNG CHARLES M. ENGEL was photographed at the Engel cabin on Expectation Mountain, where the family had a mining claim.

CHARLES MILLER ENGEL COLLECTION –
COURTESY RACHEL A. HARDWICK

THE ENGEL CABIN, on the east slope of Expectation Mountain, was a favorite place for the "good old boys" to hang out—and get away from the mundane activities of daily life. The original log cabin burned, and it was replaced by the building on this site today. Charles J. Lee (at the left) and Charles Engel, Sr., (at the right) had ridden their horses up the mountain with Charles M. Engel one summer day during the 1920's.

CHARLES MILLER ENGEL COLLECTION –
COURTESY RACHEL A. HARDWICK

CHARLES MILLER ENGEL was down at the Rio Grande Southern depot, when his father took this snapshot.

Charles snowshoed and trapped in the mountains surrounding Rico, and he was continuously involved as an avid photographer during his lifetime, recording on film the history of the town, which in later years he generously shared with researchers. For many years, he also was the local correspondent for radio station KVFC in Cortez, where Jack Hawkins regularly reported his news about the little mountain town.

During later years, Charles worked for the Rico-Argentine Mining Company, and he assisted with surveying-and-geological work before becoming warehouseman for the firm, until 1958. From 1958 until 1981, he was the official weather observer for the U.S. Weather Service, taking daily readings from the weather station located at his home.

Following surgery of the larynx in 1966, Charles turned more to writing, and he became the Rico correspondent for **The Dolores Star**, **The Dove Creek Press** and **The Cortez Sentinel,** which he supplied with weekly columns of news and views, with many photographs depicting local events. His writing often reflected his considerable sense of humor, which he displayed throughout his lifetime. This prompted many of his "Vibrations" in his weekly columns. During

CHARLES MILLER ENGEL COLLECTION –
COURTESY RACHEL A. HARDWICK

FREDA AND CHARLES ENGEL were photographed outside their Mantz Avenue home, located on Block 16, Lot 21, on the map. The young couple probably were in their mid-40's at the time.

CHARLES MILLER ENGEL COLLECTION –
COURTESY RACHEL A. HARDWICK

FREDA R. ENGEL, Charles Miller Engel's wife, was photographed about 1930, while the couple lived with Charles' parents in east Rico.

this time, he was asked to write a comprehensive résumé of the 100-year history of mining in Rico. His article was titled "Rico, Colorado: A Century of Historic Adventures in Mining," and this was published in the 1968 edition of the New Mexico Geological Society Guidebook for the San Juan, San Miguel and La Plata region.

Despite his larynx operation, he very capably served as the mayor of the Town of Rico from 1966 until 1968.

During 1972, Charles corroborated with J. A. Reid, B.S., a student at the University of Texas, in the writing of a thesis, titled "A History of Rico, Colorado: 1879–1971."

Charles Engel wrote a book on the famous Galloping Goose railbuses of the Rio Grande Southern Railroad, and he also wrote a series of historical works, titled "A Treasure-Trove of Rico History," which ran serially for several years in local newspapers.

ADVERTISEMENT IN "THE RICO ITEM,"
MARCH 12, 1915

CHARLES MILLER ENGEL PHOTO – KEN HAZEN COLLECTION

CLIMBING TO THE TOP of Expectation Mountain above 12,000 feet with heavy camera gear must have been a breathtaking adventure during the late 1920's—in more ways than one. It is believed that Freda Engel was wearing the checkered jacket, standing at the left, with two girlfriends. The panoramic camera—turning from left to right—exposed slowly enough for the trio to enter the same photograph again. The mountain range at the left includes Elliot Mountain, at

CHARLES MILLER ENGEL PHOTO – LARRY PLEASANT COLLECTION

CHARLES M. ENGEL and his father had taken the panoramic camera on a hike up Iron Draw, a gulch located northwest of Rico, to view this dramatic vista and shoot the entire south end of the valley at Rico for posterity. This picture was taken ca. 1930. Blackhawk Mountain rises in the distant background to an elevation of 12,677 feet. Numerous prospectors had claims and small mines in

12,340 feet in elevation, and Sandstone Mountain. The far-distant peaks are in the San Miguel Range, and they include Mount Wilson, Lizard Head, Vermilion Peak and San Miguel Peak. The Dolores River is in view, just beyond Sandstone Mountain, and to the right is Telescope Mountain. Silver Creek drains through the area beyond Rico, at the bottom of the scene, and Dolores Mountain rises above the town, where mine dumps were visible on Newman Hill, the lower part of the mountain.

the basin between Blackhawk Mountain and Dolores Mountain, plainly visible behind the town. The old wagon road from Rockwood (on the D&RG's Silverton Branch) made its way to Rico via Scotch Creek, located beyond Dolores Mountain. The Rio Grande Southern's station grounds and freight yard are visible at the lower right, just east of the Dolores River.

CHARLES MILLER ENGEL COLLECTION –
COURTESY RACHEL A. HARDWICK

THE INTERIOR of the Engel Mercantile Company store is pictured here as it looked in the 1940's. Charles Miller Engel was wearing white-colored slacks, and was surrounded by fellow employees and relatives. This probably was the southern portion of the store, as a connecting doorway is in this view, at the left.

TWO PHOTOS: CHARLES MILLER ENGEL COLLECTION – COURTESY RACHEL A. HARDWICK

CHARLES WAS OUTSIDE, on Glasgow Avenue, in front of his mercantile store (out of sight, to the right of this view). He was looking at the snow-sculpture effects left in the middle of the main thoroughfare by the snowplow driver. This was during the meltdown period, after the storm was over, when the snow compacts. By early spring, high-altitude snowstorms could dump five to six feet of snow on the level before it packed down. This view probably was taken during the 1940's.

Again We Are Offering

REAL BARGAINS

In Sweaters and Sweater Coats
Made by Jersild Knitting Mills

ALL WOOL
WARM and COMFORTABLE

■

Originally Priced Up To $6.00 Per Garment
Now Only $2.00
Originally Priced Up To $10.00 Per Garment
Now Only $3.50

■

Buy NOW While Sizes and Colors
Are Available

■

Phone Rico 3
for service that satisfies
Chas. Engel Mercantile Co.

FROM "THE RICO ITEM," DECEMBER 4, 1936
COURTESY MARGARET MATZICK

BUSINESS WAS THRIVING at the "Chas. Engel Mercantile Co." general-merchandise store and warehouse (next door) during the late 1930's. The entire single-story business building was operated by Charles, and the right-hand portion had been upgraded into a market, offering meat, produce and canned goods. The warehouse probably provided needed space for overstock. The automobiles lining Glasgow Avenue were manufactured during the 1930's.

UNFORTUNATELY, the Charles Engel Mercantile Company's warehouse burned during the late 1940's or early 1950's.

LARRY PLEASANT COLLECTION

CHAS. ENGEL MERCANTILE CO. GENERAL MERCHANDISE.
CHAS. ENGEL
WAREHOU
QUALITY MEATS
FRUITS-VEGETABLES

CHANDISE.
CHAS. ENGEL MERC. CO.
WAREHOUSE
FRUITS-VEGETABLES

Engel House

P.O. Box 85 • Rico, Colorado 81332 • (970) 967-2321

CHARLES MILLER ENGEL COLLECTION –
COURTESY RACHEL A. HARDWICK

THIS INTERIOR SHOT of the county clerk's office in the Dolores County Courthouse in Rico is shown in this view taken ca. 1937. County Clerk Rowena E. Snyder and her deputy, Oral C. Lutener, the brother of Freda Engel, are the subjects of the photograph. The two were married in Durango, Colorado, shortly after this picture was taken. Rachel A. Hardwick is one of their grandchildren, and she now lives in the Engel family home on Mantz Avenue in Rico.

RACHEL A. HARDWICK was standing on the front porch of the Engel house on Mantz Avenue on October 8, 1995. Rooms are now offered in the renovated home, completed by Rachel. Rachel is now the custodian of the extensive Charles Miller Engel photographic collection.

DELL A. McCOY PHOTO

CHARLES MILLER ENGEL COLLECTION – COURTESY RACHEL A. HARDWICK

THE PIONEER MINING TOWN of Dunton included a little known hot-springs resort. In this view, you are looking toward the southwest. The small settlement was photographed during 1930. Mines were located on the hill beyond Dunton.

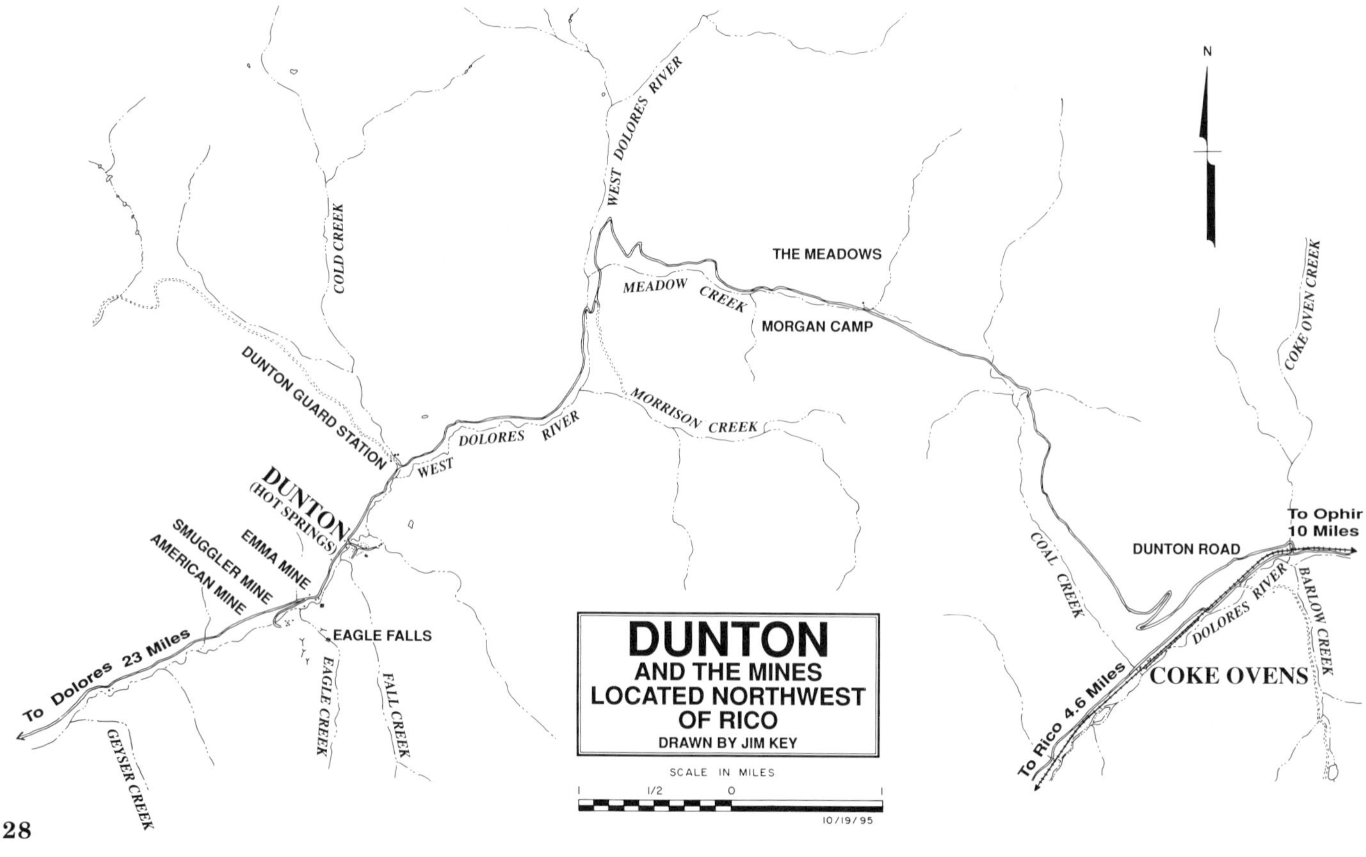

No. 2

A Mining Camp Near Rico...

THE STORY OF DUNTON

By William A. Graves

DUNTON, COLORADO, was a small unincorporated settlement in Dolores County, located about 12 miles northwest of Rico by road. The town was established in 1885, and it consisted of log structures stretching for about a mile along the West Dolores River. Early-day Dunton was a mining camp, and the mines were located about half a mile down the stream from the community. The most successful mines were the Emma, the Smuggler and the American.

As with many other mining camps, Dunton suffered from isolation and a lack of good transportation. Outgoing ore and incoming supplies had to be freighted by pack trains or wagons for very long distances, until the Rio Grande Southern was built in 1891. Even after that, the nearest rail connection was a distance of about nine miles, at Coke Ovens.

Dunton's population was less than 50 people until the Emma Mine was sold to a group of Eastern investors in 1897. The Emma was rapidly developed, and it was the mainstay of the camp. Stamp mills and other substantial mining buildings were constructed, and the town expanded. The mine's ore production and the town's population probably peaked about 1905. The greatest population has been reported to have been about 260 to 300 people. The value of the ore produced is not known, but Dunton's production probably was included in the Dolores County totals reported by the U.S. Bureau of Mines. During 1905, Dunton's probable peak year, the total county production was reported to be $172,454. During the same year, Dunton's production was estimated at about $56,000.

As in nearby Rico, the boom at Dunton did not last very long. Many people left the camp by 1910, and by 1918, the town was deserted. That year, long-time residents, Joe and Dominica Rosario, purchased the entire town and some of the patented mining claims. They operated the property as a

(Continued on Page 39)

DELL A McCOY PHOTO

THE EMMA MINE headframe was still in position beside the West Dolores River in this view taken late in September of 1978.

THIS CLOSE-UP VIEW of Dunton was shot during September of 1978. You are looking toward the east. Cabins surround an open area, much like an old town square.

DELL A McCOY PHOTO

DELL A McCOY PHOTO

LOOKING ALMOST SOUTH at the settlement of Dunton and the hot springs, the photographer took this view on October 7, 1995. The aspen leaves were late in turning gold in this scene. The Dunton mining area was beyond the buildings, and the road at the right leads to Ground Hog Reservoir. The West Dolores River is in the foreground, flowing southward.

CHARLES MILLER ENGEL COLLECTION –
COURTESY RACHEL A. HARDWICK

FREDA RICE LUTENER and Charles Miller Engel were married on September 15, 1923, and they arrived in Rico a short time after that. They had traveled up to the Dunton Hot Springs in the summer of 1925, when this picture was taken. "Duce," the dog, and Freda Engel were sunning beside the pool at these hot springs, located a short distance to the northwest of Rico.

CHARLES MILLER ENGEL COLLECTION –
COURTESY RACHEL A. HARDWICK

THE WARM SUMMERTIME in the Colorado mountains was everyone's favorite time to get outdoors after the long, cold winter. These three sisters were enjoying the Dunton Hot Springs pool. They were (from left to right): Eva Lutener, Blanche Lutener and Freda Engel. Eva and Blanche Lutener had come to Rico to visit their sister, Freda (Lutener) Engel.

CHARLES MILLER ENGEL COLLECTION –
COURTESY RACHEL A. HARDWICK

THE HOT SPRINGS POOL at Dunton was enjoyed by this group of young people from Rico during the late 1920's. The horizontal pole above the pool was used to support a ring, ball and two swings suspended above the pool.

ANN AND JAMES FERANDO PHOTO

THE FERANDO FAMILY was enjoying an outing at the Dunton Hot Springs Pool during the early 1940's. The change house consisted of two compartments in the log structure at the left.

CHARLES MILLER ENGEL COLLECTION –
COURTESY RACHEL A. HARDWICK

DUNTON'S HOT SPRINGS POOL had a log change house next to the boardwalk. The Dolores Peak in this view is a 13,290-foot mountain north of Dunton, and it is not the Dolores Mountain so prominent in the background of many of the Rico photographs.

THE OLD BUILDINGS on what was the main street of Dunton consisted of these wooden structures in earlier times. The false-front buildings probably were a grocery and market, and a two-story hotel, with a restaurant. These buildings faced toward the south.

CHARLES J. LEE PHOTO – DAVID C. BUTLER COLLECTION

MONTE BALLOUGH PHOTO – COURTESY MARGARET BALLOUGH PALMER

THE TOWN OF DUNTON was the site of this picture taken between 1899 and 1903, where the camera lens was pointed toward the northwest. The present-day gravel road is at the base of the hillside, beyond these buildings. Pack saddles contained ore from the nearby mines, which was being hauled out by this pack train.

THE EMMA MINE tunneled into the mountainside less than a mile downstream from Dunton. A rather flimsy bridge carried the ore cars across the West Dolores River. In this photograph, a train of five mine cars loaded with ore were being pulled by a single white mule. The cars had just left the mine adit and were crossing the bridge. The Emma's most productive time was after a group of Eastern developers purchased the mine's location in 1897.

MONTE BALLOUGH PHOTO – COURTESY MARGARET BALLOUGH PALMER

CHARLES J. LEE PHOTO – DAVID C. BUTLER COLLECTION

THE EMMA MINE had an enclosed head frame that was connected to the hoist house, as this view shows (at the left). It was photographed during the late 1920's, as the camera lens was aimed toward the southeast. The Emma Mine's workings followed a vein in the hillside. Eagle Creek Falls are in view, behind the Emma Mine's mill. The Smuggler's tailings dumps appear at the right.

A MINER WAS READY to head back over the bridge into the Emma Mine, in this photograph of ca. 1900. The Emma's boarding house was in view between other supporting structures (at the right). The four-wheeled cars were typical end-dumpers.

MONTE BALLOUGH PHOTO – COURTESY MARGARET BALLOUGH PALMER

MONTE BALLOUGH PHOTO – COURTESY MARGARET BALLOUGH PALMER

THIS PHOTOGRAPH of the Emma's mill and boarding house is believed to have been taken during December of 1902 by the Rico photographer, Monte Ballough. The cook and dishwasher posed outside the kitchen door. It appears that improvements had been made on the mill buildings located across the river. An enlargement had been placed to the left of the original building. A wooden flume carried water to the mill in order to produce hydro-electric power for motors and pumps.

MONTE BALLOUGH PHOTO – COURTESY MARGARET BALLOUGH PALMER

THIS SNOWY PICTURE vividly shows how difficult it could be to operate a high-altitude mine during the winter. The Smuggler-Almont Mine was 9,000 feet in altitude, and the winters were long and cold.

THE SMUGGLER-ALMONT stamp mill was located on the West Fork of the Dolores River, about one mile downstream from the settlement of Dunton. This mill was a long, narrow building, about 90 feet in length and 50 feet in height.

MONTE BALLOUGH PHOTO – COURTESY MARGARET BALLOUGH PALMER

MONTE BALLOUGH PHOTO – COURTESY MARGARET BALLOUGH PALMER

A RARE GLIMPSE inside the Smuggler Mine shows Monte Ballough (in the foreground, at the left) at the innermost workings. The miners were using candles to illuminate their work. The generous size and dry conditions inside the Smuggler made life easier for the miners.

cattle ranch, and they later increased their ranch holdings to about 500 acres. During the 1970's and the 1980's, the property was operated as a dude ranch for tourists; however, by the early 1990's, the ranch had been closed down and was for sale.

The Dunton Hot Springs townsite is now owned by a private firm, and the bridge crossing the West Dolores River has been chained off, preventing visitors from access to the site. This also cuts off access to U.S. forest trails on the other side of the bridge.

CHARLES MILLER ENGEL COLLECTION – COURTESY RACHEL A. HARDWICK

MULE POWER was in use at portal No. 6 of the American Mine at this time. The tunnel, called a drift, was driven slightly upgrade to allow water to drain from the mine.

CHARLES MILLER ENGEL COLLECTION – COURTESY RACHEL A. HARDWICK

GEYSER SPRINGS was photographed at its site southwest of Dunton. As its name implies, a heavy flow of hot water surges to the surface at intervals, filling the small basin until it overflows.

CHARLES MILLER ENGEL COLLECTION – COURTESY RACHEL A. HARDWICK

DAY'S LOG CABIN was located near Calico Peak, adjacent to Horse Creek. The miner and his wife had this prospect opening in 1930, adjacent to their rough-hewn cabin.

A HARD-ROCK MINER (at the left) was at the Day's log cabin, wearing a cap with a carbide lamp attached. All thought of work vanished when visitors arrived from Rico. This rustic scene was photographed during the summer of 1930.

CHARLES MILLER ENGEL COLLECTION – COURTESY RACHEL A. HARDWICK

Mile by Mile to...

Coke Ovens, Burns Cañon and The Rice Yard Limits

By Russ Collman

No. 3

VOLUME V of ***The R • G • S Story*** begins where Volume IV ended its coverage of the Rio Grande Southern Railroad's narrow-gauge mainline...

We ended our mile-by-mile dissertation about the RGS in Volume IV just south of **BURNS**, which was at Milepost 63.6, Elevation 9,030 feet, just north of Burns Cañon.

The last station listed was...

WINKFIELD—Milepost 64.7, Elevation 8,905 feet—was one and one-half miles north of Rico. This was little more than the site of a short spur, which served ore bins and chutes of the Mountain Spring Mine, used to load railroad cars with zinc ore. The mine owning these ore bins was located on the western slope of Telescope Mountain, above the railroad, and there was an aerial tramway that transported the ore down to the ore bunker.

At a later date, the St. Louis Tunnel was located adjacent to this point, about halfway between Winkfield and Rico. The portal for this mine entered the mountain just beneath the mainline of the RGS. This was a large mining operation, originally worked by the St. Louis Smelting & Refining Company to recover high-grade iron pyrites, used to produce sulfuric acid. This company moved the raw ore to a truck dock in Rico, where the ore was dumped into gondolas for shipment to St. Louis, Missouri, via Ridgway and Montrose.

BRIDGE 66-A—a pile trestle called the Winkfield bridge—was listed at 544 feet in length as of 1912, and it was on a 3.0-percent grade. This bridge was filled-in during 1915.

BRIDGE 66-B—another pile trestle—was just north of the switch for the Enterprise Branch. It was listed at 256 feet in length in 1912, but was filled-in as of 1915.

MONTE BALLOUGH PHOTO – COURTESY MARGARET BALLOUGH PALMER

A PACK TRAIN of mules had been loaded with wire rope (cable) measuring 3,400 feet in length and $^7/_8$ inches in diameter. It was on its way to one of the mines in the Dunton area. The RGS tank at Coke Ovens stood beside the mainline in this picture, which looked toward the northeast, in the direction of Lizard Head Pass. This was how the scene looked at Coke Ovens ca. 1900. The site of the water tank is about 1.25 miles downstream from the junction of the narrow unpaved road to Dunton and Colorado Highway 145. The mule train was taking a shortcut trail to the mining camp of Dunton.

CHARLES MILLER ENGEL COLLECTION
– COURTESY OF RACHEL A. HARDWICK

A SUDDEN SNOWSLIDE roared down the west slope of Telescope Mountain and smashed into the side of southbound rotary snowplow No. 2 and overturned its tender. The snowfighting train had just passed Burns siding, and it was about to enter Burns Cañon when the avalanche hit. Auxiliary water car W-12 was derailed, but it stayed upright. Apparently, a drag flanger had been cut in behind the lead engine, followed by two more engines and a caboose. The same slide was photographed from the opposite side of the train, and it was printed in Volume IV of this series, on Page 463. These photographs probably were taken ca. 1915.

MARGARET MATZICK COLLECTION

LOOKING UPGRADE at Burns, ca. 1920, the photographer had aimed his camera lens toward the north. If you look closely, you can see the station sign at the far left. The Dolores River had been dammed at this point about 1900. At the right, across the Dolores River, you can see the remains of this diversion dam, built to divert water from the river into the flume of the Rico Power & Light Company. This flume carried water to the head of the penstock on the east slope of Expectation Mountain. The water was used to spin the turbine in the power plant to generate electricity for Rico and the surrounding mines and mills. The mine in the foreground appears to be derelict at this time, although the cabin at the mine looks lived-in.

MARGARET MATZICK COLLECTION

THE ABANDONED DERELICT FLUME, which in earlier days supplied the water to operate the hydroelectric powerhouse of the Rico Power & Light Company, was still clinging tenaciously to the rock wall above the Dolores River in the "narrows" of Burns Cañon during the 1920's. Work had been going on intermittently in an attempt to improve the dangerous passage of the Dolores-to-Lizard Head road through Burns Cañon—a condition that persisted well into the

1940's. A recent rockslide had removed what road there was, and it covered the flume. This was a typical mountain road in the remote areas of the San Juan Region of Colorado during the 1920's. In this view you are looking down at the Rio Grande Southern's track at the south end of Bridge 64-A. The three timbers propped against the road embankment probably were used as cribbing to support the section of roadway taken out by the rockslide.

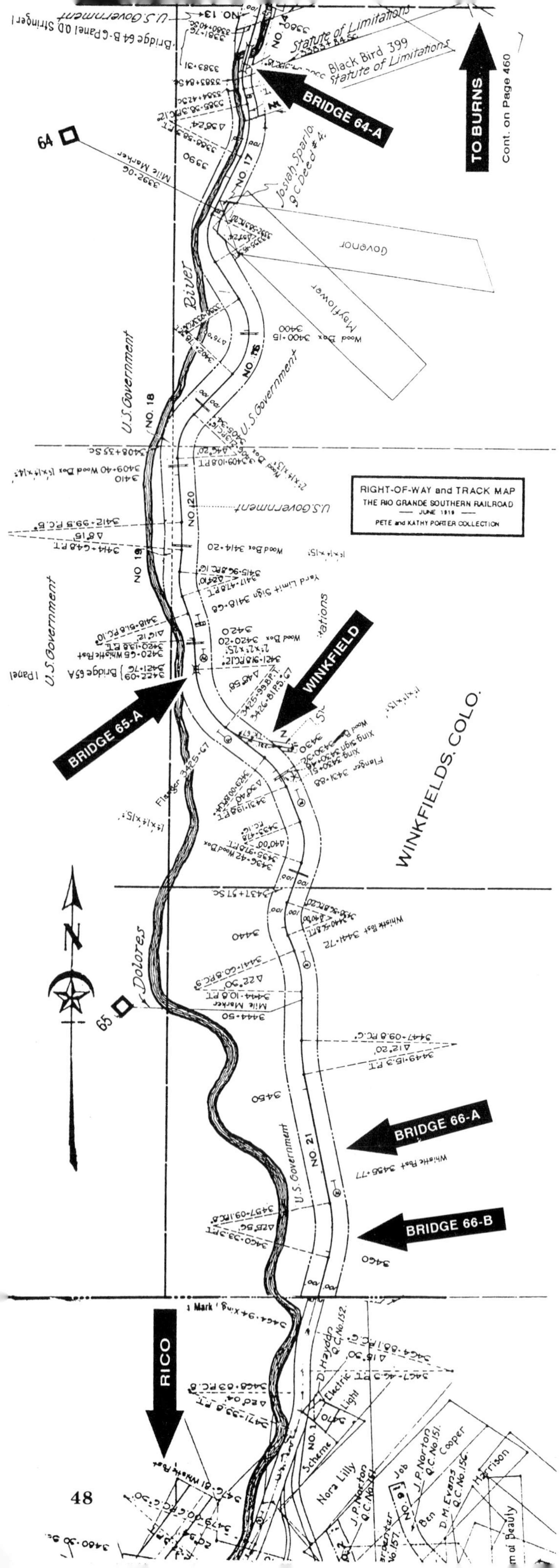

TO BURNS
Cont. on Page 460
BRIDGE 64-A
Black Bird 399
Statute of Limitations
Governor
Mayflower
River
U.S. Government
RIGHT-OF-WAY and TRACK MAP
THE RIO GRANDE SOUTHERN RAILROAD
PETE and KATHY PORTER COLLECTION
WINKFIELD
BRIDGE 65-A
WINKFIELDS, COLO.
Dolores
BRIDGE 66-A
BRIDGE 66-B
RICO

MARGARET MATZICK COLLECTION

LOCOMOTIVE No. 2 was hauling the morning passenger train, and a D&RG 3000-series boxcar was coupled behind the 2-8-0. The mixed train was rolling downgrade toward Rico when this scene was recorded ca. 1905. The Dolores River is in the foreground, and Burns siding was beyond the bluff in the background.

MARGARET MATZICK COLLECTION

RGS MILEPOST 64 was located in Burns Cañon and was photographed ca. 1905. The "R" above the number stands for the miles to Ridgway. You are looking upriver, beside the tumbling Dolores River.

SURGING WATER from the 1911 Dolores River Flood had undermined mainline trackage at this location above Rico, near Burns Cañon. This section of the track always had problems with Nature, which attempted to push out the RGS.

CHARLES MILLER ENGEL COLLECTION
– COURTESY OF RACHEL A. HARDWICK

DELL A. McCOY PHOTO

THE ASPEN TREES were almost at their height of autumn splendor in the Dolores River Valley. This scene was photographed during late September of 1993, from Colorado Highway 145, at the southern entrance to Burns Cañon. The old RGS right-of-way swung around a 16-degree curve and entered the rock-walled gorge, heading toward Bridge 64-A, at the left. Two miles farther down this beautiful mountain valley, the railroad and the highway reached the little mining town of Rico. The 13,000-foot mountain peaks—on the horizon, at the right—are approximately 25 miles to the south, in the rugged La Plata Range.

CHARLES J. LEE PHOTO – COURTESY OF DAVID C. BUTLER

ROTARY SNOWPLOW No. 2 had been withdrawn from the deep cut, and the snow-shovelers were pushing snow away from the top of the cut to a lower level, where the rotor blades could throw it from the trench. This location was near Milepost 64 in Burns Cañon. The snow-shovelers were at work on the big slide, probing for tree parts and rocks that might be in the midst of the avalanche. The rotary plow was waiting to chew through the slide and move on up the line toward Lizard Head Pass. In cases like this, the shovelers worked down through the slide in layers, to within 16 feet of the rails, to enable the rotary plow to operate properly.

CHARLES J. LEE PHOTO – COURTESY OF DAVID C. BUTLER

A SHORT DISTANCE downgrade from Burns, the rotary plow had been forced to stop because of the depth of the snow covering the track. As the rotary crew and a snow-shoveler posed for the photographer, all that could be seen of the rotary was the headlight and a small section of its wooden side. Even the stack was hidden by snow.

CHARLES MILLER ENGEL COLLECTION – COURTESY OF RACHEL A. HARDWICK

RGS PLOW-FLANGER 02 (judging by the car's length) had taken a tumble into the Dolores River one unrecorded day near the narrows of Burns Cañon, south of the siding at Burns. The roof of this rather short snowplow appears in this view, ca. 1912. The photographer was looking upriver when he shot this scene.

DELL A. McCOY PHOTO

OVERLEAF: Taken during the winter of 1995, the photographer was looking toward the south for this view. The RGS right-of-way curved to the left, around the foreground bank, where the Winkfield spur was located. The Dolores River continues past the foreground, tumbling downstream beyond Rico, in the distance.

UPPER VIEWS: ANN AND JAMES FERANDO PHOTOS

RGS PLOW-FLANGER O2 had not traveled very far north from Rico before being derailed and damaged in a solid hit by the Nigger Baby slide in 1936. Its headlight, mounted above and behind the wedge-plow blade had been nearly torn off in the blast of snow. The plow had to be re-railed and pulled back to Rico for repairs.

RGS ENGINE 22 of the snowplow train had come to a sudden halt, and the engineer on the locomotive ahead leaned from his cab window to see what had happened up ahead. The engines will have to be uncoupled and backed from the cut so section men can dig out the trapped plow and get it back on the rails.

CHARLES J. LEE PHOTO – COURTESY OF DAVID C. BUTLER

LOOKING AHEAD through one of the locomotive-cab doors toward the engine front, the wall of snow cut away by the rotary snowplow towered above the engines in this view. Of course, snow-shovelers had to reduce the height of the snow to the level of the rotary's hood before the snow-fighting train could continue its struggle toward the summit of Lizard Head Pass.

ANN AND JAMES FERANDO PHOTO

THE NIGGER BABY SLIDE had trapped Plow-Flanger 02 up ahead, and snow-shovelers were struggling to remove timbers and rocks from behind the derailed wedge plow. When the plow had been dug out, the locomotives returned to pull it back onto the rails and return with it to Rico. The rotary plow was then used to clear the line to Lizard Head Pass.

ANN AND JAMES FERANDO PHOTO

THE WET, SNOWY MONTH of February hits the high country of southwestern Colorado with a "bang" most years—and snowfighting began in earnest on the Rio Grande Southern during this month. In this case, RGS rotary snowplow No. 2 was throwing the wet stuff off the mainline near Rico. The train was heading northbound, toward Burns and beyond.

DELL A. McCOY PHOTO

OVERLEAF: This autumn scene was captured on film during 1979. The photographer was about one mile north of Rico, enjoying Colorado's dazzling Indian-summer extravaganza. You can see the long-abandoned grade of the first switchback on the Enterprise Branch of the Rio Grande Southern, which went up the mountainside beyond the Dolores River. The Enterprise Junction's turnout from the mainline was once located in the area near the left edge of this view. This was the start of this branch, which switchbacked its way up Silver Creek and onto the western slope of Dolores Mountain. The slag dump (center) belonged to the old Grand View Mining & Smelting Company. It is interesting to notice that beavers had set up housekeeping in the river bottomland, just north of Rico. Although an old shack was along the edge of the beaver pond, the industrious little creatures had built their own home (called a "beaver lodge") out of mud and sticks.

MARGARET MATZICK COLLECTION

A TRAIN CREW enjoyed this vista of the Dolores River Valley near Rico as their train drifted downgrade toward the yard. It is believed this picture of the RGS line north of Rico was taken ca. 1899. The Dolores County Courthouse is visible at the left, while a glimpse of the grade on the Enterprise Branch is in view above the town. Rico's business section is at the right, and buildings at the Atlantic Cable Mine are at the northern edge of the community.

DELL A. McCOY PHOTO

THE RGS GRADE worked its way down from Burns—located near the snowcapped mountain in this scene—as it followed the Dolores River along the right bank. From the highway bridge at Rico, it would have been possible to see RGS Bridges 66-A and 66-B during railroad days, near the right center of this picture. In 1995, a trace of the Enterprise Branch is in sight, at the right.

SNOW-CROWNED DOLORES MOUNTAIN was shining in the last rays of sunshine before darkness closed in on it during a December day in 1993. Rico is around the next curve.

WENDE PERINA STUART PHOTO

CHARLES J. LEE PHOTO – COURTESY OF DAVID C. BUTLER

THE ONE-HALF MILE beyond Milepost 65 on the RGS was one of the most difficult sections of track to keep open during the winter months. In this picture, the snow-shovelers had moved off the railroad's right-of-way to allow the rotary snowplow to advance slowly through the section cleared of debris. The rotary plow had begun this mission from the Rico yard, only a short distance downgrade from this point. This action shot along the RGS was taken during the late 1920's.

CHARLES J. LEE PHOTOS – COURTESY OF DAVID C. BUTLER

CHARLES J. LEE, an avid photographer, was living in Rico when he took these splendid action photographs of RGS snow-clearing operations, after the snowslide on Nigger Baby Hill had run. RGS rotary snowplow No. 2 had advanced up the grade, a short distance north of Rico. The plow was cutting through this infamous slide and was shooting snow off the grade into the Dolores River. The spinning rotary blades were turning at full speed, as the snow was ejected from the plow's chute.

THIS ROTARY PLOW VIEW was shot after the snowfighting train had cleared the probed section of track, and had come to a stop. At least three locomotives were lined-up behind the rotary plow to provide the motive power for this big machine. One telegraph pole had been knocked down by the avalanche, and one at the far left was leaning toward the buried right-of-way. Gabe Bradford of California came up with the following information from an RGS telegram issued ca. 1932, which was published in "The Galloping Goose" bulletin, Volume IV, Number 3: "Forest White to Section Foreman Paul at Rico—'When you start East from Rico—do not under any circumstances take any runs at hard snow or slides when power can not keep Rotary going. Use laborers to break down the face to where engs. can shove Rotary—prospect all slides carefully for rock and timber. Think if you can get a cable you could probably pull timber from slides faster than cutting it up.'"

CHARLES J. LEE PHOTO – COURTESY OF DAVID C. BUTLER

WHAT COULD BE FINER? Or... *"Ain't love grand?"* On a walk up the track out of Rico, on an early spring day, this couple seemed not to mind the snow! Anyhow, it was a pleasant way to spend a day walking along the RGS toward Burns.

CHARLES MILLER ENGEL COLLECTION – COURTESY OF RACHEL A. HARDWICK

THIS SNOWY SCENE appears to be at RGS Bridge 66-A, which had been knocked out of action by a snowslide that had run down Nigger Baby Hill. The tallest bents of this trestle were only 18 feet in height, and there were four red-spruce piles to each bent. The constant winter snowslides at this location convinced the RGS management to replace this bridge with a fill. Bridge 66-A was located in the vicinity of the Grand View Mining & Smelting Company's operation at the northern edge of Rico. It is assumed that this picture was taken in 1912.

BRIDGE-66-A

BRIDGE DRAWN TO ONE-HALF LENGTH
COMPARED WITH HEIGHT

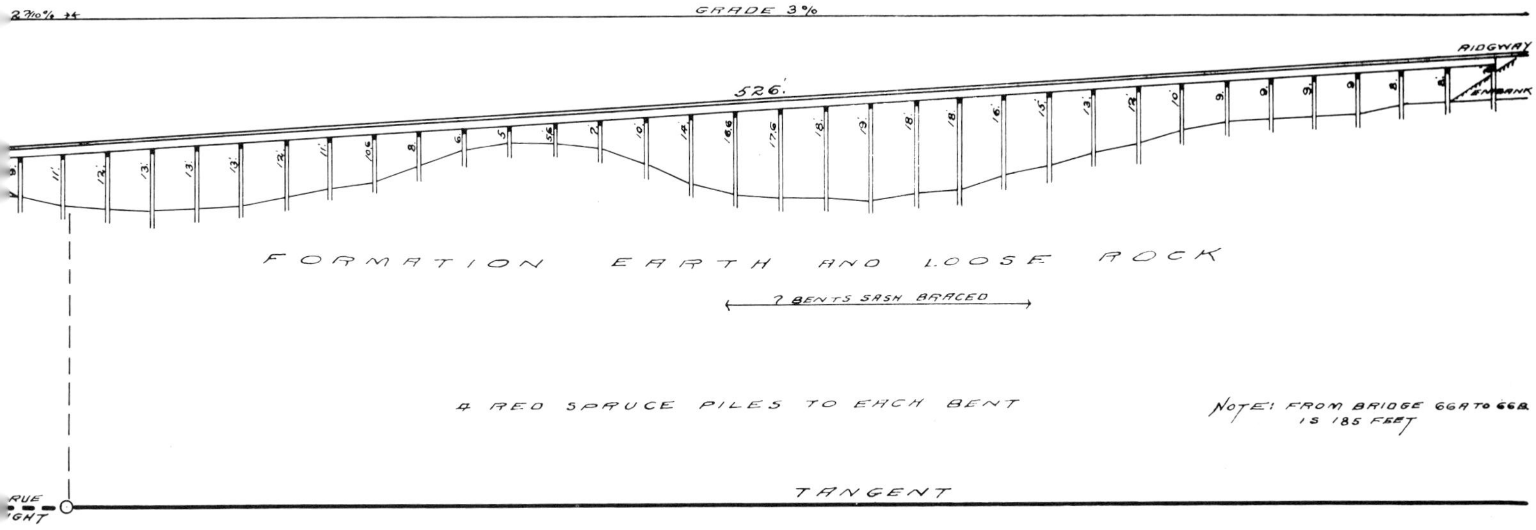

Rio Grande Southern Railroad Company

BRIDGE-66-B

BRIDGE DRAWN TO ONE-HALF LENGTH
COMPARED WITH HEIGHT

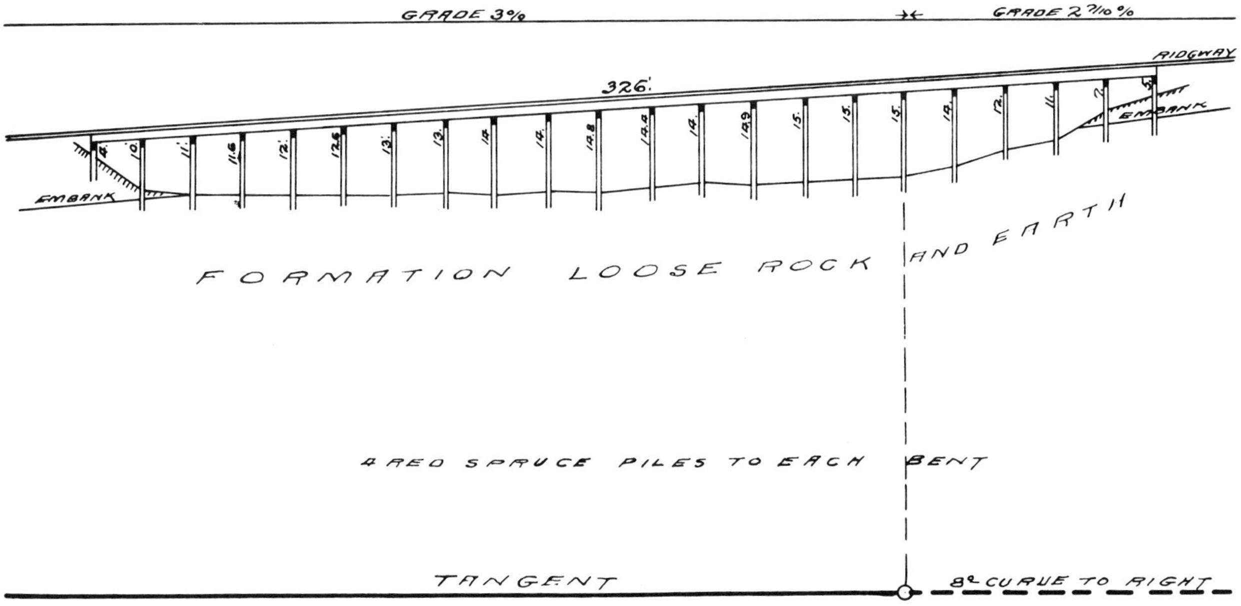

CHARLES MILLER ENGEL PHOTO – COURTESY RACHEL A. HARDWICK

BRIDGE 66-C is illustrated in this view, showing it as it crossed Silver Creek, at the lower right. This span was built as a two-panel pile trestle, and it was used until abandonment in 1951. The Congregational Church and the Dolores County Courthouse were built during the same year that the RGS arrived in Rico, 1891. The Rico schoolhouse, at the left, was built in 1892. The Atlantic Cable Mine was located next to Glasgow Avenue, adjacent to Silver Creek. Study the creek-bank deterioration in this scene from the flood of 1932, and compare it with the picture on Page 272, taken 30 years previously.

BRIDGE 66-C—a 32-foot-long pile trestle—spanned Silver Creek a short distance northeast of the Rico yard. It was on an 8.0-degree curve, located on a 1.18-percent grade. This bridge was renewed in 1909.

BRIDGE-66-C

SILVER CREEK.

BRIDGE DRAWN TO ONE-HALF LENGTH COMPARED WITH HEIGHT

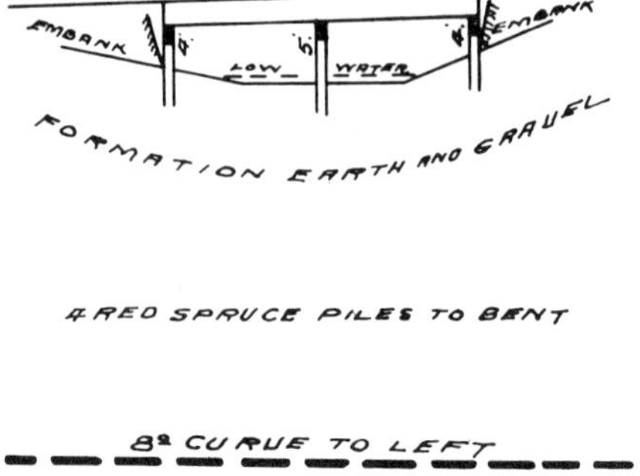

Along the Rio Dolores...

Rico's Trail Blazers

Proud Ute Indians, Early Prospectors and the Beginning of Rico

By Russ Collman

No. **4**

RICO has a history that goes back beyond the establishment of the famous silver-mining town of Aspen, and one that goes back nearly as far as that of the well-known "Cloud City" of Leadville. The mines above Rico have produced over $35 million (more like nearly $1 billion in today's inflated dollars) in precious minerals since the first prospectors arrived—such as gold and silver—as well as more mundane metals: copper, lead, zinc and iron pyrites. However, Rico is in a far more remote region than Aspen or Leadville.

In the early years, long before the arrival of the Rio Grande Southern, there was only one principal way in and out of Rico, the poor trail over the mountains to Rockwood. And since the demise of the RGS in 1951, this is true again—with Colorado Highway 145 having the distinction today. In the past, this highway was poorly maintained, and during the winter months, snow often closed this rough route over Lizard Head Pass. The isolation of Rico accounts for this mining camp's lack of fame and notoriety.

Rich mineral deposits were mined here as early as July of 1869. Two mining prospectors, Sheldon Shafer and Joseph Fearheiler, headed north from Santa Fe, New Mexico, during 1869. After they reached the so-called "Big Bend" of the Dolores River, these two men pushed on north and were impressed with indications of precious metals at Silver Creek. As a result of their findings, the Pioneer mining claim was established—the first one in the district—which covered parts of what are known today as the Shamrock, Smuggler and Riverside claims. These men also made a claim they called the "Nigger Baby," because of the abundance of black oxide of manganese at this site. Rico residents said that the men working underground at this mine came off their shifts looking like coal miners. This unusual designation was later given to part of the high peak northeast of Rico (Telescope Mountain). Nevertheless, this unusual name remains a part of Rico's colorful past. (Refer to the U.S. Geological Survey map in the first part of this volume.)

Joseph Fearheiler was killed by Ute Indians in 1870, as he was traveling toward New Mexico. Several prospectors had visited the site of Rico during 1870, but no one ventured back to this early-day mining camp on the Dolores River during the following year, presumably because of Fearheiler's untimely death.

One of the men who visited Shafer's and Fearheiler's mining claim in 1870 was R. C. Darling, who had been hired to survey the boundaries of the Ute Indian reservation at this time. He located some mining claims in the area, and then he went on with his surveying work. His name was later given to Darling Ridge, a promontory of Expectation Mountain, west of Rico.

The Ute Indians considered all of the mountain country of Colorado their homeland. They had lived in this region for hundreds of years, and they rightly believed white men were trespassers. The Treaty of 1868 set aside most of Colorado west of the high mountains for their use "forever." However, "forever" lasted only a few years. Gold and silver was discovered in the Ute Reservation in the San Juan Region, and the prospectors decided that the Utes had no need for these rich mineral deposits! So, they should be forced to move elsewhere, allowing the prospectors to acquire wealth without being menaced by "ignorant savages," as they called the native Americans.

After clashes between Indians and prospectors, Felix Brunot, chairman of the so-called "Peace Commission," was sent west from Washington, D.C., to work out a treaty that would force the Utes to give up about four-million acres in the San Juans. The resulting "Brunot Treaty" of 1873 removed the Utes from the mining region in southwestern Colorado, and a new location was established for the Southern Ute Indian Agency about 1875. However, Mr. Brunot had to call on Otto Mears to intercede with Chief Ouray before the Utes would give up their land. The Utes trusted Otto, having dealt with him before, and he was able to settle things inside a Ute lodge, sitting on a buffalo robe with the Ute leaders. Amazingly, the resulting settlement paid the Utes a mere 12 cents per acre! However, the annual payments going to the Indians were increased, and Chief Ouray was paid a salary of $1,000 per year.

During 1872—before the new treaty went into effect—R. C. Darling led a large group of men from Santa Fe to the Rico mining camp, financed by U.S. Army officers and wealthy individuals from back East. At this time, some mining work was done on diggings that were to become the Atlantic Cable, Aztec, Phoenix and Yellow Jacket mining claims. Although not much of value was produced during the 1872 "expedition," prospecting was resumed again in the Rico mining district in 1877. And by 1878, the little mining camp became quite active, particularly after Sandy Campbell, John Glasgow and David Swickhimer and others arrived.

All of this activity resulted in the location of several valuable claims, including the Atlantic Cable, Aztec, Black Hawk, Columbia, Cross, Grand View, Major, Phoenix and Yellow Jacket mines. Then, during the spring of 1879, rich silver ore was discovered on Nigger Baby Hill, which resulted in a "rush" to the Pioneer Mining District—before the mining camp even had an official name.

Before 1879, Rico was commonly known as "Carbonate City" (or the "Carbonate Camp"), and several other names were suggested for this booming camp on the Dolores River. As a compromise, a man by the name of William Weston suggested the Spanish name "Rico." Of course, the obvious reason was the early-day prospectors expected to find rich mining lodes in this section of the San Juan Region. So... "Rico" it was by the summer of 1879.

By August of 1879, Rico had a real mining boom underway. And by August 21, 1879, the first issue of **The Dolores News** was published—originally printed on the press of **The Silverton Miner** and carried over the mountains to Rico. During the same month, the first two-story house was constructed in Rico, erected by Charles A. Mantz, the first editor of the **News**.

The newspapers of Rico had considerable

influence on the town and the mining district. **The Dolores News** was the first newspaper in Rico, established in 1879, as stated above. However, other weekly newspapers followed in the footsteps of the **News**. In this volume we have chosen stories primarily from **The Rico Democrat** and its successor, **The Rico Sun**. These historic articles have come from the collection of Larry Pleasant of Dolores, who preserved these issues for the enjoyment of himself and others. (Successive owners renamed the Rico newspaper.)

During September of 1879, the first freight was hauled into Rico by wagons, bringing machinery and supplies for a sawmill. This was established about three miles north of the little mining town, near the site of what later became the siding at Burns.

In January of 1880, J. F. Wannemaker surveyed Rico and platted the streets and avenues. At this time, Rico consisted of 320 acres, a mile long (north to south) and a half-mile wide, on a terrace above the Dolores River. Later in the year, a citizen's militia was organized, established primarily to protect local residents against attacks from Ute Indians. (This group became part of the Colorado National Guard at a later date.)

On March 13 of 1880, the first president of the Grand View Mining & Smelting Company arrived in Rico. And during April of this year, the construction of the Grand View Smelter was started. Since the Denver & Rio Grande Railway had only reached Alamosa by this time, machinery for the smelter had to be hauled by wagon over the San Juan Mountains to Pagosa Springs. From this point, the wagons traveled to Animas City (just north of what became Durango), the western terminus of the D&RG line from Alamosa. The freight wagons then went west to Mancos, and then more-or-less northward over what is known as Bear Creek Hill. They were then lowered by heavy ropes to the Dolores River Valley at a site about 15 miles south of Rico. Thence, the poor struggling teams hauled their loads up along what was then known as the Rio Dolores to Rico. Good weather allowed the Grand View Smelter to be "blown in" on November 7, 1880, and this facility began to produce bullion.

Until this time, the Pioneer Mining District, Rico and the surrounding area were part of Ouray County—and a trip to Ouray must have been a very rough ride to endure by horseback or in a bouncing stagecoach. However, early in 1881, the Colorado General Assembly established Dolores County, and Rico became the county seat.

The hopes of the district's miners became rather soured during the last part of the 1880's. However, during the fall of 1887, David Swickhimer struck a rich vein of ore in the Enterprise Mine on Newman Hill (part of Dolores Mountain, east of Rico). This proved to be the turning point for Rico, and several other miners resumed work at other properties in the Rico district.

The amazing story of David Swickhimer is told elsewhere in this volume. However, it should be stated here that when the Rico State Bank was opened on February 15, 1890, he was named the bank's president. Following that, the First National Bank was opened on June 16, 1890. By the time Mr. Swickhimer sold his interest in the Enterprise Group of mining properties in 1891, a large amount of ore had been piled on the mine's dump, waiting for the arrival of the long-awaited Rio Grande Southern Railroad.

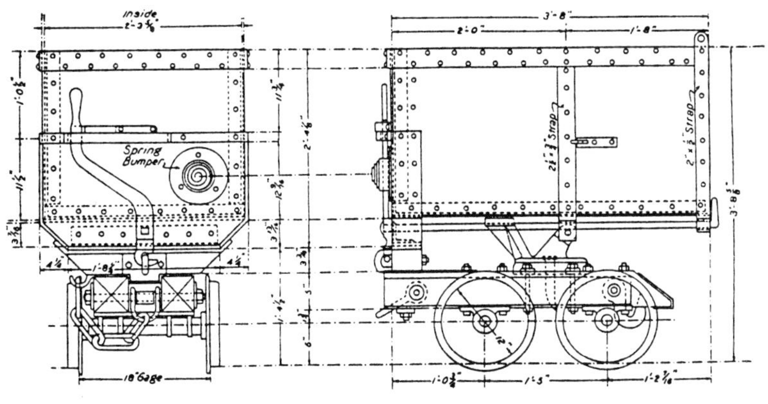

DOLORES NEWS.

VOLUME 7. RICO, COLORADO, SATURDAY, JANUARY 2, 1886. NUMBER 332.

The Dolores News.

CHAS. A. JONES,
Editor and Publisher.

RICO, COLORADO.

TERMS IN ADVANCE:

One copy one year.......................$3 00
One copy six months.......................1 75
One copy three months.......................1 00
SAMPLE COPIES TEN CENTS.

The Oldest Paper in the Dolores Country.

OFFICIAL DIRECTORY.

United States Officials.

Senators.............{ T. M. Bowen, H. M. Teller.
Member of Congress....George G. Symes,
District Judge..............Moses Hallett,
Marshal....................Walter Smith,
Clerk of Court......Edward F. Bishop,
Attorney..............E. L. Johnston.

State Officials.

Governor.......Benjamin H. Eaton,
Lieutenant-Governor...Peter W. Breene,
Secretary of State.....Melvin Edwards,
Auditor..............Hiram A. Spruance,
Treasurer...........George R Swallow,
Supt. Public Instruction...L. S. Cornell,
Attorney-General.....Theo. H. Thomas,
Adjutant-General......Frank A. Taylor,
Private Sec'y. to Gov.....Jos. T. Boyd.

SUPREME COURT.

Chief Justice..............J. C. Helm,
Associate Justices....{ Wilbur F. Stone, Wm. E. Beck.
Clerk..................Keves Danforth.

County Officials.

Sheriff..............David Swickhimer,
Clerk and Recorder.........W. L. Hull,
Treasurer...................A. H. Mundee,
Judge......................Harry C. Wyman,
School Superintendent...Helen M. Roys,
Assessor......................H. Dunton,
Surveyor...........J. F. Wannemaker,
Coroner..............Jos. N. Broughton,
Commissioners.........{ N. J. Bradley, W. G. Barnett, John Eder.
Justices of the Peace...{ F. C. Loring, J. P. Norton.
Constables.............{ Rob't. G. Brown, Dave Selby.

Town Officials.

Mayor.......................William J. Cox,
Trustees..........{ Jos. N. Broughton, Joseph Meredith, W. H. Doyle, Sr. F. Wakeman, D. A. Holmes.
Clerk.................George O. Gilbert,
Treasurer.....................A. H. Mundee,
Night Watchman.........George Hayes.

Postoffice Directory.

SOUTHERN AND EASTERN MAIL.

Arrives...6 p. m. | Departs.....7 a. m.

OURAY MAIL.

DEPARTS: Tues., Thurs., Sat. | ARRIVES: Mon., Wed., Friday.

Postoffice open from 8 a. m. to 8 p. m. Sundays from 10 to 12 a. m.

REGISTRY AND MONEY ORDERS.

Registry and money order windows open from 8 a. m. to 5. p. m.

Mail going south and east closes at 6.45 a. m. F. L. THOMPSON, P. M.

Lodge Directory.

I. O. O. F.

SILVER CRESCENT LODGE NO. 40 Holds its regular meetings at the hall on Wednesday evenings at 8 o'clock. Visiting brothers are cordially invited to attend.

D. R. CLAY, N. G.,
J. T. BEALE, V. G.
D. HAYDON, R. S.

Organizations.

JOCHMUS GUARDS.

T. H. Wagensler..............Captain.

RICO FIRE COMPANY.

Andy B. Brydon..............Foreman.

RICO CORNET BAND.

W. H. M. Bangs..............Leader.

PASADENA CORNET BAND.

Harry Iler....................Leader.

Grand Army of the Republic

Rico Post No. 63.

Meetings Second and Fourth Saturday evenings of each month

—AT—

G. A. R. HALL.

W. H. Bean, Com'd'r.
J. M. Dodd, Adjutant.

Court Sessions.

U. S. Circuit Court.—District of Colorado, Western Division at Del Norte, first Tuesday in September.

U. S. District Court.—District of Colorado, Western Division at Del Norte first Tuesday in September.

District Court, Sixth Judicial District, Charles D. Hays, Judge: Sessions, second Tuesday in May and fourth Tuesday in September of each year.

County Court.—First Monday in March, June, September and December.

County Court for Probate business, last Monday in each month.

THE

Enterprise Saw Mills,

Have a large capacity, and are now operating one Steam mill and one Water mill, cutting a great amount of Lumber. Etc.

—:o:—

All Kinds of Native Lumber,
Shingles, Lath,
Sawed Timbers, Pickets, &c.,
FURNISHED ON SHORT NOTICE.

Enterprise Saw Mills

GEO. S. BARLOW, Proprietor.

THE ST. JAMES, of Denver.

Under the management of Col. David A. Gage, this House became the popular and universally recognized home of all San Juaners, while in Denver.

THE PRESENT MANAGEMENT

Will Sustain This Popularity.

CENTRALLY LOCATED; EASY OF ACCESS;
OMNIBUSSES TO AND FROM ALL TRAINS;
RATES FULLY IN ACCORD WITH THE TIMES.

Mrs. A. H. Estes, Prop.

COLLINS & CO., HARTFORD.

Established 1826.

MAKERS OF

Picks, Hammers, Sledges, Edged Tools, &c.,

OF EXTRA QUALITY.

ANY DESIRED PATTERN OR SIZE MADE TO ORDER.

Every Tool Warranted.

For Illustrated Catalogue and Prices write to

C. A. ROBERTS & CO., Denver, Col.
ALVA ADAMS, Alamosa, "
ADAMS & BAYLEY, Durango, "
Or, COLLINS & CO., 212 Water Street, New York.

A. T. & S. F.

The Atchison, Topeka & Santa Fe R'y.

The Popular Southern Line.

FROM THE MISSOURI RIVER TO THE PACIFIC OCEAN. THE GREAT EST AND MOST LIBERAL CORPORATION ON THE AMERICAN CONTINENT, AND THE BEST MANAGED AND EQUIPPED ROAD ON EARTH. THE MAIN LINE

From Denver, Leadville, and all points in Southern Colorado, to Atchison, Kansas City, and all Eastern cities. All passenger trains equipped with Air Brakes, Miller Platforms, and all the modern improvements.

—Pullman cars on all trains between Pueblo and the Missouri River. The Only Line via Colorado Springs and Manitou. Through tickets on sale at all Principal Stations. Rates always as low as by other Lines. Baggage checked to destination.

W. F. WHITE, General Passenger Agent, Topeka, Kansas.

THE CAMP & CO. ORGAN

THE BEST!

For the MONEY to be found in the Market. VERY LOW PRICES. Send for ILLUSTRATED CIRCULAR, PRICES and TERMS. AGENTS can make Money handling these Organs.

ESTEY & CAMP,
188 & 190 State St
CHICAGO.

Protect Your Property. Save Insurance!!

Wrought Iron Fences, Cresting and Finials, Window Guards, Jail Work, Weather Vanes, Stable Fixtures, Wire Signs, Wire Cloth, Bank and Office Railing, Fire Escapes, and all kinds of Builders Iron Work, also Flower Stands and Lawn Furniture.

Manufactured by

E. T. BARNUM Wire and Iron Works,

Chicago, Ill.] DETROIT, MICH. [Windsor Ont.

ESTIMATES PROMPTLY FURNISHED.

Liberal commissions paid to any one who will give us prompt and reliable information that will secure us New Jobs of Iron or Wire Work, such as Court House and Cemetery Fences, Jail Work, Fire Escapes, and all kinds of Builders' Iron Work on public or private buildings.

Correspondence Solicited.

Address all correspondence to

E. T. BARNUM WIRE & IRON WORKS, Detroit, Mich

E. E. BURLINGAME'S

ASSAY OFFICE and CHEMICAL LABORATORY.

ESTABLISHED 1866.

All samples by mail or express will receive prompt and careful attention. Gold and silver bullion refined, melted and assayed or purchased. Write for terms. 446 Lawrence Street, Denver, Colo.

Markham Hotel { AMERICAN PLAN.

17th & Lawrence, DENVER, COLO.

Centrally located, New Management fire escapes, modern conveniences.

Rates—$2 to $4 per day.

Markham, Patterson & Thomas, Props.

Geo. R. Fisher, Manager.

SETS BEST TABLE IN THE WEST

Memphis Route South.

KANSAS CITY TO MEMPHIS! Without Change of Cars!

The attention of the people of the west and north-west is called to the Memphis extension of the Kansas City, Ft. Scott & Gulf Railroad, now completed and in operation between Kansas City and Memphis, Tenn. The completion of this short line to the south brings that heretofore to the West practically inaccessible section into close relation with the west and north-west to the very great advantage of all.

Through trains for Memphis with Pullman Palace sleeping car and elegant day coaches leave Kansas City daily, saving frequent changes and many hours time to Memphis, Jacksonville, Fla., New Orleans, and all southern cities. Tourist's tickets via this short route to Jacksonville, Mobile, New Orleans, and all the winter pleasure resorts of the South, will be on sale at all coupon offices throughout the West. A map of this new route has just been prepared and will be mailed free on application to J. E. Lockwood, G. P. & T. A., Kansas City, Mo.

The Ladies' Medical Adviser.

A Complete Medical Work for Women, handsomely bound in cloth and illustrated. Tells how to cure various diseases of the sex, with treatment at home. Worth its weight in gold. Over 10,000 sold now. Postpaid, only 50 cents. Postal note or 2 cent stamps. Address NUNDA PUBLISHING CO., Nunda, N. Y.

J. F. Wannemaker,

CIVIL AND MINING ENGINEER

U. S. Dep. Min. Surveyor.

UNDERGROUND SURVEYS MADE AND PLANS FURNISHED,

—:o:—

MINING PROPERTIES Reported Upon.

Hill Floersch,

—Proprietor of the finest—

Bath and Barber Shop In Rico.

The great rush to this popular "Emporium" has compelled the addition of a reading room, for the accommodation of such patrons as may occasionally have to wait a few moments for their turn. This is always supplied with a fresh assortment of pictorial and current literature.

A Well-Established Reputation for

Neatness, Promptness and Dispatch

Will always be merited and sustained by this shop.

DAVE SWICKHIMER,

DEALER IN CHOICE BRANDS OF

WINES, LIQUORS AND CIGARS.

RICO, COLORADO.

Plain and Fancy Drinks from Strictly First Class Goods.

BILLIARDS AND POOL.

YOU ALL KNOW THE PLACE. EVERYBODY KNOWS "SWICK."

HOME HISTORY.

The Written and Unwritten History of the Dolores, from 1869 to 1886.

Early Discoveries by Prospectors From Santa Fe in 1869.—Fort Defiance Officers and Others work the Atlantic Cable, "Dolores" and Poison Tunnels 17 Years Ago.—Crude Attempts at Smelting.—Many Subsequent Years of Idleness.—Important Events of Each Year, Socially, Business and Chronological.

Yesterday ushered in the year 1886 and in the morning of the new year, which promises such great things for Rico, its citizens will take an interest in turning back and hastily reviewing the events of the preceding few years, during which the region that but a short time ago was a wilderness has been converted into a prosperous mining district, with Rico as its centre. This review is simply a brief statement of occurrences as they have happened, and it is intended to chronicle every important or interesting event which has occurred. The work has been laborious and hurried and there may be errors and omissions. We hope and believe they are few.

To "begin at the beginning," in 1869, the following party started out from Santa Fé on a prospecting trip: Wm. Hill, Jack Eccles,—Miller, Joe Fearheller, Tom Sager, "Tinker" Brown, and "Pony" Whitmore. In July of that year they penetrated into this region, then the Ute Reservation, coming by way of La Plata, Baker's Park (now Silverton) and the Trout Lakes. In August they went south to Fort Defiance, Arizona, and reported the result of their discoveries. E. N. Darling and party, a government surveying outfit, had just reached Defiance from the east and became interested in their accounts from far-off Dolores. A company of fifteen was formed, including Capt. Bennett, Lieut. Ford, R. C. Darling and J. Carey French, and seven of the company came to where Rico now is and spent the winter of 1869-70 in working upon the Atlantic Cable, the Dolores (now the Aztec) and the "poison tunnels," just across the river. As a matter of incidental interest we may state that during that winter a cow and calf, horse and burro, wintered upon the slopes of Nigger Baby Hill and came out fat in the spring. They had nothing to eat but the natural grasses to be nipped from the hill-side.

In May 1870, R. C. Darling and Falstina Caldrene, a Mexican, put up a Mexican adobe furnace just across Silver creek and endeavored to treat by the Mexican process the ores from the Atlantic Cable and the Dolores. They melted the ore, but the furnace would always fall in at the critical moment. The remains of this adobe concern are still quite plainly to be seen a few steps beyond the Silver creek bridge. Late in the summer the whole outfit left the place except Joe Fearheller and Sheldon Shafer, the latter being a cousin of Mrs. J. P. Landon, now of Rico.

From this time until 1877, occasional wandering prospectors spent a few days here, but in the year named John Glasgow, Bob Snyder, Ed Robinson and others came and remained. In 1878, several prospected with good results and in the spring of 1879, quite a little boom was started. The purchase by Senator Jones and important discoveries attracted many people and our chronicle will therefore begin with the summer of

1879.

July—17, The miners hold a meeting to draft local regulations, with R. C. Darling as president.

August.—Early in this month the owners of mill-sites covering the ground where Rico now stands concluded to survey their mill sites into lots. These owners were Humaston and Campbell, Brown and Buchanan (who sold to Stuart and Kingsmill, of Silverton,) Doyle and Elliott and E. A. Robinson. In addition to this was the claim of R. C. Darling to 160 acres, his cabin standing just east of the road where it now crosses Silver creek. These lots were to be free to actual settlers, the mill-site people holding for themselves every alternate lot. The rush to the "carbonate camp" has now fairly begun and neighboring camps begin to send delegations to prospect....3, The name of Rico is selected....4, A petition is forwarded to Postmaster General asking that a postal route be established from Silverton via Howard Fork (Ophir).... 11, Senator John P. Jones and party arrive from Nevada and visit several mines ....13, This afternoon Senator Jones and others purchased from Sandy Campbell and Chas. Humaston the Grand View and other lodes for $60,000, and from John Glasgow the Yellow Jacket and Phœnix lodes. Consideration, secret, on account of litigation....15, Finds a large number of people in the town, flushed with excitement and hunting "carbonates"....20, A meeting for purpose of effecting a town organization is held; trustees chosen—M. Winfield, James ("Tendye") Brown, A. C. Brainard, Jos. Castello and D. Crowley....21, DOLORES NEWS appears, issued from *Miner* office at Silverton....31, "Kid" McGoldrick kills "Frenchy," being Rico's first murder.

September.—1, A count shows 105 new log cabins in Rico; Aug. 1 there were none....The first wagons enter Rico containing machinery for McJunkin's saw mill, coming around from Silverton via the Mancos and Bear Creek Hill....2, Rico's first trial commences, the prisoner being the "Kid." Col. Chas. A. Mantz prosecutes; "Beaver Aleck" and A. O. Terry for the defense, before Justices Ross, of Ophir, and Baldwin, of Rico ....20, The first mail arrives....23, Wm. G. Morrow, a saloon keeper, accidentally shoots and kills himself. Chas. A. Mantz, then editor of the NEWS, delivers a funeral oration....24, The Indians hear of Rico and send a delegation to find out what all the fuss is about....The "Dolores Jockey Club" commences a series of races....28, Rev. H. P. Roberts, of Silverton, preaches the first sermon in Rico. It was an out doors affair—the collection not being omitted. Immediately after the sermon a large purse was made up for a horse race, and all adjourned to the race track after the benediction.

October.—7, Rico's first election is held ....26, An Episcopal church organization is formed.

November.—5, A meeting of citizens is held to consider the question of squatter sovereignty, owing to great trouble from lot jumpers, and adopted resolutions compelling owners to erect cabins in order to hold lots....16, New mail route in force—compare the route with present facilities—from Alamosa, via Silverton, Animas City and Parrott City; tri weekly ....27, Rico enjoys her first ball on this Thanksgiving in Day, "Theo. Barlow's large store house." The ladies present were "Mrs. Walter Higgin, Mrs. Alderman and daughter, Miss Collins, Mrs. Hubbard, Mrs. Snyder and daughter and Mrs. Robison."

December.—4, Another meeting to discuss lot question. Resolutions by L. Alderman, A. S. Goodrich, John Gault, J. M. Ohlhausen and A. A. Waggoner.... 22, First election (special) resulting in the election of Frank Raymond, mayor, over E. P. Kent by a few votes. D. A. McGraw, recorder; trustees—H. E. Snyder, M. A. Bean, E. P. Kent and M. C. Marston.... 25, Christmas ball at H. E. Snyder's Pioneer Hotel.

1880.

January.—1, Whiskey and Tobacco famine—Dilute alcohol and boot-heels in great demand....10, A number of Rico boys, Pat Cain, Sam Gray, Lige Ledford, Joe Charest, Waggoner and Dave Swickhimer, start for the Arizona excitement ....17, A Literary Society is formed with L. Alderman, C. Enderich, A. K. Prescott, D. Hayden, E. C. Miller, J. W. Park, A. O. Terry, J. W. Randall, Frank Hartman, A. L. Raplee, G. W. Spencer and M. C. Marston as members.

February.—2, First school, Miss Alice A. Snyder (now Mrs, Chas. Middaugh) teacher....15, Staple articles of food begin to vanish from stores and a bread famine becomes imminent....18, O. T. Tyler, a mail carrier coming from Parrott City to Rico, is frozen to death.

March.—7, A party starts to open up a trail to get beef and other luxuries to the hungry multitude....News arrives of the killing by Indians of Merrick and Mitchell in the Navajo country.

April.—10, John Foote's pack train arrives with several thousand pounds of merchandise; sells flour from his jacks at $35 a hundred....The result of the spring election was another victory for Raymond over Kent for mayor and for D. McGraw over J. W. Lamon for clerk; trustees elected were Jos. Costello, W. L. Hull, Amos Lane and L. Alderman....17, J. Dorsey Haines has his leg broken by a falling rock in Marguerite Basin....30, John Curry sells NEWS to Jones & Hartman.

May.—2, A regularly organized barber shop puts in an appearance and "cleanly shaved faces on our streets" is chronicled ....10, The spring boom is fairly on—everybody has money, and real estate deals are very active at fancy figures.... The trail to Ouray is shoveled out at large expense, and some provisions arrive over that route....17, Puzzle cabin burned ....24, Piedmont, the now flourishing city of the west side, effects a town organization, with Charlie Forsberg, President and Joe Lacome, secretary....26, John Bissell, with his safe and banking fixtures endeavors to ford the Dolores and drowns two mules.

June.—3, Al McKenzie and J. G. Goodrich, on their way to Rico are fired on by a party of settlers in "The Park;" one horse killed, and narrow escapes all round ....4, Daily mail ordered by postal department from Ouray to Rico, to commence June 14....15, Theo Belle starts a stage and express line from Animas City to foot of Hermosa range, with saddle service into Rico....16, Machinery for Grand View smelter being "snubbed down" Bear creek hill....17, A party by the name of Todd, from Animas, shoots off his own finger, while drunk....A pack train brings in light machinery for Elgin smelter....John Foote's dance hall has a "grand opening," with a thousand visitors....Skeleton found on West Dolores, with pick by its side....18, Daily mail from Cascade.

July.—4, About 11, a. m. a long string of wagons bring in the machinery for the Grand View smelter....5, About one o'clock in the morning, John B. Grigsby was shot and instantly killed in the Bon Ton saloon by Jim Neumeyer; the latter received a severe wound in the shoulder ....13, Forsberg & Downs receive the first billiard table ever brought into Rico ....14, "Big Jim Dunnigan," the 6 foot, 10½ inch giant, who lately left Rico, kills a fellow policeman and on the 19th was lynched by fifty Mexicans....20, Dr. Landon breaks one of his ankle bones.... Wm. Bell breaks a rib by being thrown from a horse....24, Mabel King is severely kicked by a fractious horse ....John B. Geisel, a prospector, packs his burro and starts off for Mount Wilson; and neither man nor animal have ever since been heard of....25, Richard Gentry resigns his position as manager of the Grand View and C. D. Porter, of the Black Hills, is appointed....John Hilden robs Mrs. Gus Talbot, and escapes; is captured later by L. Alderman, constable ....The Miners' Union is formed....28, "Kid" McGoldrick is sentenced at Ouray to 6 years in the pen. for the murder of "Frenchy" at Rico....The grand jury at Ouray returns an indictment for murder against Jim Neumeyer, and Deputy Sheriff Sullivan comes after him.

[The disappearance of John Geisel in this month vividly recalls a poem by a Colorado author, published several years ago. It so clearly indicates the spirit which impels the prospector to take great personal risks, and shows their not infrequent fate, in such beautiful verse, that we reproduce it here in connection with the disappearance of the Rico prospector and dedicate it to his memory:]

THE PROSPECTOR.

"The night is near, the trail is dim,
With chapparal overgrown;
Till morning stay, go not that way
Alone."

"Day after day, year after year,
I'm like a dry leaf blown
By the shifting gales, o'er hills and vales
Alone.

"And wintry storms have o'er me lower'd;
And snowflakes round me flown;
As night I lie beneath the sky
Alone.

"And summer days have seen me oft
Where the sun has fiercest shone—
O'er burning sands, in desert lands,
Alone.

"Far in the mountains gleams a light
That guides me on and on;
I follow its gleam, through forest and stream,
Alone.

"'Tis Fortune's beacon leads me on,
O'er yonder bluff it shone;
Ere falls the night I scale the height,
Alone."

* * * * * * * *

Over the bluff in a deep ravine,
With bushes all o'ergrown,
Lie a pick and a pan, and the bones of a man,
Unknown.

August.—10, The Johnny Bull strike is made....11, Some miscreant cuts down the Garfield and Arthur stringer across Glasgow avenue....15, Parson Hoge establishes "St. Luke's Chapel," corner of Glasgow and Campbell....16, Bank of Rico commences business with a grand flourish....21, C. D. Porter resigns as Grand View manager and F. M. Ennlicli is appointed—Hugo Arnolds, as metallurgist, completes as fine a combination as ever used champagne and beer for flux ....22, The Odd Fellows hold a preliminary meeting and effect an organization as "Silver Star Lodge" (now Silver Crescent); E. B. Cushing, Chas. Munn and Jacob Summa are named as officers....28, Chief Ouray died at Southern Ute Agency; of Bright's disease of the Kidneys.

September.—4, The office of J. W. Park is riddled with bullets, the attorney being inside; supposed object, to frighten him into giving up papers and books in a certain administration suit....10, H. J. Caldwell had a fine horse stolen from him; while in Silverton....16, The Grand View smelter blows its whistle for the first time ....The town of Durango is located, two miles below Animas, four blocks being surveyed....W. H. Bennetts made foreman of Grand View and Rico mining companies....19, Jim Belford addresses an audience; H. M. Teller did not arrive in time to deliver an address; Tom Patterson and John S. Hough deliver addresses on the evening of the 24th:

October.—1, Barlow and Sanderson put on stage line from Animas City to the end of the D. & R. G. track....3, Young Jackson is killed by Utes on the Cimarron; near Ouray....7, Like a meteor of light O'Brien Moore enters the limits of Rico ....14, Citizens hold an indignation meeting concerning the Ute outrages....16, Division of Ouray county is agitated.... 28, C. D. Knappen shoots Pat McDonald in the abdomen at the door of the Rob Roy saloon, the ball penetrating the bladder and intestines and causing death 28 hours later....59, The body of a dead Indian is found near Big Bend....30; Knappen is secretly taken to Ouray to avoid a lynching.

November.—8, Election is held—A. A. Waggoner is elected commissioner; Worden Grigsby and Will C. Gilmer, justices; Jim Cart and Dave Swickhimer, constables; A. W. Hudson is elected representative for Ouray and San Juan counties ....9, A Mormon named Ericson suicides on the ranch of Sam and Jud Burghardt ....17, The Grand View smelter blows in for the first time, and at 11, p. m. the

Continued on next page.

Commercial

—AND ALL OTHER KINDS OF—

Job Printing

—AT THE—

Dolores News Office.

The Finest Stock and Most Experienced Workmen Employed.

Orders at home or from abroad, attended to with promptness

MARGARET MATZICK COLLECTION

THE DOLORES NEWS office at Rico was located on River Street when this view was produced in 1882. Notations on the back of this old picture give the individual's names as follows (left to right): Old Man Hartman, Charles A. Jones, Sherman Waggoner, Virginia (Jennie) Bartlett (later, Mrs. Charles A. Jones), Lillian Hartman (later, Mrs. Lillian Hartman Johnson of Chicago) and Willie, the "printer's devil" (an apprentice). Expectation Mountain is in the background.

FROM THE DOLORES NEWS
Rico, Colorado, January 2, 1886

HOME HISTORY

The Written and Unwritten History of the Dolores Region from 1869 to 1886

Early Discoveries by Prospectors From Santa Fé in 1869—Fort Defiance Officers and Others Work the Atlantic Cable, Dolores and Poison Tunnels 17 Years Ago—Crude Attempts at Smelting—Many Subsequent Years of Idleness—Important Events...

YESTERDAY ushered in the year 1886, and in the morning of the new year, which promises such great things for Rico, its citizens will take an interest in turning back and hastily reviewing the events of the preceding few years, during which the region that but a short time ago was a wilderness has been converted into a prosperous mining district, with Rico as its center. This review is simply a brief statement of occurrences as they have happened, and it is intended to chronicle every important or interesting event which has occurred. The work has been laborious and hurried, and there may be errors and omissions. We hope and believe they are few.

To "begin at the beginning," in 1869, the following party started out from Santa Fé on a prospecting trip: William Hill, Jack Eccles, a Mr. Miller, Joe Fearheller, Tom Sager, "Tinker" Brown and "Pony" Whitmore. In July of that year, they penetrated into this region, then the Ute Reservation, coming by way of La Plata, Bakers Park (now Silverton) and... Trout Lake. In August, they went south to Fort Defiance, Arizona, and reported the result of their discoveries. E. N. Darling and party, a government-surveying outfit, had just reached Defiance from the East and became interested in their accounts from far-off Dolores. A company of 15 was formed, including Capt. Bennett, Lieut. Ford, R. C. Darling and J. Carey French. And seven of the company came to where Rico now is, and [they] spent the winter of 1869–1870 in working upon the Atlantic Cable, the Dolores (now the Aztec), and the Poison tunnels, just across the river.

As a matter of incidental interest, we may state that during the winter a cow and calf, horse and burro wintered upon the slopes of Nigger Baby Hill, and came out fat in the spring. They had

COURTESY THE PLEASANT PRESS – PUBLISHED 1977

THE DOLORES NEWS, published in Rico on January 2, 1886, contained a capsule account of the previous years of history that had taken place in Rico, starting with 1869. We are greatly indebted to Larry Pleasant of THE PLEASANT PRESS in Dolores, for his efforts in saving this history, which he, in turn, now shares with all of us.

nothing to eat but the natural grasses... from the hillside.

In May, 1870, R. C. Darling and Falatina Caldrone, a Mexican, put up a Mexican adobe furnace just across Silver Creek, and endeavored to treat by the Mexican process the ores from the Atlantic Cable and the Dolores. They melted the ore, but the furnace would always fall in at the critical moment. The remains of this adobe concern are still quite plainly to be seen a few steps beyond the Silver Creek bridge. Late in the summer, the whole outfit left the place, except Joe Fearheller and Sheldon Shafer, the latter being a cousin of Mrs. J. P. Landon, now of Rico.

From this time until 1877, occasional wandering prospectors spent a few days here, but in the year named [1877], John Glasgow, Bob Snyder, Ed Robinson and others came and remained. In 1878, several prospected with good results, and in the spring of 1879, quite a little boom was started. The purchase by Senator Jones and important discoveries attracted many people, and our chronicle will therefore begin with the summer of...

1879.

JULY 17—The miners hold a meeting to draft local regulations, with R. C. Darling as president.

AUGUST—Early in this month, the owners of mill sites covering the ground where Rico now stands concluded to survey their mill sites into lots. These owners were Humaston and Campbell, Brown and Buchanan (who sold to Stuart and Kingsmill of Silverton), Doyle and Elliott and E. A. Robinson. In addition to this was the claim of R. C. Darling of 160 acres, his cabin standing just east of the road, where it now crosses Silver Creek. These lots were to be free to actual settlers, the mill-site people holding for themselves every alternate lot. The rush to the "carbonate camp" [Rico] has now fairly begun, and neighboring camps begin to send delegations to prospect....

AUGUST 3—The name of "Rico" is selected [for the town].

AUGUST 4—A petition is forwarded to the Postmaster General, asking that a postal route be established from Silverton, via Howard Fork (over Ophir Pass).

AUGUST 11—Senator John P. Jones and party arrive from Nevada and visit several mines.

AUGUST 13—This afternoon, Senator Jones and others purchased from Sandy Campbell and Charles Humaston the Grand View and other lodes for $60,000, and from John Glasgow the Yellow Jacket and Phoenix lodes. Consideration, secret, on account of litigation.

AUGUST 15—Finds a large number of people in the town, flushed with excitement and hunting "carbonates."

AUGUST 20—A meeting for the purpose of effecting a town organization is held. Trustees chosen: M. Winfield, James ("Tendye") Brown, A. C. Brainard, Joseph Castello and D. Crowley.

AUGUST 21—**The Dolores News** appears [for the first time], issued from **The Silverton Miner** office at Silverton.

AUGUST 31—"Kid" McGoldrick kills [a man by the name of] "Frenchy," being Rico's first murder.

SEPTEMBER 1—A count shows 105 new log cabins in Rico; on August 1, there were none.... The first wagons enter Rico, containing machinery for McJunkin's new mill, coming around from Silverton via the Mancos and Bear Creek Hill.

SEPTEMBER 2—Rico's first trial commences, the prisoner being the "Kid." Col. Charles A. Mantz being the prosecutor; "Beaver Aleck" and A. O. Terry for the defense, before Justices [from both Ophir and Rico].

SEPTEMBER 20—The first mail arrives [in Rico].

SEPTEMBER 23—William G. Morrow, a saloon keeper, accidentally shoots himself. Charles A. Mantz, the editor of the **News**, delivers a funeral oration.

SEPTEMBER 24—The "Dolores Jockey Club" commences a series of races.

SEPTEMBER 28—Rev. H. P. Roberts, of Silverton, preaches the first sermon in Rico. It was an outdoors affair—the collection not being omitted. Immediately after the sermon, a large purse was made up for a horse race, and all adjourned to the race track after the benediction.

OCTOBER 7—Rico's first election is held.

OCTOBER 26—An Episcopal church organization is formed.

NOVEMBER 5—A meeting of citizens is held to consider the question of squatter sovereignty, owing to great trouble from lot jumpers, and adopted resolutions compelling owners to erect cabins in order to hold lots.

NOVEMBER 16—New mail route in force—compare the route with present facilities—from Alamosa, via Silverton, Animas City and Parrott City; tri-weekly.

NOVEMBER 27—Rico enjoys her first ball on this Thanksgiving Day in "Theo. Barlow's large store house." The ladies present were Mrs. Walter Higgin, Mrs. Alderman and daughter, Miss Collins, Mrs. Hubbard, Mrs. Snyder and daughter and Mrs. Robison.

DECEMBER 4—Another meeting to discuss the lot question. Resolutions by L. Alderman, A. S. Goodrich, John Gault, J. M. Ohlhausen and A. A. Waggoner.

DECEMBER 22—First election (special), resulting in the election of Frank Raymond, mayor, over E. P. Kent by a few votes; D. A. McGraw, recorder; trustees: H. E. Snyder, M. A. Bean, E. P. Kent and M. C. Marston.

DECEMBER 25—Christmas ball at H. E. Snyder's Pioneer Hotel.

1880.

JANUARY 1—Whiskey and Tobacco famine—Dilute alcohol and boot-heels in great demand.

JANUARY 10—A number of Rico boys, Pat Cain, Sam Gray, Lige Ledford, Joe Charest, Waggoner and Dave Swickhimer start for the Arizona excitement.

JANUARY 17—A Literary Society is formed,

with L. Alderman, C. Enderich, A. K. Prescott, D. Hayden, E. C. Miller, J. W. Park, A. O. Terry, J. W. Randall, Frank Hartman, A. L. Raplee, G. W. Spencer and M. C. Marston as members.

FEBRUARY 2—First school, Miss Alice A. Snyder (now Mrs. Charles Middaugh), teacher.

FEBRUARY 18—O. T. Tyler, a mail carrier coming from Parrott City to Rico, is frozen to death.

MARCH 7—A party starts to open up a trail to get beef and other luxuries to the hungry multitude.... News arrives of the killing by Indians of Merrick and Mitchell in the Navajo country.

APRIL 10—John Foote's pack train arrives with several-thousand pounds of merchandise; sells flour from his jacks at $35 a hundred [pounds].... The result of the spring election was another victory for Raymond over Kent for mayor and for D. McDraw over J. W. Lamon for clerk; trustees elected were: Joseph Costello. W. L. Hull, Amos Lane and L. Alderman.

APRIL 17—J. Dorsey Haines has his leg broken by a falling rock in Marguerite Basin.

APRIL 30—John Curry sells **The Dolores News** to Jones & Hartman.

MAY 2—The spring boom is fairly on—everybody has money, and real estate deals are very active at fancy figures.... The trail to Ouray is shoveled-out at large expense, and some provisions arrive over that route.

MAY 17—Puzzle cabin burned.

MAY 24—Piedmont, the now-flourishing city of the west side, effects a town organization, with Charlie Forsberg, president, and Joe Lacome, secretary.

MAY 26—John Bissell, with his safe and banking fixtures, endeavors to ford the Dolores [River] and drowns two mules.

JUNE 3—Al McKenzie and J. G. Goodrich, on their way to Rico, are fired on by a party of settlers in "The Park." One horse killed and narrow escapes all around.

JUNE 4—Daily mail ordered by postal department Ouray to Rico, to commence June 14.

JUNE 15—Theo Belle starts a stagecoach-and-express line from Animas City to foot of Hermosa Range, with saddle service into Rico.

JUNE 16—Machinery for the Grand View smelter being "snubbed down" Bear Creek hill.

JUNE 17—A party by the name of Todd, from Animas City, shoots off his own finger, while drunk.... A pack train brings in light machinery for the Elgin smelter.... John Foote's dance-hall has a "grand opening," with a thousand visitors.... Skeleton found on West Dolores, with pick by its side.

JUNE 18—Daily mail from Cascade.

JULY 4—About 11:00 a.m., a long string of wagons bring in the machinery for the Grand View smelter.

JULY 5—About 1:00 a.m. John B. Grigsby was shot and instantly killed in the Bon Ton Saloon by [Policeman] Jim Dunnigan; the latter receiving a severe wound in the shoulder.

JULY 13—Forsberg & Downs receive the first billiard table ever brought into Rico.

JULY 14—"Big Jim" Dunnigan, the 6-foot, 10 3/4-inch giant, who lately left Rico, kills a fellow policeman and on the 19th, was lynched by 50 Mexicans.

JULY 20—Dr. Landon breaks one of his ankle bones.... William Bullock fractures a rib by being thrown from a horse.

JULY 24—Mabel King is severely kicked by a fractious horse.... John B. Geisel, a prospector, packs his burro and starts off for Mount Wilson; and neither man nor animal have ever been heard of since.

JULY 25—Richard Gentry resigns his position as manager of the Grand View [smelter], and C. D. Porter of the Black Hills, is appointed .... John Hilden robs Mrs. Gus Talbet, and escapes; is captured later by L. Alderman, constable.... The Miners' Union is formed.

JULY 28—"Kid" McGoldrick is sentenced at Ouray to six years in the penitentiary for the murder of "Frenchy" at Rico.... The grand jury at Ouray returns an indictment for murder against Jim Neumeyer, and Deputy Sheriff Sullivan comes after him.

AUGUST 10—The Johnny Bull strike is made.

AUGUST 11—Some miscreant cuts down the Garfield-and-Arthur stringer [banner] across Glasgow Avenue.

AUGUST 15—Parson Hoge establishes "St. Luke's Chapel," [at the] corner of Glasgow and Campbell.

AUGUST 16—Bank of Rico commences business with a grand flourish.

AUGUST 21—C. D. Porter resigns as Grand View [smelter] manager, and F. M. Ennlicli is appointed—with Hugo Arnolds as metallurgist.

AUGUST 22—The Odd Fellows hold a preliminary meeting and effect an organization as "Silver Star Lodge" (now Silver Crescent); E. B. Cushing, Charles Munn and Jacob Summa are named as officers.

AUGUST 28—Chief Ouray died at Southern Ute Agency, of Bright's disease of the kidneys.

SEPTEMBER 4—The office of J. W. Park is riddled with bullets, the attorney being inside; [the] supposed object, to frighten him into giving up papers and books in a certain administration suit.

SEPTEMBER 10—H. J. Caldwell had a fine horse stolen from him while in Silverton.

SEPTEMBER 16—The Grand View smelter blows its whistle for the first time.... The town of Durango is located, two miles below [south of] Animas City, four blocks being surveyed.... W. H. Bennetts made foreman of Grand View and Rico mining companies.

SEPTEMBER 19—Jim Belford addresses an audience; H. M. Teller did not arrive in time to deliver an address; Tom Patterson and John S. Hough deliver addresses on the evening of the 24th.

OCTOBER 1—Barlow & Sanderson put on a

stagecoach line from Animas City to the end of the [new] Denver & Rio Grande track.

OCTOBER 3—Young Jackson is killed by Utes on the Cimarron, near Ouray.

OCTOBER 7—Like a meteor of light, O'Brien Moore enters the limits of Rico.

OCTOBER 14—Citizens hold an indignation meeting concerning the Ute outrages.

OCTOBER 16—Division of Ouray County is agitated.

OCTOBER 28—C. D. Knappen shoots Pat McDonald in the abdomen at the door of the Rob Roy Saloon... causing death 28 hours later.

OCTOBER 29—The body of a dead Indian is found near Big Bend.

OCTOBER 30—Knappen is secretly taken to Ouray to avoid a lynching.

NOVEMBER 8—Election is held—A. A. Waggoner is elected commissioner; Worden Grigsby and Will C. Gilmer, justices; Jim Cart and Dave Swickhimer, constables; A. W. Hudson is elected representative for Ouray and San Juan counties.

NOVEMBER 9—A Mormon named Ericson suicides on the ranch of Sam and Jud Burghart.

NOVEMBER 17—The Grand View smelter blows in for the first time, and at 11:00 p.m. the furnace was charged and in a short time afterward, bars of bullion were piled up, marked: "Grand View, Lot No. 1."

DECEMBER 17—Rico's first and only fatal snowslide occurs at the Rob Roy mine, near the head of Horse Gulch.... Herman Kelfer, a Prussian, the victim.

DECEMBER 24—Oscar Pruett was killed at a Christmas Eve dance at the house of F. M. Hamblatt, on the San Juan, by one of the Hamblatts.

DECEMBER 25—Bert Wilkinson killed "Comanche Bill" at Durango.

1881.

JANUARY 10—The lower country has another killing. Porter Stockton is called upon by 18 men and killed, being riddled with bullets.

JANUARY 22—Representative Hudson introduced the bill asking that a county, to be called "Dolores," be cut from the territory of Ouray County.

FEBRUARY 3—O'Brien Moore, alias J. Barlow Luddy, drops the mask and leaves Rico, after swindling numerous persons.... a fatal fall of Fred Herman in the Little Pittsburg [mine] shaft—overcome by foul air, and fell from the bucket within two feet of the surface.

FEBRUARY 6—A New York gentleman commences suit against Hon. Steve Elkins, Senator John P. Jones, Ex-Senator J. B. Chaffee and D. H. Moffat, Jr., for misrepresentation and fraud in the title of the Grand View property at Rico. These gentlemen, it was claimed, had bought the property for $100,000 and had let friends in on "the ground floor," but cinched them for $400,000.

FEBRUARY 10—The county of Dolores is created by the passage of House Bill 111, which went through, as Cushing telegraphed to Julius Thompson "with a whoop." Everybody is happy over the news.

FEBRUARY 11—John W. Bailey obtains a $54,429.41 verdict against the Grand View.

FEBRUARY 17—Orlando Davis "does up" an editor by the name of Jones.

FEBRUARY 25—A company is organized to build a toll-road from the mouth of Horse Creek to the Hot Springs on the West Dolores via Horse Gulch.

FEBRUARY 28—Governor Pitkin appoints a few officers for the new county of Dolores....

MARCH 1—An armed force, under Tom Carroll, proceeds to the Johnny Bull mine to take and hold it for himself and Joel B. Erhardt. They stop the day shift from going to work, and Superintendent Hollbaugh comes to town for a force of men. Eugene B. Cushing arms 68 men and it looked as though there would be bitter war. Fabulous prices were paid for all kinds of service, and on both sides at least 125 men were employed. Officers Cart and Swickhimer took possession for the Hollbaugh side by process of law, and the battle was bloodless.

MARCH 8—Tom Greatorex killed in the Coliseum Auditorium (Durango) by Jack Roberts of Rico.

MARCH 9—Additional Dolores County appointments are: county commissioners, Sam H. Burghardt and A. K. Prescott (Prescott being postmaster, A. A. Waggoner was substituted); treasurer, Eugene B. Cushing; assessor, J. P. Norton; surveyor, F. W. Gove; coroner, A. H. Mundee.

MARCH 17—First meeting of Dolores County commissioners. S. W. McCormack is admitted as commissioner.... Jack Roberts, murderer of Greatorex, lynched on the Ute Reservation.

MARCH 19—Report that W. H. Sockrider has been murdered by the "Farmington Gang." Sockrider, however, turned up smiling.

MARCH 20—C. D. Knappen escapes from jail at Ouray, but is recaptured at Cow Creek.

MARCH 28—Militia company organized [legally, to provide for law and order].

APRIL 1—Work is begun on the Rockwood-to-Dolores wagon road, under contract to W. S. Poindexter.

APRIL 5—Spring election—J. W. Winkfield, with 177 votes, elected over J. F. Wannemaker for mayor... Charlie Middaugh, clerk; trustees, A. M. Friend, Worden Grigsby, Sam H. Burghardt and S. W. McCormack.

APRIL 10—The celebrated fight at Durango between the Farmington Gang and the Stockton–Eskridge outfit takes place.

APRIL 12—Mormon Charlie lynched in Durango for killing J. Polk Prindle.

APRIL 22—A Denver & Rio Grande [passenger] car dashed 200 feet down the precipice at Toltec Gorge—eight persons killed.

MAY 2—The Indians attack and kill John

Thurman and Dick May at a ranch on the Dolores [River], below Big Bend. They secured a large number of horses and left.

MAY 12—I. W. Lacy, the cattle king, killed at a dug-out near Fort Lewis by Dan Howland.

MAY 15—Harg Eskridge killed "Kid" White near Amargo.

MAY 23—J. B. Horten drowned in the first crossing [of the Dolores River] below town.

MAY 28—Billy May comes to Rico to raise men to go to the Blue Mountains and recover livestock stolen by the Indians.

JUNE 15—The cowboys and Rico delegation have a fight with the Indians in a spur of Little Castle Valley, in the Sierra la Sal Mountains of Utah, which resulted very disastrously for the whites. Of the Rico boys, H. H. Melvin, Jack Galloway, Wiley and Hard Tartar, James A. Heston and T. C. Taylor were killed, and Jim Hall and Harg Eskridge wounded. Dave Wills, of Piute Springs, Tom Click, of the Dolores and Wilson brothers of Grand Valley were also killed, and Jordan Bean wounded.

JUNE 21—A rescuing delegation starts for Utah on hearing partial news of the Indian fight of the 15th and 16th by a special courier.

JUNE 30—The Wabash employees present Ed Suydam with a horse.

JULY 1—Judge Burris arrives to open first term of District Court in Dolores County.

JULY 20—H. J. Hardess arrives and takes charge of the metallurgical department of the Grand View.

JULY 24—Jasper Bean kills Issac Denby, at the Beidler ranch down the river.

JULY 25—Work begins on Durango-to-Rico toll road.

JULY 31—Joseph Lovett sold the Pussie mine to H. D. Dumont for $70,000.

AUGUST 5—Durango celebrates the arrival of the Denver & Rio Grande Railway. Rico sent down a delegation to assist in the "blow out."... Another skeleton found on West Dolores by J. A. Ross.

AUGUST 24—Bert Wilkinson, Dyson Eskridge and "The Kid" kill Clate Ogsbury, marshal of Silverton.

AUGUST 26—C. D. Knappen, murderer of McDonald, escapes from the Rico jail—has since been captured.

AUGUST 30—Ike Stockton betrays Wilkinson for a monied consideration.... A party of Indians camped out have all their horses stolen. They go to the [Indian] agency on foot to make complaint.

SEPTEMBER 2—The Rico–Durango toll road is completed.

SEPTEMBER 7—The first stagecoach drives into Rico, coming over the Tabor & Wason line.

SEPTEMBER 14—A number of witnesses go to Pueblo to attend the Johnny Bull Hawk trial.

SEPTEMBER 20—Roudebush is held up and has a hole shot through his hat on his way to the Wabash [mine]....

SEPTEMBER 27—Jim Sullivan kills the notorious Ike Stockton in Durango, while Stockton was resisting arrest.

OCTOBER 14—George Wetherby has his leg amputated as a result of an accidental pistol shot on September 7.... A petition is sent asking for the establishment of a military post at Cross Cañon.

OCTOBER 25—Contract is let by W. G. Waring for preliminary work on Rico Reduction Works.

NOVEMBER 1—Arthur Shepherd arrives to write booming articles for Rico at $300 a month.

NOVEMBER 2—George Grupe, a well-known Rico citizen, drowned in [Colorado] River.

NOVEMBER 8—Fall election—successful candidates Rosser for judge over Levis; Raymond for clerk over Middaugh; Dawson for sheriff over Hayes and Summa.

NOVEMBER 21—Rico *Electric Light* [newspaper] appears, with Eber C. Smith as editor.... J. K. Walker attempts suicide at the Pioneer Hotel by shooting.

NOVEMBER 23—Gus Talbot adjudged insane and sent to asylum.

DECEMBER 1—Tom McCulston is charged with the murder of Hiram Hyssell at "555," a notorious Durango resort.

DECEMBER 26—George Frink died at Bob McGrew's ranch on the Mancos [River], from the effects of a shot fired on the 15th last by Mr. Rush, Sr.

1882.

JANUARY 2—Trains begin running to Rockwood [on the D&RG's Silverton Branch] to accommodate Rico traffic.... Joe Tremble's log cabin burned.... Rico Mining Exchange formed.

JANUARY 10—The Santa Clara Company orders lumber for their mill.

JANUARY 11—Jim Neumeyer recaptured at Trinidad [Colorado].

JANUARY 14—H. B. Jenkins arrives to take charge of the Wabash [mine].

JANUARY 19—The DuPonte building on Glasgow Avenue burns.

JANUARY 23—County commissioner McCormick takes Neumeyer to Colorado Springs for safe keeping.

FEBRUARY 4—The McJunkin sawmill burns to the ground.

FEBRUARY 6—O'Brien Moore, who lately telegraphed his own obituary in sensational dispatch, bobs up serenely in old Mexico.

FEBRUARY 10—Jack Lyon nearly killed by rolling timber at the Zulu Chief claim. Dan Taylor calls to Jack to "look out," which he did, with disastrous results.

MARCH 7—The mill-site claimants gain a decision in their favor before Land Comissioner McFarland at Washington, D.C.

MARCH 28—Rico's first piano arrives to be manipulated at the Exposition Dancehall by Jerome Johns.... Sheriff Bowman, of Gunnison, attempts to arrest George Howard (well known in Rico) and other cattle thieves. In the melee, Howard was killed.

MAY 2—Miners' mass meeting adopts resolutions pledging not to work underground for less than $4.00 a day.

MAY 6—At midnight all the Chinese in the town were raided, beaten and robbed. Jig Hume, one of the participants, received a severe wound at the hands of one of his own party.

MAY 10—Sam McClemmends, alias "Sourdough Sam," a Rico stagecoach driver, was arrested for complicity in cattle stealing in Utah—and afterwards, sentenced to 20 years in the Utah penitentiary, which sentence he is now serving.

MAY 13—In a reign of petty thieving, William Byles, Beverly R. Keim and John Quincy Adams are robbed. Mrs. C. F. Middaugh is assaulted by a ruffian in her own home.... Frank Bard appointed commissioner, to serve with Harry Cahn and A. A. Waggoner.

MAY 15—Jesse Love resigns and George A. Smith goes on as marshal.

MAY 18—George N. Woods murders M. G. Buchanan at Durango.... Henry Park has a frightful fall on the cliffs up Silver Creek—several ribs broken.

MAY 20—Harg Eskridge adds another murder to his list, shooting John Bennett at Antonito, Colorado.

MAY 31—Andy Bigger kills Harry Diehl at Ophir [the old mining camp].

JUNE 3—At 11:00 o'clock, Marshal Smith and Sheriff Dawson attempted to arrest "Trinidad" and Wall for stealing saddles from Schueler & Bang [store]. Smith's brain is pierced by a bullet and his body riddled. The murderers escape, hotly pursued by citizens, but they cut the traces of horses hitched to an incoming wagon and get away.

JUNE 4—The unfortunate marshal is buried.

JUNE 5—Vigilantes order Dick Simms and "Coyote Bill" to leave town.... Pat Cain is appointed marshal.

JUNE 12—William Bingham "skips" with the funds of "The Restaurant."... Hank Steinman arrested for stealing horses.... News comes that the murderers of Marshal Smith had been captured near the Navajo Indian Agency by the Indians who were hired by Barney Watson, sheriff of La Plata County.

JUNE 13—The murderers are brought in by Sheriff Watson, who gets his $1,500 reward.

JUNE 15—At an early hour this morning, the bodies of Tom Wall and Charlie Cummings ("Trinidad") are found hanging in John Gault's stable, near the jail.

JUNE 26—Rico wins the famous case against the mill-site claimants before Secretary Teller.

JUNE 27—Rico's first telephone in operation—from the Grand View smelter to the mine.

JUNE 28—The first legal hanging in the San Juan [region] takes place at Durango—that of George Woods.

JUNE 29—The Rico smelter commences operations at noon today.... [Miner] smothered by a landslide in an open cut on the West Dolores. ... D. A. McGraw is appointed postmaster....

JULY 1—A new stagecoach company, consisting of J. L. Sanderson, George Mescrole and George Blake, commences operations between Rico and Rockwood [on the D&RG].

JULY 8—Neumeyer's trial for the killing of John Grigsby resulted today in a verdict of manslaughter, and he was sentenced to 18 months imprisonment at Cañon City.

JULY 18—Neumeyer escapes from jail, but is recaptured by Jim Hall next day at Trout Lake.... Dave Munn discovers a skeleton between the headwaters of the San Miguel and Mineral Creek—some of the bones now at the **News** office; owner can have same by identifying and paying advertising charges.

JULY 24—Gus Talbot, Rico's insane man, drowned in Cottonwood Creek, near Buena Vista, Colorado.

JULY 26—Governor Pitkin pays us a visit.

AUGUST 1—Toll road to Ames agitated.

AUGUST 7—The Santa Clara mill turns out its first silver brick.

AUGUST 8—Robert F. Patton killed by an explosion of giant powder in the Little Carrie [mine]—John R. Brown badly injured at the same time.

AUGUST 27—Father of A. V. Goria injured by upsetting of stagecoach near Dallas [adjacent to what is now Ridgway].

SEPTEMBER 14—Mike Hellwarth has his arm broken in two places by a premature explosion in Aurora tunnel.

SEPTEMBER 23—W. B. Norton severely injured by falling tree at General Sheridan mine.

SEPTEMBER 24—Jim Sullivan arrested at Pagosa Springs for stealing cattle.

OCTOBER 1—Durango Land Office opens for business.

OCTOBER 24—J. C. Grierson succeeds Amos Lane as manager at the Grand View, and other charges are made.

NOVEMBER 6—*The Rico Clipper* appears—W. L. Stevens, publisher.... First patent received—the Puzzle.

NOVEMBER 7—Election—Gerry, district judge; Galloway, state senator; Letcher, representative.

NOVEMBER 8—An unknown assassin shoots and kills John Berg at Trout Lake.... R. H. Ptolemy slips under his wagon and crushes his arm.

DECEMBER 14—Mrs. J. A. Robinson robbed of $117.60.

DECEMBER 25—John Kring's horse fell, breaking Kring's leg.

1883.

JANUARY 11—Stephens attempts to bribe Representative Letcher to vote for Tabor for senator—that is, he told Letcher he would give him $2,000 or $3,000 for his vote.

JANUARY 13—E. G. Segner & Company failed at noon.

JANUARY 17—Richard Pohl, cashier of the Bank of Rico, leaves town under peculiar circumstances and creates a great sensation; he is arrested on a snowbound D&RG train near Cumbres Pass.

JANUARY 22—Mabel Coulter, a *nymph du pave*, endeavors to suicide by taking morphine, but fails.

JANUARY 27—Announcement of new bank by Newman, Pittman & Company.

FEBRUARY 5—Several citizens of Rico hold a meeting to form a town company for the West Dolores, which results in the founding of Bowen.

FEBRUARY 8—J. H. Seck, that metallurgical cove oyster, commences suit against the Santa Clara company for $1,900.... Captain Davidson, Al Rogers and Charlie Kelley tumbled about by a snowslide near the C.B.&Q. mine in Horse Gulch.

FEBRUARY 20—George Messerote buys Blake's interest in stage line.

MARCH 2—Fred Dressler, a stagecoach driver, kicked by his horses, while alone on top of the range, and is seriously hurt.... John W. Grindle found dead on the West Dolores trail.

MARCH 19—Joe Hale has his leg badly broken while at work in the Atlantic Cable shaft.

APRIL 8—Spring election—Landon, mayor; Gilbert, clerk; Hull, treasurer.

APRIL 18—George Canning falls down the Swansea shaft—dies a few days later.

APRIL 27—Jack Sears arrested on Mantz Avenue by Sheriff Jim Bishop of San Miguel County for the murder of John Berg.

APRIL 31—Sears has his preliminary trial and is discharged.

MAY 19—Thompson Dean, owner of the Grand View smelter, makes his first and only visit to Rico.

JUNE 2—Rico Board of Trade organized.

JUNE 11—M. A. Bean, one of Rico's pioneer citizens, and the brother of W. H. Bean, suddenly disappears. He left Silverton in the afternoon for his mines in Cascade Basin, and no trace of him has ever been discovered.

JUNE 25—Rico Reduction Works commence running.

JULY 4—Grand celebration, races, shooting matches, etc.

JULY 16—Harry Bohrer, a workman at the Rico Reduction Works, is caught in the belts and hangs suspended some time before discovered, but few bones broken.

JULY 20—The South Park contact boom—20 shafts are started.

AUGUST 11—Plaster of Paris molds of silver dollars and halves are found near an old cable in Aztec Gulch—evidently used by counterfeiters.

SEPTEMBER 1—The Grand View has a strike of large proportions, caused by reduction of wages at mine and mill.

SEPTEMBER 15—Hugh Ray and party on their way to work at the Grand View mine are turned back by a party of armed men.

OCTOBER 11—The Santa Clara mill shuts down for the winter.

NOVEMBER 6—Fall elections; Swickhimer, sheriff; Hull, clerk...; Barnett, commissioner; Mundee, treasurer; Robinson, judge.... About 5:00 o'clock this afternoon, a wagon load of ore overturns near Big Bend....

NOVEMBER 25—A sensational shooting at the Miner's Arms—nobody hurt.

DECEMBER 16—Bruce Hunt killed by bank burglars at Durango.

DECEMBER 21—A stuck stagecoach in the snowdrifts—snowshoe service—21, 22, 23, 24 days of terribly fatal snowslides all over the San Juan country—Ferg Slater of Rico lost in the Mendota slide in Marshall Basin.

DECEMBER 29—Dave Selby and a companion have their feet badly frozen between Rico and Trout Lake.

1884.

JANUARY 1—O.B.B. gives a grand reception for its lady friends.... Numerous gentlemen receive....

FEBRUARY 1—The long-to-be remembered blockade begins today.... Controversy between 79'ers as to comparative depth of snow.

FEBRUARY 7—Pierce and Clegg started from Hope & Cross mine without snowshoes and are unable to get to town or get back—help reaches them at dark.

FEBRUARY 18—The unprecedented snows continue—no mail, no trains, no news, "no nothing" over the line of the railroad from Durango all the way to Alamosa. There is just enough of the telegraph poles sticking out to give the appearance of crucifixes in a Catholic cemetery.

FEBRUARY 27—Fresh meat supply low.

MARCH 29—Provisions playing out, and snowshoe rabbits begin to look sick.

APRIL 4—People begin to worry over the prospects of a continued blockade, and insane talk of shoveling out a trail is indulged in, but a protracted diet of rabbit tracks, snowballs and pepper is not conducive to muscular development, and lungs are the only things exercised.

APRIL 9—The Pasadena smelter people give notice that they will locate their smelter at Rico.

APRIL 15—Provisions hauled in from Ames on sleds... flour is divided by the pound.

APRIL 24—The blockade is broken on the D&RG railroad, and the first train reaches Durango since February 2nd.

APRIL 29—A party of ladies arrive in Rico, after enduring... [terrible] hardships for four days in the snow between Rockwood and Rico. [Men were also with the ladies for this ordeal.]

MAY 20—Discovery of the robbery of the St. James Hotel of silver plate, wine, cigars, etc., by Tom Hume and W. A. Shoot. They deny charges [against them], but the goods are found in their possession, and they leave the country.

MAY 23—Great floods down the river—all the ranchmen floating around on their chicken coops.

MAY 26—Telegram announcing the appointment on the 21st inst. of Filer L. Thompson as postmaster is received.

JUNE 5—Panic on... [the main street of Rico], caused by the war of prices of flour, which lower from $15.00 to $5.00 in a few hours. The hungry are jubilant.

JUNE 22—Killing of Billy Wilson by Pat Cain, near Silverton.

JULY 1—The first stagecoach arrived in Rico since February 1st.

JULY 15—The cowmen and soldiers who followed the Indians nearly to the Colorado River are ambushed and lose two men....

JULY 24—Durango has a little Indian scare, because of a conflict between town and government authorities.

JULY 30—James H. Burns is drowned in the Dolores, seven miles below the mouth of the San Miguel. Jim Hall is with him, but unable to give him any assistance.

AUGUST 15—Barry Cahn turns over stock to McCord-Nave Mercantile Company.

AUGUST 26—Bill Solomon, alias B. S. Bill, killed at Animas City [just north of Durango] by Doc Sears.

SEPTEMBER 10—A little son of Mr. and Mrs. James West falls from a burro and badly breaks an arm.

SEPTEMBER 11—Waggoner and Barnett find another skeleton on the West Dolores.

SEPTEMBER 18—Pasadena smelter charges the Neville reverberatory furnace.

SEPTEMBER 15—The C.H.C. begins to ship ore [from Rico].

SEPTEMBER 30—The county [of Dolores] purchases the Rico & Durango and Coal Banks toll roads.

OCTOBER 16—Warden Grigsby takes charge of the stagecoach line under lease.... Judge Getty orders the Cahn stock sold at sheriff's sale.

OCTOBER 17—Ed Quigley is caught in the belts at the Santa Clara [mill], and narrowly escapes being crushed to death.

OCTOBER 21—Lee Cottle arrested on the West Dolores by Sheriff Swickhimer, under charge of larceny.

NOVEMBER 25—The Little Maggie complications create a sensation.

NOVEMBER 26—Tom Cain killed at Ash Fork, Arizona.

DECEMBER 19—The promoters of the Expectation tunnel scheme hold a meeting.

DECEMBER 26—Three snowslides in one day at the Honduras [mine].

1885.

JANUARY 1—Horse-shoeing match between George Tabor and Tom Taney—won by the former.

JANUARY 2—Rumor that the D&RG... declared that they would not clear the snow blockade on Conejos Range.

JANUARY 17—Town council passes the ordinance to compel saloon keepers to keep their places free from minors.

FEBRUARY 11—Mrs. E. E. Rust breaks an arm at the skating rink.

FEBRUARY 14—Announcement that Samuel James, Jr. would erect a magnificent skating rink [in Rico].... Rico Cornet Band organized.

MARCH 7—Dolores County made a part of the Sixth Judicial District.

MARCH 18—The Pasadena [smelter] offices burned.

APRIL 5—Andy Brydon reopens the St. James Hotel.

APRIL 17—Sneak thieves rob M. E. Field & Company.

APRIL 24—Tom Williamson somewhat demoralized by a blast in the Swansea [mine].... A stagecoach-load of passengers upset in Hermosa Park—Mrs. Stein injured.

MAY 12—Landslide at the Newman [mine].

MAY 22—The Waring–Iler foundry made its initial effort.

MAY 29—John Bissell announces that he will close the Bank of Dolores.

MAY 31—Decoration Day [now known as Memorial Day] is appropriately observed.

JUNE 11—W. H. Benneta leases skating rink and Manager Newton steps down and out.... Rico Sunday School has a present of a library from Miss May Tucker of Northampton, Mass.

JUNE 15—The Chicago Comedy Company pays us a visit.

JUNE 19—Unknown parties attack an Indian camp on Beaver Creek and kill eight of them.

JUNE 29—County Judge E. A. Robinson found dead in his chair—apparently suffered by the use of morphine.

JULY 4—A grand celebration.

JULY 17—Reported disappearance of John Atkinson from Rockwood a sensation.

JULY 23—The town board conclude to enforce the "habitual drunkard" law.

AUGUST 1—Application forwarded for charter for a post of the Grand Army [of the Republic—a Civil War veterans organization].

AUGUST 27—Mason Jackson and Mark Munyan held up near Rockwood.

SEPTEMBER 14—Parker, the pen man, delivers a series of orations on the street corner.

SEPTEMBER 22—Judge Hayt opens his first term of court [in Rico].

SEPTEMBER 25—Mock trial of C. M. Wyman for stealing Hill Floersch's barber pole.

SEPTEMBER 28—Munt Grigsby kidnapped from the stagecoach at Perley Wason's and held on a charge of seduction—is released next day.

OCTOBER 24—Grand Army [of the Republic] Post mustered in.

OCTOBER 27—Post Office Inspector Waterbury discontinues Trout Lake post office.

NOVEMBER 18—The McIntyre Building burns—a complete loss.... No more insurance for Rico.

DECEMBER 7—Sudden death of George W. Bemis near Big Bend.

DECEMBER 11—Meeting between Swickhimer brothers for the first time in 14 years.

— Of Interest to Rico —

Saturday, January 2, 1886

THE BULLION PRODUCT of Rico this year, we understand, is about 135 cars. The Pasadena smelter, which has been running steadily this season, has shipped 85 carloads of bullion. The Grand View, which started later, we understand has shipped about 50 carloads of bullion. This would indicate that Rico is doing some work.

— ***The Durango Idea***

Local Brevities

THE STAGECOACH now comes in on runners.

JOHN JAMES, the Grand View [smelter] coke burner, returned from Trinidad [Colorado] this week and is at his station at the coke ovens.

THE PARTICIPANTS in the race at the [ice-skating] rink Thursday evening were Bob Ray and Clint Dutcher; distance, 150 laps, or five miles. Prize (offered by Manager Bennetta) a fine gold watch and chain. Bob Ray won by five laps and skated the distance in 28 minutes. A forfeit was up on a bet that the winner could not skate 100 laps in 20 minutes, but as Bob covered 112 laps in that time during the race, the bet will no doubt be forfeited. Willie Ray and George Enderich skated a 10-lap race, which was won by the former by a fourth of a lap.

CHARLES MILLER ENGEL COLLECTION – COURTESY RACHEL A. HARDWICK

SANDSTONE MOUNTAIN is the more prominent formation at the left, on the skyline beyond Rico, in this photograph of the town probably taken during the mid-1880's. Glasgow Avenue runs directly north with most of the major wooden business buildings lining this street through the town. Rico was incorporated in the State of Colorado on October 25, 1880.

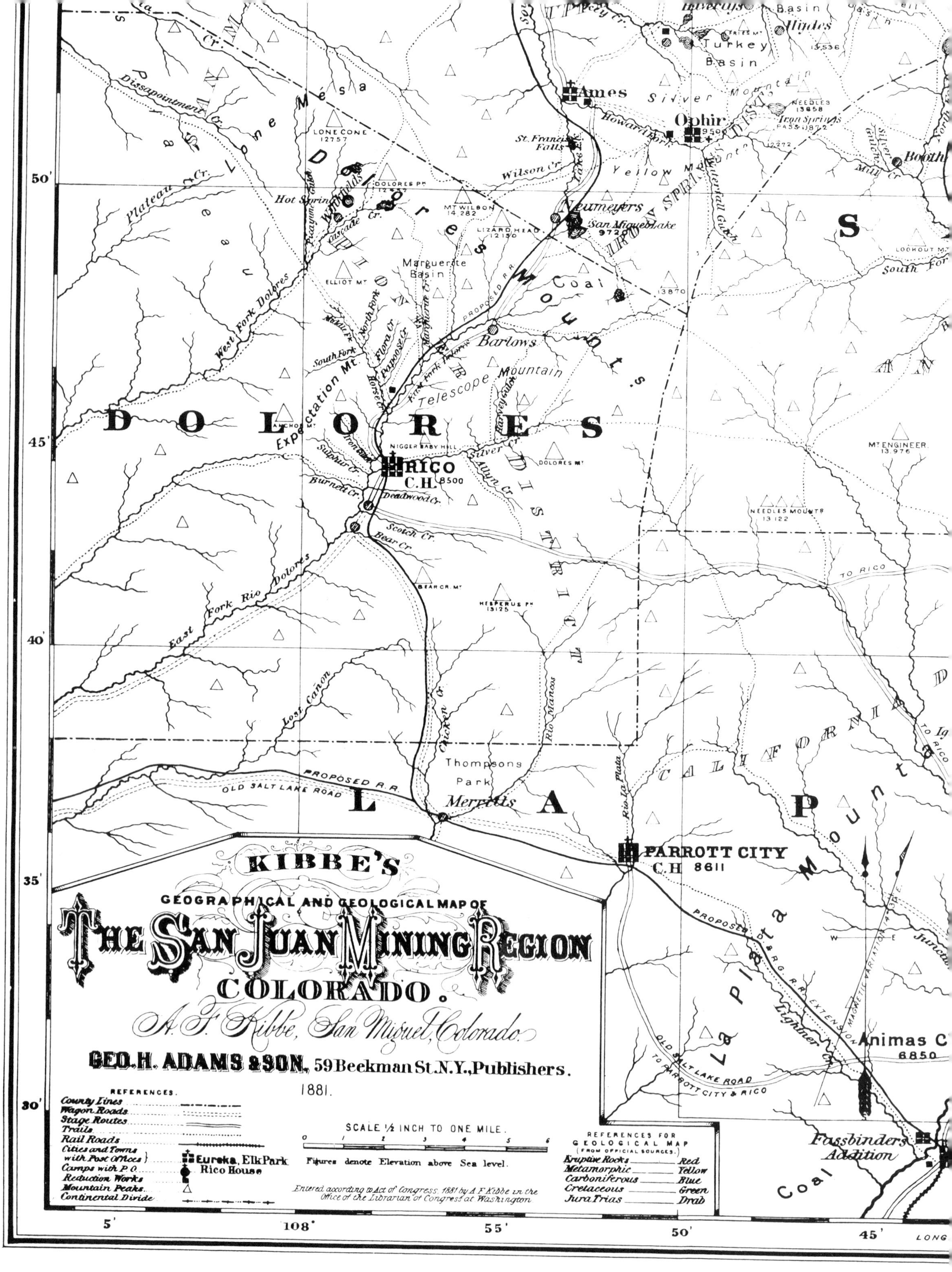
KIBBE'S
GEOGRAPHICAL AND GEOLOGICAL MAP OF
THE SAN JUAN MINING REGION
COLORADO.
A. F. Kibbe, San Miguel, Colorado.
GEO. H. ADAMS & SON, 59 Beekman St. N.Y., Publishers.
1881.
REFERENCES.
County Lines
Wagon Roads
Stage Routes
Trails
Rail Roads
Cities and Towns with Post Offices
Camps with P.O.
Reduction Works
Mountain Peaks
Continental Divide
Eureka, Elk Park
Rico House
SCALE 1/2 INCH TO ONE MILE.
0 1 2 3 4 5 6
Figures denote Elevation above Sea level.
REFERENCES FOR GEOLOGICAL MAP (FROM OFFICIAL SOURCES.)
Eruptive Rocks Red
Metamorphic Yellow
Carboniferous Blue
Cretaceous Green
Jura Trias Drab
Entered according to Act of Congress, 1881 by A. F. Kibbe in the Office of the Librarian of Congress at Washington
DOLORES
L A P
RICO C.H. 8500
PARROTT CITY C.H. 8611
Animas C
6850
Ames
Ophir
Neumeyers
San Miguel Lake 9720
Barlows
Merritts
Thompsons Park
Fassbinders Addition
Turkey Basin
Silver Mountain
Yellow Mt.
Telescope Mountain
Expectation Mt.
Dolores Mounts.
La Plata Mount
Lone Mesa
Disappointment Cr.
Plateau Cr.
West Fork Dolores
East Fork Rio Dolores
Lost Canon
Rio La Plata
Rio Mancos
Chicken Cr.
Scotch Cr.
Bear Cr.
Silver Cr.
Burnett Cr.
Deadwood Cr.
Sulphur Cr.
Wilson Cr.
Howard Fork
Hot Springs
LONE CONE 12757
DOLORES PK 12457
MT WILSON 14282
LIZARD HEAD 12150
Marguerite Basin
ELLIOT MT
DOLORES MT
BEAR CR. MT
HESPERUS PK 13125
NEEDLES MOUNTS 13122
MT ENGINEER 13976
NEEDLES 13858
13870
NIGGER BABY HILL
PROPOSED R.R.
OLD SALT LAKE ROAD
TO RICO
OLD SALT LAKE ROAD TO PARROTT CITY & RICO
PROPOSED D&RG R.R. EXTENSION
MAGNETIC VARIATION
Coal
50'
45'
40'
35'
30'
5'
108°
55'
50'
45'

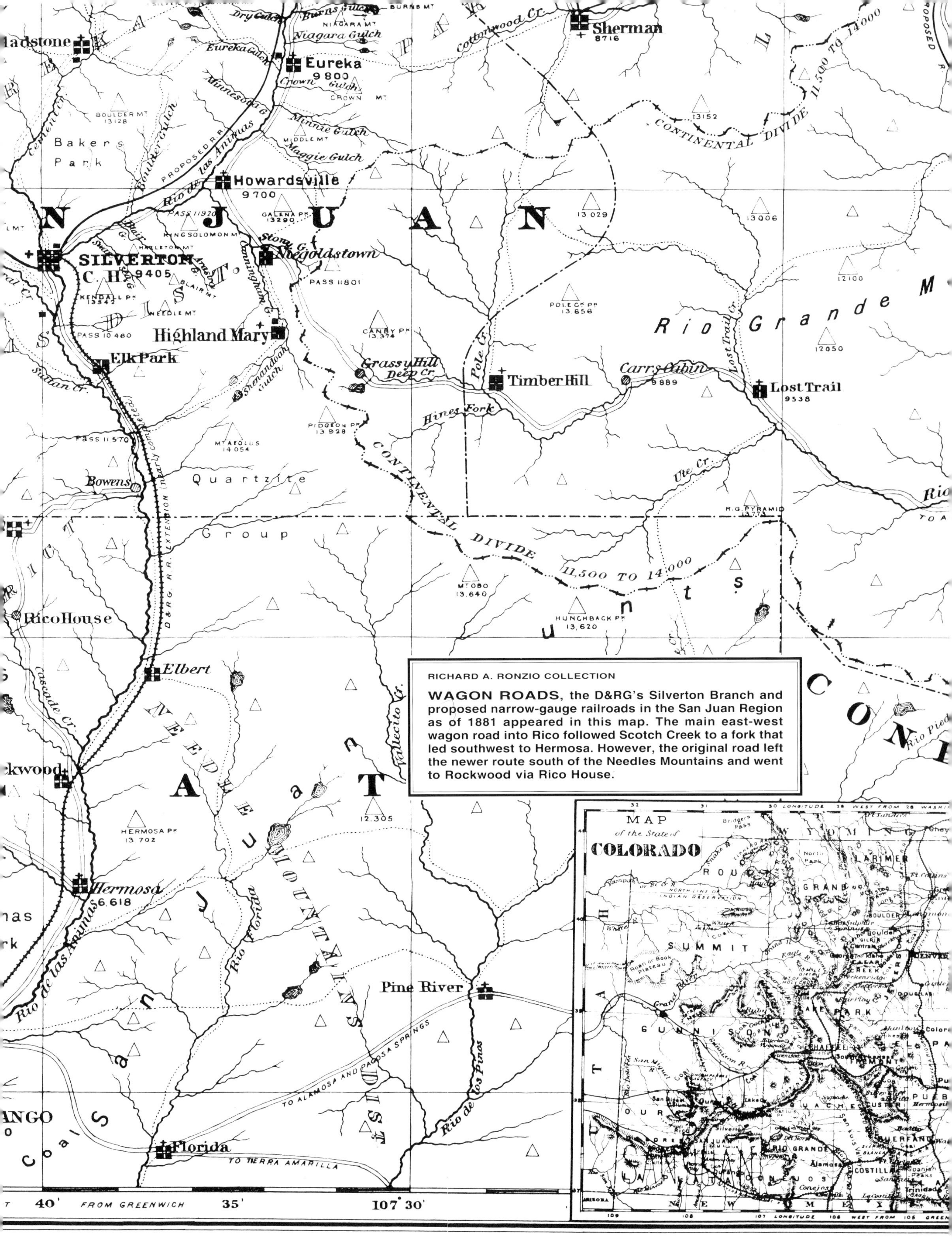

RICHARD A. RONZIO COLLECTION

WAGON ROADS, the D&RG's Silverton Branch and proposed narrow-gauge railroads in the San Juan Region as of 1881 appeared in this map. The main east-west wagon road into Rico followed Scotch Creek to a fork that led southwest to Hermosa. However, the original road left the newer route south of the Needles Mountains and went to Rockwood via Rico House.

CHARLES J. LEE PHOTO – COURTESY OF DAVID C. BUTLER

THE ENTIRE COMMUNITY would become alarmed when someone was late returning to Rico from one of the nearby mines. A man named Rooney had not returned from a walk while heading up the RGS grade during the 1920's. Over 60 men went out to search for him, armed with long poles, or rods, which could easily be pushed down through the snow to make contact with objects in the avalanche. This was the site of the infamous Nigger Baby Slide, on the mainline and Enterprise Branch. A glimpse of the town of Rico can be seen at the upper right, where the head frame of the Atlantic Cable Mine is barely visible, at the northern end of Rico's main street.

THE GRAND VIEW Mining & Smelting Company had built their plant at the north end of Rico in 1882, at the left edge of the photograph. A plume of smoke is visible at the south end of Rico where the Pasadena Smelter and Rico Reduction Works had set up shop. This was about 1885, and the town teemed with new structures, both dwellings and business buildings. THE RICO DEMOCRAT, dated March 6, 1891, stated that the bullion product of Rico in 1885 was about 135 wagons. You are looking down the Dolores River Valley.

ORVAL L. JAHNKE COLLECTION

CHARLES MILLER ENGEL COLLECTION –
COURTESY RACHEL A. HARDWICK

THE PASADENA SMELTER, the second one to be built in Rico, began operating in 1884. It was located near the final wye on the Rio Grande Southern Railroad, near the south end of town. In this scene, you are looking up the Silver Creek drainage. Captain William A. Adams was the principal owner. This was usually called the Adams Smelter in newspaper articles. The Pasadena operation was mentioned in THE RICO DEMOCRAT on August 21, 1891, in an article stating that J. H. Goldsworthy had secured a contract to build a wagon road to the mines on Newman Hill. This road started at a point near the Pasadena Smelter. Newman Hill was directly behind this structure. Reportedly, the Pasadena Smelter shipped 85 wagon loads of bullion during 1885.

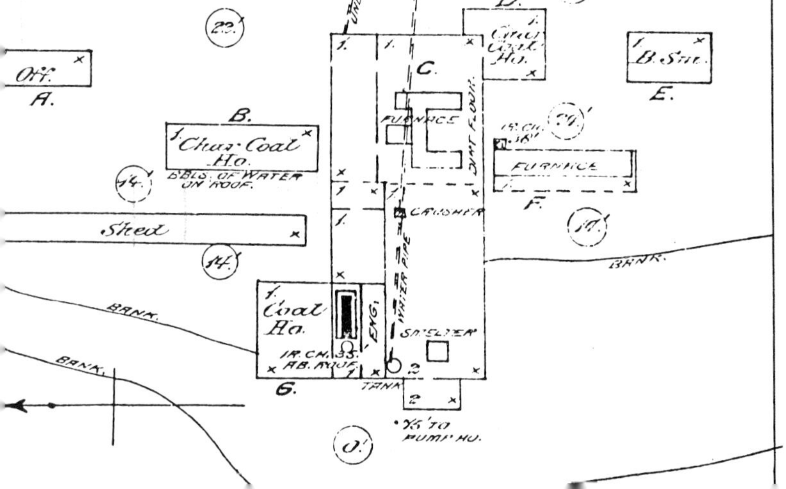

THE STORY OF RICO AND THE PIONEER MINING DISTRICT

A Chronology of Events

By William A. Graves

THE HISTORY OF RICO, Colorado, is the account of a small, remote mountain town—a mining camp—in a beautiful setting along the Dolores River, in the heart of the San Juan Region. The town has had a turbulent history of good times and bad times, so typical of many Western mining camps. However, Rico has been spared the fate of becoming just another abandoned ghost town. Just the opposite is true—Rico is a live-and-growing community in the 1990's. This chronology will identify many of the main events that occurred during Rico's existence as the principal mining town in the Pioneer Mining District, as well as its importance as the main commercial center in the northern part of the Dolores River Valley.

1776 Juan Maria de Rivera, from the Spanish territory that is now the state of New

(Continued on Page 88)

LARRY PLEASANT COLLECTION

LOOKING SOUTHWEST across the town of Rico, it is apparent that the town was booming in about 1885. However, nearly all of the buildings were made of wood, which would prove to be a fire hazard in times to come. So often, the buildings would burn due to a fire starting in a stove flue, often in the dead of night. Rico's little white schoolhouse was in the foreground, described as "...that neat and cozy little schoolhouse, with its bright paint and its steeple and surroundings." False-front business buildings lined Glasgow Avenue, seen from left to right (south to north) in the distance. Smoke was rising from the Adams Smelter, beyond the buildings at the left. Farther south, down the Dolores River, smoke rises from coke ovens.

RICO
— 1886 —

SANBORN
MAPPING AND GEOGRAPHIC INFORMATION SERVICE

Mexico, explored the Dolores River Valley and passed through the present-day site of Rico. Very likely, similar expeditions passed through this area from time to time, part of the vast region the royal Spanish ***conquistadors*** (conquerors) claimed as "New Spain."

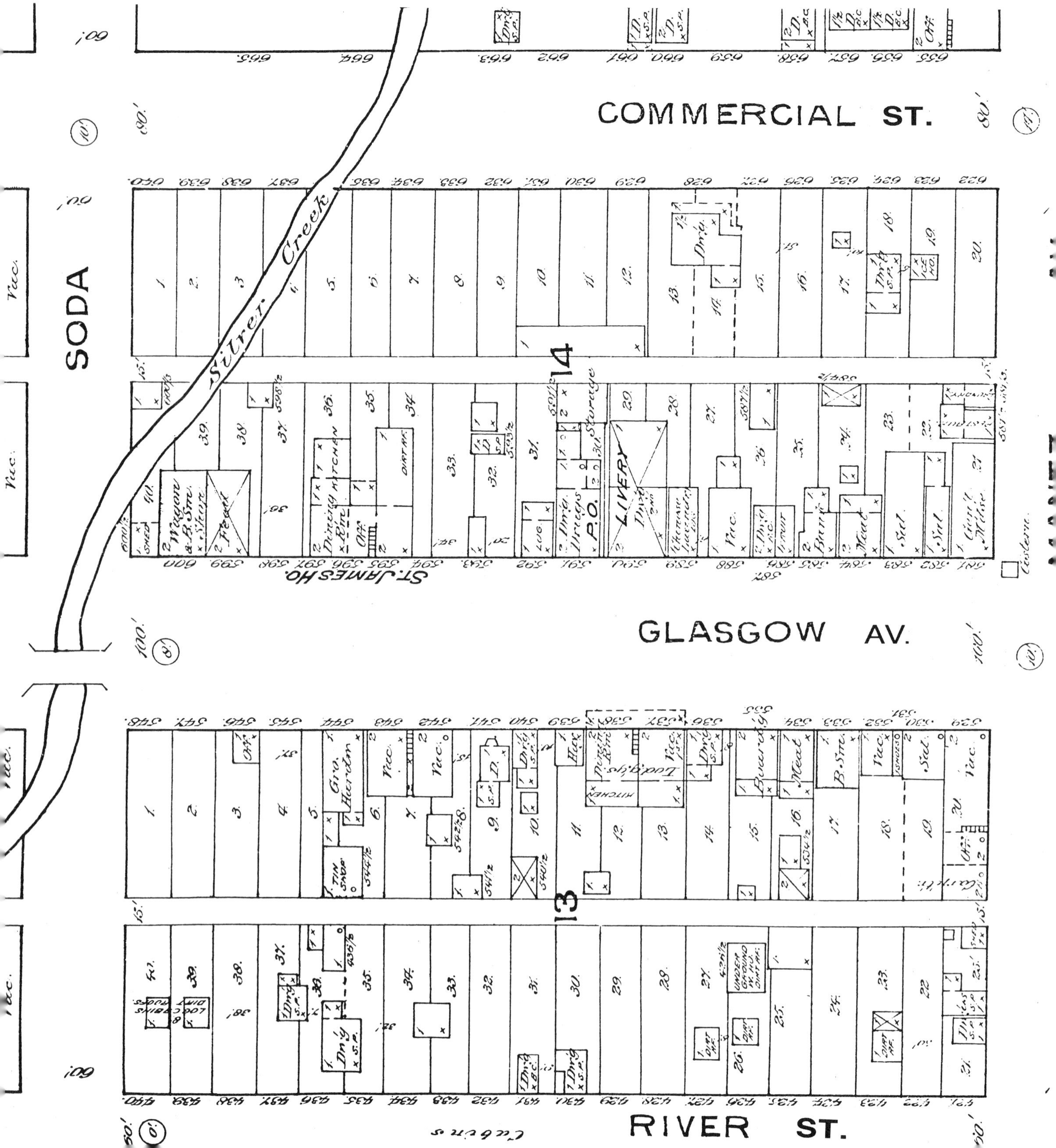

1833 Colonel William G. Walton led a party of about 60 American fur trappers from the St. Louis Fur Company into the Dolores River area, and the men spent the summer of 1833 along the river and at Trout Lake. They reported many Indians in the area, and they found the remains of some old Spanish smelters.

1861 Lieutenant Howard (first name unknown) led a group of prospectors from Bakers Park (where Silverton is now located) over the mountains at Ophir Pass to prospect along the Dolores River.

1869 Mining prospects in the Rico area

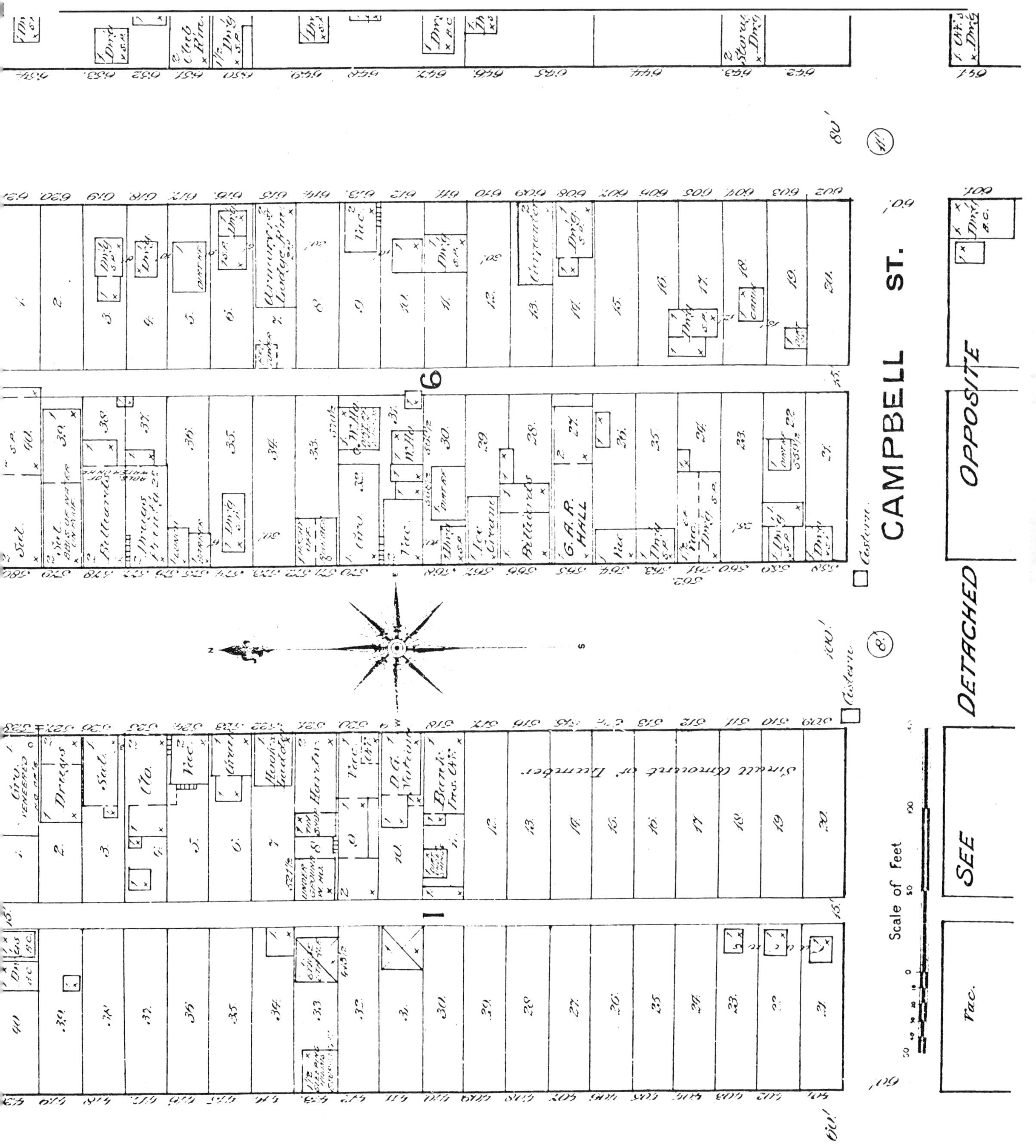

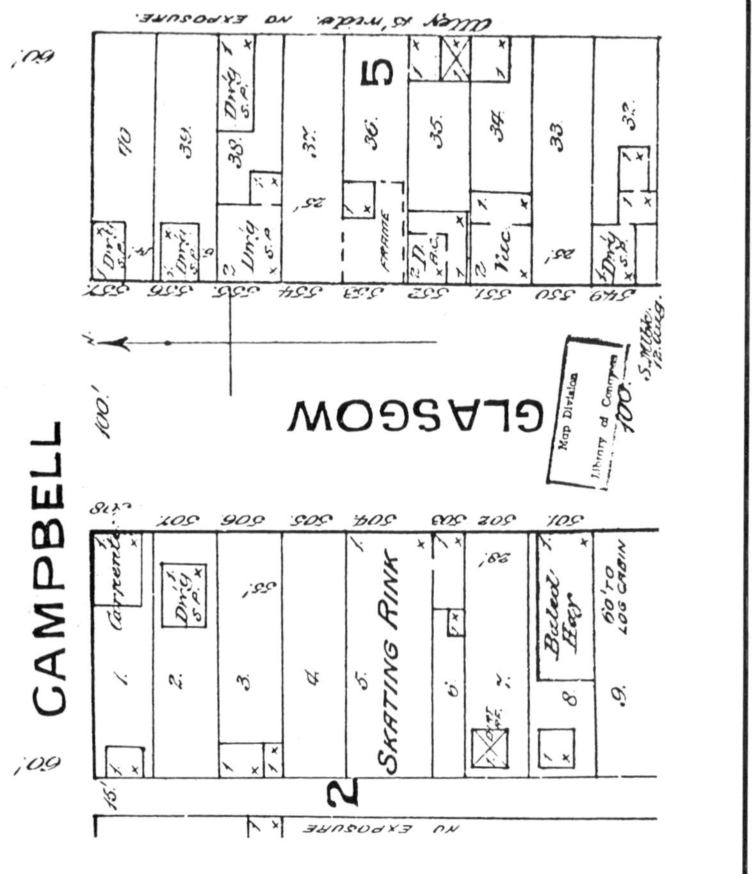

looked very promising to two prospectors, Sheldon Shafer and Joe Fearheiler, and they decided to explore the area as thoroughly as possible. The two men built a log cabin on the banks of Silver Creek, near the site of the future South Park Mine, and they located a claim in July of 1869, which they called the "Pioneer." Their claim formed parts of the future Riverside, Shamrock and Smuggler claims. The Pioneer claim also gave its name to the official designation for the mining district. In the fall, they built a better cabin and worked their Pioneer claim through the winter.

1873 The Brunot Treaty with the Indians made prospecting legal in the Rico area. However, the Ute Indians did not want to give up their ancient mountain land, rightfully theirs for hundreds of years. A few renegades harassed, stole livestock

DENVER PUBLIC LIBRARY, WESTERN HISTORY DEPARTMENT

FREIGHT WAGONS hauled by oxen were arriving in great numbers at Rico during the early 1880's, when this picture was taken. Fortunately, the town fathers had a surveyor lay out a 100-foot width for Glasgow Avenue, Rico's main street. This enabled wagons to turn around in the middle of the block with ease. The buildings in this scene lined Glasgow Avenue, facing west in Block 6, and they can be identified by using the Sanborn Map of 1886. Beginning at the far left, at the intersection with Mantz Avenue, the first white building was a saloon, to the right of the covered wagon. The two darker colored buildings housed a saloon and a billiard parlor. A two-story drugstore building was next in line, with three small commercial buildings following. Two more one-story structures appear next, embracing an assay office, barber shop and a grocery store. The two-story false-front building, with awnings on the second floor, contained a drugstore. A small drugstore was in line, after the vacant lot, and it was neighbor of an ice-cream store. The Bon Ton was a notorious saloon that reportedly had several shoot-outs. Nigger Baby Hill is centered on the background. (Note: "Nigger Baby Hill" is used in its historical context throughout this volume. This name dates back to 1870, when prospectors found large deposits of black oxide of manganese on this hill.) THE RICO DEMOCRAT newspaper carried an article during 1891, stating that Beeman's bull train and that of Jones' came in over the Rockwood Road with freight, and his outfit was "loaded out" with ore for the return trip.

DAVID C. BUTLER COLLECTION

CITY SHOEING SHOP was the name on this structure, built on the east side of Glasgow Avenue, in Block 14, close to where it crossed Silver Creek. This view was taken on July 1, 1886, when a pack train of burros was on the avenue. On many occasions, these pack trains carried sacks of ore out of Rico and returned with provisions for the stores in town. This was a popular place to have horses shod, and repairs also could be made on wagons. The Dashaway Stables was their competitor in Rico at this time.

from, and/or even killed white settlers and prospectors.

1874 Professor Ferdinand V. Hayden's survey party made the first official survey of the Rico area, and Hayden named most of the nearby peaks and accepted the musical Spanish names bestowed on rivers and mountains by Spanish explorers many years earlier.

1877 Ouray County was created on January 18, 1877, and at that time, this political subdivision was a very large county, which included the Dolores River and the Rico area.

1877–1878 During 1877, the first significant mineral discoveries were made at Rico. Promising silver strikes were made by a prospector named Sandy Campbell, who located the Grand View lode. The news of Campbell's discoveries quickly spread, and the mad stampede was on! Mining claims were staked for the Atlantic Cable, Aztec, Black Hawk, Columbia, Cross, Hope, Major, Pelican, Phoenix and Yellow Jacket in 1877 and 1878 by such men as Professor Hayden, John Glasgow, Charles Hummiston and David Swickhimer. These claims were located in three segments of the mining district: Nigger Baby Hill, on Telescope Mountain; Newman Hill, on Dolores Mountain; and along the stream banks of Silver Creek.

1879 Word of the rich silver lodes on Nigger Baby Hill spread quickly during the spring of 1879. The first "outside money" needed to develop the ore bodies at Rico came to the region later that year. A group of well-known Nevada Comstock investors—which included Senator John P. Jones—bought the Grand View, Yellow Jacket and Phoenix claims from

E. H. WILDER PHOTO – BILL ELOE COLLECTION

NEW FACES APPEARED on Glasgow Avenue almost daily during the early 1880's. These were get-rich-quick hopefuls, attracted to Rico by the news of gold or silver strikes. Many of them had nothing more than a few personal items in their pack and a burro to ride. The businesses were certainly an attraction at Rico, with their brightly painted signboards placed out over the boardwalk. Near the center of this photograph was the establishment of Hunt & McGraw–Druggists, and King's Meat Market was located farther down the avenue. This was the only town of any kind for a great distance, located in what had been the heart of the Ute Indian Nation. The Ute (or "Utah") territory originally included all of western Colorado until 1874. The buildings in this scene faced toward the east on Glasgow Avenue, and you can see them listed on the Sanborn Map of 1886. The tail building at the right, one lot from the corner, at the intersection of Mantz Avenue (with a vacant lot on this side of it), still stands today.

A CLOSER VIEW of two commercial establishments on the west side of Rico's Glasgow Avenue shows Davis & Raymond's general store, alongside the Gold Room, no doubt a saloon—of which there were many. This picture was taken in 1884 or 1885, and it shows a number of burros outfitted with pack saddles for hauling goods. An 1891 newspaper article in THE RICO DEMOCRAT complained that "...there is a statute against permitting jacks to run at large, which should be enforced in this town. Several complaints have been made lately of depredations committed by them. In the trail they are useful, but in town they are a nuisance."

L. PALENSKE & CO. PHOTOGRAPH – BILL ELOE COLLECTION

ORVAL L. JAHNKE COLLECTION

THESE STORE FRONTS lined Glasgow Avenue in Rico ca. 1887, facing toward the east. The log structure at the left had a projecting sign above its entrance that read "Horse Herd," which probably indicated the building housed a stable or blacksmith shop. It was located near the middle of Block 1, on the west side of Glasgow Avenue. The local boys gathered there to discuss politics, mining strikes and probably the lack of marriageable women in the community. It was a common sight to find freight wagons coming into Rico loaded with provisions to stock the merchants. The heavier-laden wagons were pulled by oxen—slow, but reliable. Mules, with ears longer than those of horses, were favored as good pulling power. Notice the large oil lamp on the boardwalk to attract attention to a shop. Sandstone Mountain is the prominent ridge on the skyline at right. A portion of Block 13 appears at the far right.

Glasgow, Campbell and Hummiston for $60,000 (or $124,000, depending on which report you choose to believe). The Nevada Comstock men made plans to create a corporation with $10 million in capital. Rico became a boom town! From one log cabin in August, Rico had grown to 105 buildings only a month later! And the progression grew to 240 structures in December and 400 by February of 1880. However, many of the buildings consisted of only log walls without roofs, because milled lumber was not available.

Two sawmills were erected near Rico during the later part of 1879. These mills

were water-powered, and their machinery had to be freighted-in from Alamosa, which was the nearest railhead (on the Denver & Rio Grande) at that time. No wagon roads had yet been constructed, and it was a tremendous feat to cut a trail through the mountains from Durango to Rico. The cost of freighting was greater than the cost of the milling machinery. The first sawmill was set up by James R. McJunkin, about four miles north of Rico, while George Barlow erected the second sawmill about a mile upstream from McJunkin's mill. Neither mill produced much sawn lumber in 1879 before winter set in, but both operated night and day during the following year, in an attempt to keep up with the building boom at Rico.

The first meeting to organize the town was held on August 18, 1879. The town was incorporated during December, and Frank W. Raymond was elected Rico's first mayor on December 22, 1879. The year of 1879 was the period for many "firsts" for Rico. The first edition of **THE DOLORES NEWS,** Rico's first newspaper, appeared on August 21, 1879. The editor was Charles A. Mantz, and he soon became well known for his colorful writ-

(Continued on Page 98)

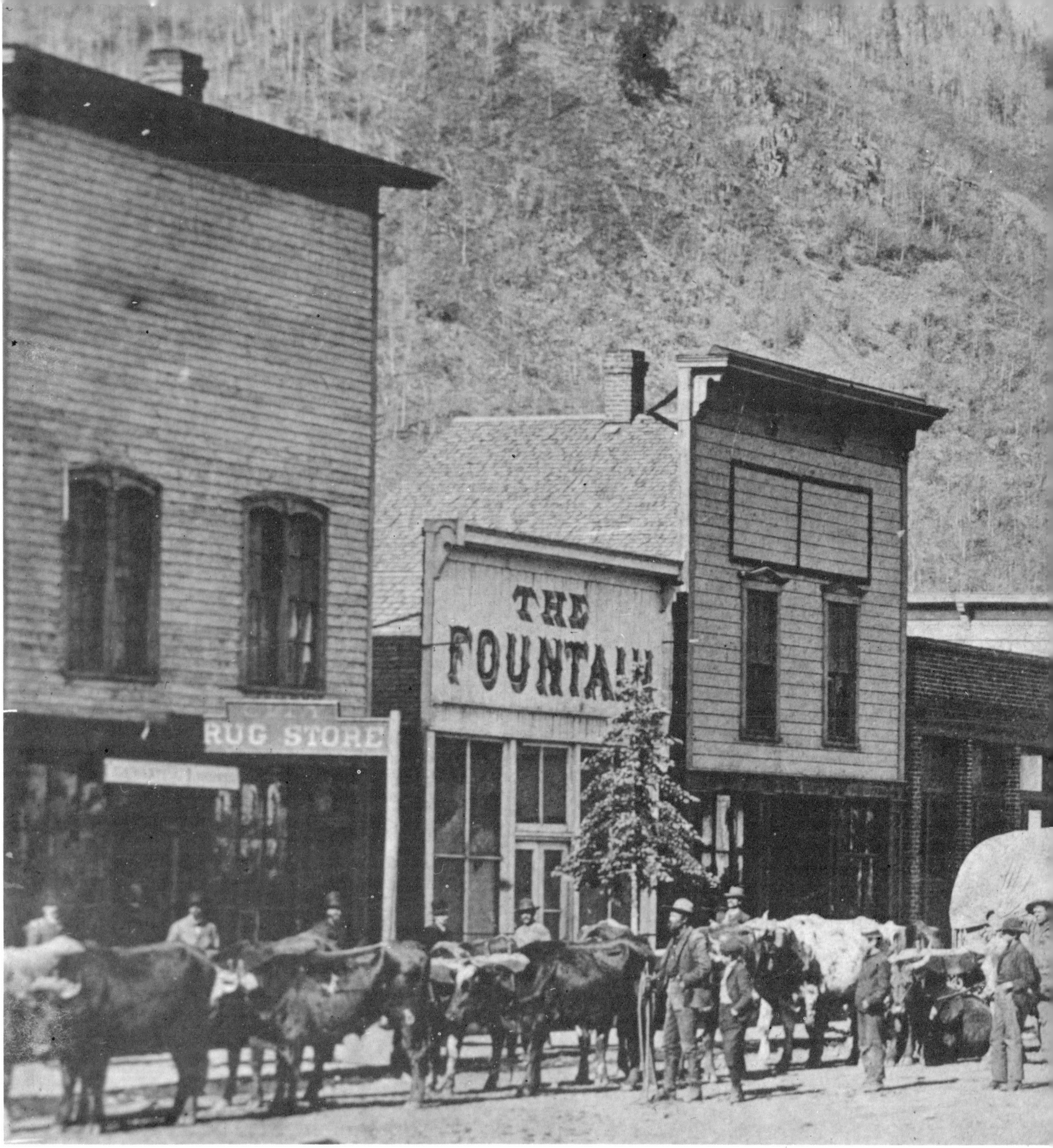

MONTE BALLOUGH PHOTO – COURTESY MARGARET BALLOUGH PALMER

MONTE GEORGE BALLOUGH was a resident photographer who resided in Rico from the late 1880's until November of 1903. His daughter, Margaret, was the source of information pertaining to his photographs printed in this volume. (These views were included in the book, *Beyond the Third Rail,* by Doris B. Osterwald.) This picture probably was taken early in 1889, perhaps one of Monte's first views. What a change has occurred along the west side of Glasgow Avenue since the 1880's! An unusually large gathering of "bull trains" (wagons hauled by oxen) had just reached Rico. At the intersection of Mantz and Glasgow avenues (near the center of this view) you can see Rico's first building constructed of brick. By this date, the two-story First National Bank had been built next to Mantz Avenue. Another two-story structure was next door (to the north), which housed a barber shop and meat market. The large wooden building at the northern end of the picture (at the far right) was at the site of the structure that became the Hotel Enterprise. It was either razed or remodeled in order to make room for the larger hotel.

From The Rico Democrat:

GEORGE JONES' BULL TRAIN, loaded with plunder for Rico's merchants, came in on Tuesday from Rockwood. The bull train will, in a short time, exist only in the memory of our people.

—The lumber yards of Rico are well stocked with all grades of lumber. This is as it should be. There will be a great many buildings constructed in Rico before the close of '91. Mark the prediction.

—The finest assortment of fruit we have seen in a long time was received on Tuesday by J. F. Kelly. Among the lot were several crates of California peaches, that for size and flavor beat anything yet brought to Rico.

—Our streets this week presented a scene of unusual activity, owing to the presence of several bull and pack trains, loaded in with merchandise and out with the ore shipments of the Enterprise Mining Company.

RICHARD A. RONZIO COLLECTION

FROM AN ANGLE farther up Dolores Mountain, an unknown photographer took this view of Rico, believed to have been taken during the late 1880's. Commercial Street is in sight, at the far right, located parallel with Glasgow Avenue, which is situated a block down the hillside. Snowcapped Elliott Mountain is on the distant horizon. Rico boasted of having a skating rink during 1885. The rink apparently was located in an open field in Block 2, Lot 5.

ing style. The first organization in town was the Miners Union. However, the Miners Union was a social club and not a labor union. It should be pointed out that Rico never had the severe labor strife and violence that embattled Telluride, only 20 miles away—although there were labor disputes with management, as might be expected.

Business establishments began to develop along Glasgow Avenue, Rico's main commercial thoroughfare. During September, the U.S. Post Office Department opened an agency in a drug store.

Rico's worst Indian scare took place on the night of October 5, 1879. The Meeker Massacre was still on everyone's mind, and all precautions were taken. The women and children were housed in a strong, windowless cabin, and the men converted Raymon's General Store on Glasgow Avenue into a fortress. However, the Indians never appeared, and the only casualty of the night was an unfortunate burro, who was the victim of a "spooked" sentry's bullet.

1880 Although Rico was snowbound during January of 1880, J. F. Wannemaker was able to get into the place to survey the new townsite and divide it into lots and streets. The lots were 25 feet wide, while Glasgow Avenue was 100 feet wide, the town's main street. It ran north and south and eventually became the route for the highway through town. The townsite also ran north and south, and it measured one mile in length by one-half mile in width.

The winter of 1880 was an extremely difficult time for Rico. Everything was in short supply, and food was the most critically deficient item. Merchants and residents had attempted to lay-in food and other supplies for the winter. However, the town was blockaded by snow for a much longer time than was anticipated. No pack trains were able to get through with provisions from November until April of the following year. Food shortages were the most critical, and any kind of food sold at unbelievable prices. The ordeal was not forgotten, and it was not repeated the following winter. (See the newspaper accounts of this snow blockade elsewhere in this volume.)

The building boom that began in 1879 continued right on through 1880. The official 1880 census listed Rico's population as 894, and the town was growing rapidly. By summertime, Rico had six general stores, four meat markets, three hotels, two dancehalls, three blacksmith shops, two banks, one doctor, two lawyers, many saloons and only one tent

(Continued on Page 102)

DENVER PUBLIC LIBRARY – WESTERN HISTORY DEPARTMENT

THE ST. JAMES HOTEL in Rico appears on the Sanborn Map & Publishing Company map of 1886. This hotel was located in Block 14 and faced west on Glasgow Avenue, not far south of the Silver Creek bridge. The frame structure contained a dining room and kitchen on the main floor, in the left half, which also had an office. The staircase to the rooms was through the central door. Boarding here cost $7.00 per week. The right third of the main floor probably was leased as a retail store, with living quarters above. This side had a dirt roof over the rear of the structure. A silver plate, wine and cigars were stolen from the hotel on May 20, 1884. The culprits left the county when discovered. On April 10, 1891, THE RICO DEMOCRAT reported that Mr. Fred Archambault, proprietor of the St. James Hotel, was raising nine hogs, which had an average weight of 200 pounds each.

LOUIS HART,
ONE PRICE CLOTHING HOUSE
HATS, BOOTS, SHOES &c.

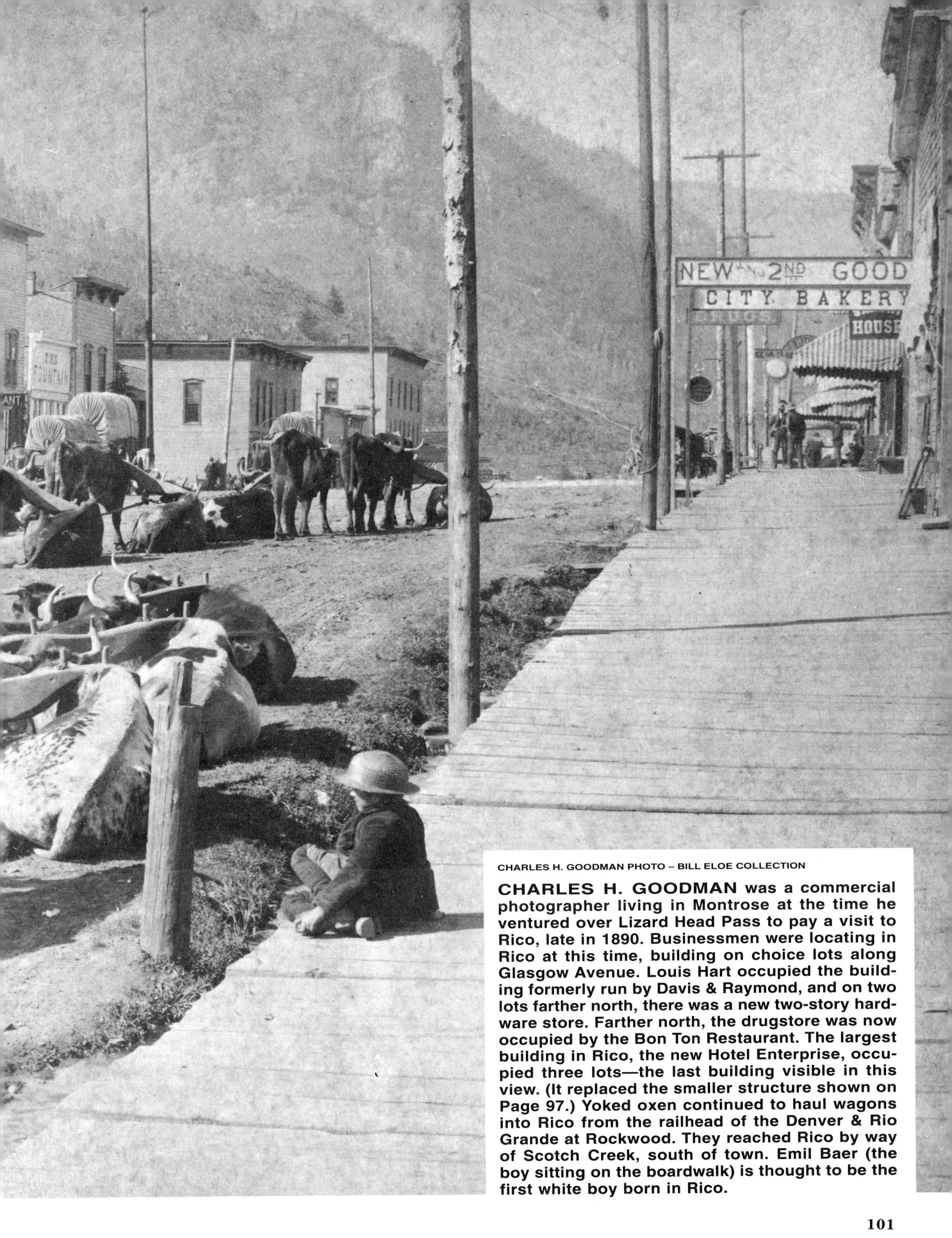

CHARLES H. GOODMAN PHOTO – BILL ELOE COLLECTION

CHARLES H. GOODMAN was a commercial photographer living in Montrose at the time he ventured over Lizard Head Pass to pay a visit to Rico, late in 1890. Businessmen were locating in Rico at this time, building on choice lots along Glasgow Avenue. Louis Hart occupied the building formerly run by Davis & Raymond, and on two lots farther north, there was a new two-story hardware store. Farther north, the drugstore was now occupied by the Bon Ton Restaurant. The largest building in Rico, the new Hotel Enterprise, occupied three lots—the last building visible in this view. (It replaced the smaller structure shown on Page 97.) Yoked oxen continued to haul wagons into Rico from the railhead of the Denver & Rio Grande at Rockwood. They reached Rico by way of Scotch Creek, south of town. Emil Baer (the boy sitting on the boardwalk) is thought to be the first white boy born in Rico.

building for a church. (Does the ratio of saloons to church buildings bespeak the moral values of the boom period?)

Two additional sawmills were constructed near Rico, and lumber was voraciously consumed to build hundreds of houses, as well as for improving existing structures—utilized for better roofs, doors and windows. Nearby coal was available to townspeople for the taking; however it was of poor quality. A public school board was formed, but there was not yet a school building. A school term of several months was established; however, the students were forced to meet in very poor quarters.

The story of the Grand View Smelter began in 1880. Rico needed two things in order to prosper in the economy of 1880: It needed a smelter and better, less-expensive transportation. Of course, a railroad would provide cheaper transportation in the years to come. However, in 1880, a railroad to Rico was hardly even a dream. The nearest "end-of-track" (as railroaders called the point at the end of a rail line) was somewhere between Alamosa, Colorado, and Chama, New Mexico, en route to Durango, Colorado, by way of remote Toltec Gorge and Cumbres Pass. This was General William J. Palmer's famous narrow-gauge "San Juan Extension," being built to tap the mineral riches of the San Juan Region. Realistically, Rico dreamed of a passable road for heavy freight wagons, so that wagons could replace mule trains that had to use many miles of poor pack-train trails.

The outlook for a smelter was more promising. Most of the ore in Rico's Pioneer Mining District was high in lead content, which was both a blessing and a curse. On the one hand, the composition of the silver-lead ore made it ideal for smelting. And smelting would greatly reduce the volume and weight of the refined product that had to be transported to the outside world. On the other hand, the value of lead was only about 4 cents per pound, while the cost of freighting ore from Rico to Alamosa was about 12 cents per pound. Only a concentrated product from a smelter, which was rich in silver or gold, could be handled at a profit. During February of 1880, Senator John P. Jones announced the creation of the Grand View Mining & Smelting Company. He stated that this firm would construct a smelter in Rico. The town was overjoyed at this announcement!

Every new rumor about the Grand View Smelter produced a new wave of excitement in the town. The machinery for the new smelter arrived in Alamosa during late April of 1880. However, moving the heavy machinery to Rico was an incredible task. Eighty-five mules were required to create a wagon train for the machinery. Of course, Rico did not yet have a wagon road, and when the wagons came down Bear Creek Hill, south of Rico, the wagons had to be lowered by "snubbing" (reeling) them down the hill by using ropes and chains that encircled large trees. The wagon train entered Rico on the Fourth of July amid wild celebration. The site for the new smelter was about 500 feet north of Silver Creek, near the outskirts of town. Four months were needed to erect the smelter and test it.

The first production run of 12 hours was on November 17, 1880, which produced 25 bars of silver bullion, and this proved to everyone's great relief that the silver-lead ore was of a type that could be smelted. Although the smelter had produced silver bullion, the smelting machinery had some operating problems typical of any new plant. However, by December, the smelter was running steadily around the clock. Twenty-four men worked two 12-hour shifts.

1881 Although Ute Indian reservations had been established in the 1870's, violent incidents between Indians and whites were still common, both on and off the reservations. Sometimes, prospectors could not resist the temptation to enter the reservations in their search for minerals. And the Indians still harassed white settlers. During June of 1881, two ranchers in the Dolores Valley were killed by Indians, and a large group of whites was

MARGARET MATZICK COLLECTION

THE SILVER LEAGUE THEATRE was known as one of the finest entertainment centers in southwestern Colorado. It occupied Lots 5, 6 and 7, in Block 2, and it faced toward the east. This fine theater building was photographed in 1892. However, the structure burned to the ground in 1898. This happened in spite of the efforts of both the M.P.S. and the A. P. Adams volunteer fire brigades. Charlie Rohde (with the arrow over his head), at the far right, was the owner of the theater at this time. It occupied the lot previously used as an ice-skating rink.

organized to chase and punish the Indians. The chase led to the La Sal Mountains in Utah, where the Indians barricaded themselves among the rocks. In the inevitable gun battle that followed, men on both sides were killed. Six of the whites that were killed were from Rico.

Civilization was gradually coming to Rico, though. Dolores County was created from part of Ouray County in 1881. Ministers of different faiths visited Rico from time to time and held religious services in several buildings; however, no church congregation was formally organized until later. Two additional sawmills were erected near Rico, and frame buildings of improved construction appeared.

Rico needed a wagon road very badly to replace the pack-train trails. After two years of false starts, a successful company was incorporated in 1881. The firm was financed by $6,000, raised by the residents of Rico, in order for it to build a toll road. The new toll road was built south along the Dolores River to its confluence with Scotch Creek, where it turned east and climbed up along Scotch Creek into the La Plata Mountains. It crossed the mountain ridge and dropped down along Hermosa Creek to the town of Hermosa on the Animas River. Hermosa was 11 miles north of Durango, and a wagon road had already been built to connect the two towns. The new road was completed in September of 1881. It was a success and greatly lowered freight rates.

1882 When construction of the Denver & Rio Grande's Silverton Branch reached Rockwood during January of 1882, a new and better wagon road was built to connect with the railroad at Rockwood. The new road ran south from Rico along the Dolores River, to the confluence with Scotch Creek. The road then headed east up steep slopes along Scotch Creek, over the La Plata Mountains and down to Rockwood, for a total distance of 35 miles. Two stage companies began to provide connecting Concord stagecoach service from Rockwood to Rico, using the new wagon road. **The Dolores News**, published in Rico, reported that the Concord coaches were "...immense and gorgeous." The stagecoaches were pulled by four-horse teams, and the scheduled time for the trip was eight hours. The one-way fare was $7.00, plus an additional $1.70 train fare to Durango. The D&RG and the stagecoaches also carried mail and express shipments.

Rico struggled for several years to provide a public school building. Finally, in 1882, donations, a loan and a two-mill property tax provided the money required to construct the new schoolhouse—$1,239.00—and hire a teacher. Mrs. Robert Darling, the wife of Robert Darling, a Rico pioneer, was hired as the new school teacher. The new school building was described by **The Dolores News** as "...that neat and cozy little school house, with its bright paint—its steeple and surroundings." The little building was filled to capacity, and the 1882–1883 school term ran for five months. A closing-day school program for parents and friends, which featured the children, was well received.

The year 1882 was the "year of the smelter" in Rico. The new Grand View Smelter had been in operation for a year and a half. The hands-on operating experience was invaluable, and the output and efficiency of the smelter was raised to a very satisfactory level. The success inspired the management to plan a large expansion program. The new smelter addition would have a concentrator that would make smelting of low-grade ore profitable. A large new building was constructed, and new machinery was installed. Trial runs were successful, and the smelter was soon running smoothly and producing about two tons of bullion every 24 hours.

Successful smelting at the Grand View plant brought competition. Three new smelters were built in Rico in 1882. The Rico Smelting Company's smelter was built on the south side of town, near the Dolores River. The Santa Clara Mining & Milling Company and the Rico Chemical Reduction Company constructed new smelters north of town, near the head of Horse Gulch. The new smelters had some mechanical problems, and the Rico Smelting Company had financial problems, but the smelters operated well enough to prove that Rico's ore could be efficiently and profitably processed.

RICO BUSINESS ENTERPRISES as of 1882 (from THE DOLORES NEWS):

4	General merchandise stores	4	Billiard halls
1	Hardware house	12	Saloons
2	Banks	2	Wholesale liquor houses
2	Hotels	2	Drug stores
3	Sawmills	1	Cobbler shop
1	Smelter	2	Barbers
2	Brickyards	2	Jewelers
3	Livery stables	1	Gunsmith
3	Blacksmith shops	2	Dressmaker shops
1	Bakery	5	Laundries (1 Chinese)
2	Restaurants	3	Assay offices
2	Meat markets	1	Photograph gallery
1	Boarding house	1	Stage and Express office
1	Bath house	1	Wagon shop
2	Confectionary stores	3	Painters
3	Physicians	5	Ore cabinets
6	Lawyers		and several carpenters
3	Tobacco and cigar stands	?	Grocery stores

1883 With the new smelters on line, Rico's ore production doubled in 1883. Production was geared almost entirely for silver. Gold, copper and lead were also produced, but 88 percent of the total value was silver. Rico was truly a silver camp!

RICO AREA MINE LOCATIONS in 1883 (from THE DOLORES NEWS):

	MINE LOCATIONS	NUMBER OF MINES
1	Dolores Mountain	23
2	Telescope Mountain	18
3	Horse Gulch	8
4	Elliot Mountain	12
5	Expectation Mountain	15
6	Printer Boy Hill (on Expectation Mountain)	12
	Total..	88

1884 The candidates in the 1884 presidential election were Grover Cleveland and James Blaine. Democrat Cleveland was seen as a "gold bug," so a silver camp like Rico voted for his opponent. This was no surprise, because the people of Rico felt that the Republicans were more sympathetic to the mining industry, and for many years voted accordingly.

The winter of 1883–1884 produced snow eight feet deep in Rico. The wagon roads were blockaded from February to June, and Rico ran short of food and coal. Flour sold for as much as $50 for a 100-pound sack. And mail delivery was very erratic.

1885 A wagon road was built along the Middle Fork of Horse Creek north of Rico to a non-incorporated settlement named "Dunton," on the West Fork of the Dolores River (as the West Dolores River was called at this time). Dunton's economy primarily depended on ranching, but there was also a little mining activity in the area. (See "The Story of Dunton" on Page 29 in this volume.)

1887 David Swickhimer uncovered an incredibly rich blanket of gold and silver at his Enterprise Mine. The Enterprise became the most productive mine in Rico's Pioneer Mining District. (See "David Swickhimer and the Enterprise Mine" on Page 219 in this volume.)

1889 The phenomenal production of silver ore at the Enterprise Mine rejuvenated mine development, and encouraged the discovery of new ore bodies throughout the district. Ore production quadrupled in one year. Almost 80 percent of the total ore value was silver.

1890 The Sherman Silver Purchase Act of 1890 brought about the biggest mining boom that Rico ever knew. This act called for the U.S. government to purchase 4,500,000 ounces of silver each month, and silver was Rico's main product. The purpose of the silver-purchase act was to stabilize the Nation's currency and to maintain the public confidence in paper money by backing it with silver. Theoretically at least, paper money could be redeemed at any time for silver coinage. The amount of silver that the government was to purchase was about equal to the country's production, and the act specified that the silver was to be bought at market price. The result of such large purchases was to increase the market price of silver from 94 cents an ounce to $1.05 an ounce in 1890. Rico could not have been in a better position—the opportunity to sell large amounts of silver at a high price. Mine owners and managers had correctly forecasted the boom, and the Rico area's silver production increased from 123,852 ounces in 1888 to 618,615 ounces in 1889, and to 848,785 ounces in 1890. All the mines in the Rico area were in a square about three miles on each side. About 2,000 mining claims were recorded in this area.

Rico was the only incorporated town in the Pioneer Mining District. The official 1890 census showed that the population of Rico had grown to 1,134 people. An additional estimated 400 people lived outside the town limits, but within the mining district. With the population soaring, there was not an empty building to be found. Earlier wooden buildings were replaced with brick and stone structures. Some of these buildings are still standing today (as of 1995). The first church building in Rico was the Congregational Church. It was organized by the Rev. T. C. Dickinson, with financial help from the national organization of the Congregational Church.

(Continued on Page 129)

From THE DOLORES NEWS, January 2, 1886

OF INTEREST TO RICO

One of the first results of re-organization of the Denver & Rio Grande Railway will be the construction of a branch by way of the Big Bend to Rico. The company could hardly make a more profitable inventment.

— *The Durango Herald*

OVERLEAF: An unknown man had climbed a short distance up Expectation Mountain in order to photograph this view of Rico. This presumably was done about 1890, as the Hotel Enterprise was in its original small size. This hotel underwent extensive rebuilding, beginning on March 20, 1891. It is seen as the largest building in Block 13, Lots 11, 12 and 13. The First National Bank building had been erected near the center of this view. Most of the buildings are marked or drawn on the Sanborn Map of 1886. Later, the bottomland along the Dolores River (in the foreground) was sold to the Rio Grande Southern Railroad for its station grounds. In this view you are looking toward the east, with Mantz Avenue in the center and the Silver Creek drainage at the left. The white front of the St. James Hotel can be seen across the street from the Hotel Enterprise.

CHARLES MILLER ENGEL COLLECTION
– COURTESY RACHEL A. HARDWICK

FROM THE RICO DEMOCRAT
Published at Rico, Colorado, 1891

— RICO —

"THE PRIDE OF THE SAN JUANS!" OUTPUT OF MINES FOR 1890: $1,250,000

WHERE IS RICO?

RICO, the business center and seat of government for Dolores County, is situated on the Dolores River, at an altitude of 8,737 feet. At present, the distance to the railroad on the north is 26 miles, at Ilium [Vance Junction, on the Rio Grande Southern], and on the south [and east] 35 miles to Rockwood [on the Silverton Branch]... of the Denver & Rio Grande.

TO REACH this Land of Promise, two routes are open from Pueblo [Colorado], by either of which the traveler is greeted on every hand by a view of the grandest scenery conceivable to the mind of man. By one route the Royal Gorge of the Arkansas Cañon, the sublime views of Marshall Pass, the Black Cañon of the Gunnison, and the broad Valley of the Uncompaghre, like a panoramic view, fills one with wonderful delight. The other route, to the south, is via Alamosa, the Conejos Range [over Cumbres Pass in the San Juans, and up]... the Valley of the Animas and Durango, is no less attractive in point of grandeur.

A STAGECOACH RIDE of 30 miles through pleasant valleys and wild gorges brings you to the Valley of the Dolores; thence, up the river for five miles, and you can exclaim "EUREKA!"

BUSINESS POINT

BEING THE ONLY TOWN of importance within a radius of 50 miles, the immense mining interests of the [Pioneer Mining] District, together with cattle, coal and timber industries being tributary thereto, has enabled the merchants and other businessmen to realize handsomely on business ventures.

WITH THE COMPLETION of the Rio Grande Southern early this season [1891], every indication points to the doubling of her population. With proper facilities for the shipment of the product of her mines, industries here that have lain dormant, will again re-echo through the valley, giving lucrative employment to thousands of restless-and-willing hands.

TO THE CAPITALIST, we say, "This is your opportunity!" To [the person]... of limited means, we say, "Come!" To all we say, "Join in making this what the Almighty intended it should be—the home of a wealthy-and-contented people, in the enjoyment of the fruits of their industry, to which they are justly entitled."

RICO'S MINES

MINERALS were first found here in 1869, by a surveying party, and a few claims were located. Not until 1879 was any considerable mining done. In that year, rich minerals were found in what is now known as Newman Hill, and from which colossal fortunes have since been taken, with no perceptible diminution in either quantity or grade of ore.

RICO'S STATUS as one of the leading mining district's of Colorado was firmly established in 1889, when the Enterprise Mine was opened, disclosing bodies of minerals of almost-fabulous richness—from which princely fortunes have been amassed, and there are now on its dumps ore to the value of more than $100,000. Among other prominent mines are the Jumbo, Swansea, Laura, Golden Age, Pigeon, Logan, C.H.C., Butler and Cobler, all of which are netting their owners handsome returns.

RICO'S MINING DISTRICT is not alone confined to the precious metals. Immense beds of bituminous... coal crop out for a distance of over 15 miles. The output of Rico's coal mines in 1890 was over $15,000, all of which was consumed at home.

HEALTH RESORT

ONLY A FEW MILES from Rico, on another branch of the Dolores River, are the Hot Springs [at Dunton]. To the healing properties of the waters of which many pain-racked mortals can bear testimony. The cure of rheumatics in countless numbers has here been effected during the past few years.

NEAR THESE SPRINGS a [resort] hotel has been erected, and those who have partaken of its fare and enjoyed the hospitality of Mine-Host Dunton, take leave of this delightful spot with regret.

TROUT FISHING is also to be found near these springs. Those who enjoy the chase can find on the Table Lands near the hot springs, blacktail deer in countless numbers, which fall an easy prey to the hunter. Mountain grouse are also plentiful, while in every thicket and on every hillside mountain quail abound in great number. To the hunter who is intent on more noble game, an opportunity is offered, as his search for the black and grizzly bear will be rewarded with a view of "Bruin" on his native heath.

RICO'S HOTELS

THE TRAVELER, tourists and others, who prefer "hotel life," will find all the conveniences here to be found in any first-class hotel. The Hotel Enterprise, built this last season at a cost of $32,000 [a great sum in 1890], is first class in all its appointments, has attentive-and-polite servants, and a culinary department to which none in the state is superior. The Hotel St. James also commends itself to the favorable consideration of the traveling public.

CHARLES MILLER ENGEL COLLECTION – COURTESY RACHEL A. HARDWICK

THE RICO DEMOCRAT, dated August 21. 1891, carried an article stating that..."J. H. Goldworthy has secured a contract to build a wagon road to the Newman mines. The road starts at a point near the Pasadena Smelter, below town. This would seem to indicate that the facilities heretofore enjoyed by the Newman people are not sufficient to meet the requirements of the company in future operations." This photograph, taken about 1890, shows the Newman Group, located at the far right. Dolores Mountain is at the left, while Newman Hill is at its base. There were about 50 miles of tunnels inside Newman Hill, the lower part of Dolores Mountain. Note that many of the mine names shown on this photograph are incorrect. The Vestal Mine actually was at the place marked "Star," above the "Montezuma" lettering (at the upper right). Please note that the Enterprise Tunnel is shown at the wrong location. It should have been marked farther to the right. Only a barn and cleared bottomland, used to grow hay, occupied the flood plain beside the Dolores River.

MINING IN RICO

PRIOR TO COMPLETING the RGS into Rico and constructing the Enterprise Branch, THE RICO DEMOCRAT'S editor, T. J. Tarsney, encouraged mine owners by reporting their news in a positive way. This, in turn, helped to attract Otto Mears' attention regarding the need to haul ore by railroad out of Rico.

THE ASPEN MINE

JANUARY 15, 1891—M. A. Wood, of Rico, and Messrs. Fitzgerald and Van Atta, of Telluride, have for some time been actively engaged in sinking a shaft on this property. Work was brought to a standstill about three weeks ago, their shaft house having been destroyed by fire, which resulted in a total loss of their machinery. With commendable energy they have erected other buildings and placed in position new and improved machinery, and work will at once be resumed. As this property is located in line with

(Continued next page)

the Enterprise, Newman, Jumbo and other great properties, we may soon hear that they have reaped the reward which their energy and enterprise merits.

JANUARY 30, 1891—Active operations are going on at the Pro Patria Tunnel. The work is under the direction of Thomas Young. The tunnel has been run in a distance of 40 feet.

MINES AND MINING

PAYING MINES AND PROMISING PROSPECTS

Few Shipments, But Much Development Work, the Present Condition of Our Camp

JANUARY 15, 1891—Many things have conspired to bring about the present lull in mining matters, as well as in Rico as elsewhere throughout the West.

On Newman Hill the big producers are not shipping to any extent, and while they are employing about their usual number of men, they confine themselves to the work of developing their respective properties and are breaking little or no ore. The Jumbo is gradually proving the value of its owners' investments; the Enterprise grows more valuable day by day; the Montezuma keeps pace with its already established popularity, and the Newman is still exposing to view the hoarded treasures of the celebrated Chestnut vein.

The other properties on Dolores Mountain and throughout our camp are not *non plussed* by the present unpromising outlook, but are daily driving their tunnels and sinking their shafts to the sure success of Rico and their respective interests.

A few days yet, when Rico gets her railway, and receives the attention of capitalists; and when the silver bill, which is inevitable, shall have passed, Rico will be the land of promise and of fortune of the San Juan country.

Mining Notes

The Union-Carbonate people are still sinking their shaft, which is now in the lime-shale formation and at a depth of 275 feet.

Considerable work is being done on the Telegraph [mine] by W. W. Parshall. A vein of high-grade mineral has been opened up, and they are now driving for air.

W. W. Parshall and John Clarke are sinking a shaft on the Grand Prize, which adjoins the Nebraska and Enterprise. This shaft is down about 300 feet, where they have already cut the contact.

Work on the Aspen on Newman Hill still continues, and with better success in the sinking of the shaft, which is now sunk to a depth of 330 feet. Water, which has been their great obstacle, is now decreasing.

The Nebraska is shipping as much ore as any mine in camp. During the past week two carloads of high-grade ore, identical with that found in the contact vein of the Enterprise, have been sent to the smelter.

The Stephenite tunnel in Deadwood Gulch, which is owned and worked by the proprietors of the Snowflake, is now in a little over 50 feet. It is expected that this tunnel will cut whatever leads may be found in the Snowflake tunnel 300 further in depth.

In the Silver Age tunnel on Newman Hill, owned by the McLaughlin Brothers, and under lease to Hay & Houghton, a streak of very high-grade ore has been discovered, which gives promise of a pay lead in the near future. There is no reason why this should not be as good a property as its neighbors, with sufficient development.

Under the management of N. H. Cone, the shaft on the Vestal has already reach a depth of 450 feet, and every day adds at least three feet to its depth. The delay occasioned last week by the breaking of the rocker on the station pump has been overcome, and the work of sinking proceeds at the same rapid pace which has heretofore marked the work on this claim.

The shaft on the Ætna group is now 160 feet deep. Mr. L. F. Hill has reason to congratulate himself on his success in sinking this shaft to this depth without any use of machinery. The water, which has occasioned so much anxiety in the past month, has in a measure ceased, and it is probable that the work of sinking may be continued for the next two months without hindrance in this direction, for the water in this, as in the other shafts on Newman Hill, will now steadily decrease until spring.

The Grand View on Nigger Baby Hill is being worked under lease by Joseph Spurlock. He is exploring the Discovery tunnel, where two defined veins have been opened; one perpendicular, and the other an inclined vein. High-grade ore has been found in both, most of which will assay 200 ounces.

The Snowflake group's tunnel on Dolores Mountain, owned by John Eder, Mat Woods and

(Continued on Page 112)

IT IS POSSIBLE that this was the head frame and hoist house for the Enterprise Mine. Sandstone Mountain rises near center of the view. A wood-fired steam boiler provided power for the hoist mechanism that raised and lowered a bucket into the shaft below the head frame. Generally speaking, mining work began higher on the mountain, with a shaft that worked straight down. If sufficient ore was found, a tunnel was drilled into the shaft from a point lower on the mountain, allowing for the use of a nearly level narrow-gauge rail network (a mining tramway) and providing a passageway from the outside world. This tunnel also was used to drain water from the mine since it sloped toward the adit (portal).

ORVAL L. JAHNKE COLLECTION

CHARLES MILLER ENGEL COLLECTION – COURTESY RACHEL A. HARDWICK

DURING JUNE of 1890, Enterprise Mine bullion had been sacked and stored in stacks nearly six feet tall, waiting for the bull (oxen) train to arrive. Metals smelted from the processed ore contained 80-percent silver. Trees cut into specified lengths were stacked outside the smelter to be burned as fuel in a crude attempt at separating out the gold and silver ore. The smelter furnace was inside the building with the large chimney, at the right. A couple of new buildings were being roofed, back in the trees.

others, is at present writing, in 1,150 feet. The hope and expectation of cutting some of the big leads on this hill now impels the steel drill at a lively pace. Scientific examination and survey predicts a success for this property at an early date.

Work on the Logan group on Telescope Mountain, adjoining the C.H.C., was resumed last Monday. This group of claims present a most promising prospect. A fissure vein has been opened in places which is 40 feet between walls, and a contact vein has been discovered more than 15 feet thick at point of discovery. This property is owned by Louis Haberman, Charles Rohde and others, who are full of hope for the future fortune of their property. The ore is of a sand carbonate nature, similar to the C.H.C. adjoining.

Benjamin M. Picker, who represents Adam Herzinger & Co.'s Return tunnel, on Dolores Mountain, expresses himself as well satisfied with the work of development so far done on this group. At 165 feet in the tunnel a lead of carbonates carrying 19 ounces in silver and 66 percent lead, and the sulpherets 17 ounces in silver and 75 percent in lead. After cutting this vein, which was four feet in width, the tunnel was driven 300 feet further, and a shaft sunk to the contact, where a vein of mineral was uncovered, which will no doubt assay 100 ounces in silver alone. This adds another to the list of Rico mines, and bids fair to hold its own with the older properties on this hill.

FEBRUARY 13, 1891—A new ore house is being erected at the Silver Age.

The Golden Age shipped a carload of high-grade ore the first of the week.

(Continued on Page 114)

ORVAL L. JAHNKE COLLECTION

THIS MINING CREW was at the Enterprise Mine in 1890. They were taking time off their shift for a group photograph. They were posing beside and on top of sacked silver ore, each of which weighed between 50 and 80 pounds. This shipment was destined for Swansea, Wales, and it was said that each sack contained $70 to $80 worth of silver. The blacksmith shop was at the right.

FORT LEWIS COLLEGE CENTER OF SOUTHWEST STUDIES

THE OFFICE BUILDING of the Enterprise Mine was the structure at the left, and the blacksmith shop, at the right, could handle mine-car repairs. The entrance to the mine tunnel was beyond the cars, at the right. No structures remain at this location today (as of 1996).

COLORADO HISTORICAL SOCIETY

A NUMBER OF BULL TRAINS had reached the Enterprise Mine tunnel in this view of 1891. Bunkhouses and boarding houses were among the buildings beyond the ore smelter. The wagons had arrived loaded with provisions and would leave with heavy metal bullion. The oxen would have to struggle over the mountain range east of Rico by way of Scotch Creek to reach Rockwood, located on the Silverton Branch of the Denver & Rio Grande Railway.

The Montezuma shipped a car of high-grade ore the first of the week.

The Jumbo is keeping up its reputation—shipping three cars of high-grade ore during the week.

The C.H.C. ore chute, which pinched some time ago has opened up again richer than ever. The ore is being shipped to the Brace sampler.

David Fliesher and Andy Emmerson, who have been running a cross-cut tunnel on the Blue Lode in Horse Gulch, came down Saturday for a few days rest, and returned Tuesday.

The Stanley No. 2 shaft is down 250 feet, and is still being pushed energetically. The black shale, which overlies the contact from 60 to 75 feet, was encountered during the early part of the week.

At the Enterprise, development work is being carried on with the usual force. No ore is being broken, except where necessary in the course of the development work. Nearly 300 tons of high-grade ore [is] in the ore house.

At the Newman group everything is moving along as smoothly as usual, under the management of Mr. H. Klingender. Mostly development work is being done. The shipments amount to about two cars per week of first-class ore.

Work of sinking the shaft on the Ætna still continues, but with considerable opposition from water. The shaft is now about 180 feet deep, and it is to be hoped that fortune will be kind enough to permit these gentlemen to sink to contact ere spring, and thereby avoid being drowned out.

T. J. Lewis, S. E. Curtis, J. W. Picket and F. M. Bryant are the owners of the Outlaw, a claim with a bright outlook, located on Expectation Mountain. They have been at work during the past week, and have opened a vein in the cross-cut that may be a producer in the near future.

The South Park mine is still being worked under lease by Louis Haberman and others. In 1883 and 1884 considerable rich ore was shipped from this property, some running as high as 900 ounces in silver to the ton. Thirty-thousand dollars was offered and refused about that time for a half interest.

Considerable annoyance and delay had been occasioned during the past week to the Vestal by the wearing out of valves; caused by the severe action of the sand matter drawn up by the pumps. The shaft is now down 475 feet, and by our next publication we expect to report that they have struck contact.

Work of sinking a shaft has been commenced on the Iron Dollar, on Nigger Baby Hill. "Sandy" Campbell, one of the owners, is engaged in doing this work and reports marked improvement as they gain in depth. Recent work has developed a well-defined vertical, which a few weeks before was reported as a flat vein.

The dissatisfaction among the parties interested in the present lease of the Butler on Nigger Baby Hill, has caused a suspension of work for a few days. Efforts are being made to harmonize matters, which if successful, will be of advantage to everybody interested; for the Butler is a good property, and if properly operated by the lessees, cannot fail to be successful.

At the Argonaut but little is being done. This claim has been worked in a half-hearted way for a year past by lessees. Some good ore has been taken out and shipped, but the work lacks the vim and energy that goes to make a mine. 'Tis a pity, too, for the Argonaut is one of the best prospects in the district, and with proper development will undoubtedly become a large producer.

The Pontiac Lode on the northern slope of Expectation Mountain, the property of Frazier & Foote, is being worked steadily under the supervision of Mr. Frazier. A tunnel 300 feet in

length has been driven. For over 200 feet the tunnel has driven through a body of iron carbonate ore that assays from 12 to 20 ounces of silver per ton. The tunnel is still being driven ahead in the hope of finding a better grade of ore.

Mr. A. G. Dunning wears a black eye and several other bruises on various parts of his body, the result of a tussle with one of his mules. Last Wednesday, while packing ore from the Golden Age mine to the wagon road, one of the animals in his train became frightened and plunged off the trail, landing in a snowbank some distance below. While trying to free the animal from his pack and dig him out, Mr. Dunning was severely kicked about the face and arms.

FEBRUARY 20, 1891—From T. C. Johnson, a miner employed on the Nebraska, we learn the particulars of a snowslide which occurred on Thursday morning, near that mine. Mr. Johnson left the boarding house to go to his work about 7:00 o'clock, and while walking along he heard a slight noise, which his experienced ear told him was the starting of a slide. Throwing himself on the snow at full length, it was only a moment until the huge wave of snow was upon him, carrying him some 300 feet. Mr. Johnson thinks that had he remained in an upright position, he would not be alive to give his experience. The foreman of the mine, Mr. Charles Cannum, Edward Fosbeck and Joseph Hendershot were only a few moments behind Mr. Johnson when the slide came.

About 3:00 o'clock yesterday afternoon an other snowslide occurred on Dolores Mountain. Three men in the employ of H. Roys, wood contractor for the Enterprise mine, were supposed to be buried in the snow, but later two of the men were reported as not in the slide. A large relief party are now searching the slide for the missing man.

MAY 15, 1891—The Black Hawk will be worked as soon as the snow has sufficiently disappeared from Nigger Baby Hill to permit access to the mine. Considerable ore has been shipped from this property, some of it averaging 600 ounces

MARGARET MATZICK COLLECTION

CHIEF OURAY signed a treaty giving up the Ute's summer hunting grounds in the San Juan Region in 1874. However, it was still considered newsworthy to report on Indians coming into town during the 1890's, although both black men and Navajo Indians commonly worked as cowboys in the Southwest, and the sight of Indian cattle wranglers was nothing new. On September 4, 1891, the Rico newspaper reported that... "A party of five Navajo Indians, with horses, ponies and blankets for sale, was an object of curiosity to many... newcomers...." On one 1890's Fourth of July celebration, these native Americans had come to town, Ute Indians proudly dressed in their finest ceremonial costumes.

of silver; and there is a vein of carbonates of lead and iron, carrying about 35 ounces in silver exposed in this property, which varies from 4 to 10 feet in thickness between walls.

JUNE 12, 1891—Prospecting is being done extensively on Expectation Mountain and along Silver Creek.

The enterprising lessees of the Butler report satisfactory results from their work on this property.

The Adams concentrator has for weeks been pounding away, and the result is a large quantity of ore awaiting shipment, most, if not all, from the Jumbo.

The new owners of the Enterprise [mine] and other properties recently purchased, have been in town during the week looking over their property. We are informed that they intend increasing the force on these mines at once.

Mr. John Farrish, the mining expert, who made the examination of the Enterprise and other mines recently sold to Red Mountain parties, among whom were Messrs. Posey & Crawford, was a passenger on the Ilium stagecoach Tuesday. It is reported that Mr. Farrish will have charge of these rich properties under the new management.

Work on the Black Hawk is progressing rapidly. Mr. Jordan, formerly of the Enterprise, is in charge.

The Enterprise Mining Company have shipped 14 cars of ore since Monday, most of which has been high grade.

Lee Wood, who has a lease on the Grand View, reports that property looking up. Will commence shipment at once.

The Intrinsic, owned by A. P. Adams, is showing a fair body of ore, which runs high in lead and about 20 ounces in silver.

A shaft has been built on the Iron Dollar. This property is being worked by Alex Campbell, et al., who have some good mineral.

Chamberlin & Company have 200 tons of ore at the concentrating works awaiting shipment. This ore is mainly from the C.H.C. [mine].

The C.H.C. and Ethelena are working good forces and have considerable ore in bins. Shipment will commence as soon as cars can be had.

The Adams concentrating works have about 80 tons of Enterprise concentrates sacked and ready for shipment that will run $300 per ton.

OCTOBER 30, 1891—Ore shipments are continuing as fast as cars can be had. Enterprise is sending out ore at the rate of 20 cars per week.

NOVEMBER 28, 1891—Two cars of high-grade ore daily is the present output of the Enterprise. This class of ore runs from $400 to $1,000 per ton. Figure is up.

There is now at the Rico Ore Market about 800 tons of medium-grade ore, which will be shipped as soon as the switch from the Rio Grande Southern is completed.

FROM THE RICO DEMOCRAT
Rico, Colorado, February 20, 1891

RICO'S NEW CHURCH

REV. S. O. DICKINSON IS ORDAINED AS PASTOR

The Rev. G. M. Sanders, Assisted by the Revs. Reid and Chase, Dedicated the People's Congregational Church to the Services of Our Lord

THE HEAVY SNOW that fell during Monday and the evening of that day, did not deter the people from attending the ordination ceremony that had been announced to take place at the [new] People's Congregational Church, and by 8:00 o'clock, that beautiful structure was filled by the people of Rico, who take a just pride in the long looked-for consummation of their desire to see such a building erected.

The exercises began by the rendering of the following beautiful hymns: "Nearer My God to Thee" and "What a Friend We Have in Jesus," followed by the Rev. J. H. Reid of Telluride, after which "Coronation" was sung by the congregation.

A discourse was then delivered by the Rev. Reid; subject, "The Fatherhood of God," which was delivered in a forcible and sound, argumentative manner, and listened to with rapt attention by the immense audience.

Then, the Rev. C. M. Sanders, superintendent of Home Missions, lifted up his voice in supplication to the Throne of Grace for approval on their work; following which, the Rev. A. L. Chase offered up the ordination prayer, during which all the ministers assisted in the solemn-and-impressive ceremony of consecrating the life of him who had been chosen to lead the flock, to the service of the Master.

The Rev. Sanders then gave the charge to the new minister, which if observed, cannot fail to do good work in this part of the moral vineyard, and give us a higher moral standing.

At the close of the ordination service, the "Avé! Maria!" ["Hail! Miriam! Mother of the Messiah!"] was rendered by Mrs. F. L. Thompson, who was assisted by Mrs. E. L. Davis at the organ.

After the benediction, a great many pressed forward to shake the hand of the newly ordained minister, and [they] were introduced to the visiting clergymen.

All went to their homes, feeling that the evening had been profitably spent.

On Tuesday evening, the exercises attending the dedication of Rico's new church building was held in that beautiful structure on Mantz Avenue. A large attendance greeted the new pastor, who was assisted by the Rev. Sanders, superintendent of Home Missions for Colorado and Wyoming, the Rev. Reid of Telluride, and the Rev. Chase of the Montrose parish.

The [dedication] exercises began at 7:30 p.m. by the singing of an anthem by the choir, led by Mrs. F. L. Thompson, followed with prayer by the Rev. Chase.

After the rendering of some appropriate music by the choir, responsive readings were had, the Rev. Reid taking the part of the pastor.

The Rev. Sanders then delivered the dedication sermon, which was a fervent appeal to the people to rally around the pastor of the church, and to aid and assist him in the upbuilding of the good work in Rico.

At the close of the sermon, the "Avê Maria" was rendered [as it was during the Monday service], and the benediction [was] pronounced by the pastor, after which the congregation dispersed.

CHURCH SERVICE DISRUPTED

Rico, Colorado, April 24, 1891

THERE are a number of young hoodlums in Rico whose conduct at church on Sunday evening was such that the minister felt called upon to eject them. If they have no regard for the feelings of others, or reverence for the place, they should be brought to a knowledge of the fact that there is a severe penalty attached to disturbing religious worship.

NEW FENCE AT CHURCH

Rico, Colorado, August 7, 1891

A. O. HEADLY, largely interested here in mining and milling, departed for his home in Newark, New Jersey, on Monday. Before his departure, he made a donation to the church, by which the church grounds will be properly graded and a fence built to enclose... [the grounds].

CHARLES MILLER ENGEL COLLECTION – COURTESY RACHEL A. HARDWICK

LOCATED ON THE NORTHWEST CORNER of the intersection of Mantz Avenue and Silver Street, the People's Congregational Church was in very good condition at the time this photograph was taken. This church probably was erected during 1890, and it was the first church building in Rico. THE RICO DEMOCRAT reported that the fence was being built during the summer of 1891. As of the mid-1950's, Rico's Presbyterian congregation had assumed the ownership of this building.

CHARLES MILLER ENGEL COLLECTION – COURTESY RACHEL A. HARDWICK

A FEW YEARS LATER, a new two-story brick building had replaced the small wooden house that once stood behind the church (to the north). To the northwest, 12,337-foot snow-shrouded Elliot Mountain looms up in the left background. THE RICO DEMOCRAT, dated March 27, 1891, reported: "About a dozen men contributed work on new seats for our church with 15 new seats; 150 people can be accommodated."

FROM THE RICO DEMOCRAT:

JULY 31, 1891—The Rev. S. C. Dickinson of the People's Congregational church was agreeably surprised on opening that sanctuary last Sunday morning, to find that someone had been there before him and had placed in position for his use a large upholstered chair, and had also spread upon the platform a fine Brussels carpet of neat design. It was subsequently ascertained that these neat and useful additions to the church furniture was the gift of Silver Crescent Lodge No. 40, [of the] I. O. O. F.

Rev. Father Pickle of Silverton visited Rico last week and conducted Catholic worship on Friday, Saturday and Sunday. The services conducted here are known to the faithful of that persuasion as a "retreat." During the stay of the Rev. gentleman here, considerable interest was manifest by the Catholic people, and it is more than likely they will have purchased or built a place of worship in a short time, a committee being at present working to that end.

August 28, 1891—The Rev. Father Pickle of Silverton, will celebrate mass next Sunday morning at G. A. R. hall and also conduct service in the evening. All invited.

December 12, 1891—The new bell for the church has arrived, and will be used tomorrow to call out Rev. Dickinson's flock.

LARRY PLEASANT COLLECTION

WE DO NOT KNOW the circumstances concerning this gathering on the front steps of the People's Congregational Church in Rico, but it is possible the ladies were wearing Easter bonnets ca. 1900.

DURING THE 1920's, women's hats changed dramatically, and, of course, it was nearly considered sinful not to wear one at church, especially for Easter services. The kids in this group were probably eagerly awaiting their chance to frolic in the snow. Apparently, the photographer was using a Kodak postcard camera from the size of the leather carrying case hung on one of the fence pickets.

CHARLES E. LEE PHOTO, NORMA L. BUTLER COLLECTION

THE RICO DEMOCRAT.

VOL. 1. RICO, DOLORES COUNTY, COLORADO, FRIDAY, MARCH 6, 1891. NO. 8

OFFICIAL DIRECTORY.

Rico Town Officials.

Mayor	F. L. Thompson
Clerk	F. R. Lewis
Treasurer	W. W. Parshall
Trustees	J. C. Clark, J. N. Brough on, E. H. Hackett, J. O. Campbell, J. P. Landon, F. L. Thompson.
Marsha.	J. J. Hefferman
Police Justice	C. A. McLaughli

Dolores County Officials.

Judge	Albert E. Kennedy
Sheriff	John C. Clark
Clerk and Recorder	Frank R. Lewis
Treasurer	W. W. P rshall
Assessor	Charles F. Burger
School Superintendent	Fernando Roys
Surveyor	Albert E. Armo
Commissioners	E. H. Hackett, J. M. Dodd, S. R. Robinson

Sixth Judicial District Officials.

District Judge	Geo. T. Sumner
" Attorney	Chas. A. Johnson
Clerk of Court (Dolores Co.)	F. R. Lewis

State Officials.

Governor	John L. Routt
Lieutenant Governor	William Story
Secretary of State	Edwin J. Eaton
Auditor	John N. Henderson
Treasurer	James N. Carlisle
Sup. Pub. Ins	Nathan B. Coy
Attorney General	Joseph H. Mauphin
State Senator 21st District	Lou N. White
Representative for the counties of Ouray San Miguel and Dolores	Alonzo P. Adams

United States Officials.

Senators	H. M. Teller, E. O. Wolcot
Representative in Cong	Hosea Townsend

Mail and Travelers' Directory

Arrival and Departure of Mails.

Postoffice open from 8 a. m. to 8 p. m. On Sundays from 10 to 12 a. m.

Registry and money order department open from 8 a. m. to 5 p. m.

Mail arrives at Rico, and departs according to the following schedule:

ARRIVES.

Durango mail....................7:30 p m
Eastern mail, via Telluride....................2:00 p m

DEPARTS.

Durango mail....................7:00 a m
Eastern mail, via Telluride....................9:30 a m

JOSEPH MEREDITH, P. M.

Rico Stage and Express Line.

TIME SCHEDULE.

Stages depart from Rico for the East via Rockwood at....................7 30 a m
Stages depart from Rico for the East via north Hard ... at....................9 30 a m

Local Organizations and Societies.

PEOPLE'S CONGREGATION CHURCH, Corner Mantz avenue and Silver street Service every Sabbath at 11 a. m. Sabbath School, 2:30 p. m. Evening service, 7:30 p. m. S. O. Dickinson, Pastor.

HAZEN POST, G. A. R., meets the second and fourth Saturday of each month. S. H. Burghart, Commander.

McPHERSON CAMP, S. of V No. 9. Meetings every first and third Saturday of each month at G. A. R. hall. A. H. Mundee, Captain.

SILVER CRESCENT LODGE, I. O. O. F., No 40, holds it regular meetings at the hall on Wednesday evenings at 8 o'clock. Visiting brothers are cordially invited to attend. Geo. Sanborn, N. G.; R. C. Jones, Secretary.

RICO ENCAMPMENT, I. O. O. F., No. 19. Regular meetings the first and third Saturday in each Month, at 7:30 p. m. at Odd Fellow's hall. James Mea.s, C. P.; W. J. Watts, Scribe.

PROFESSIONAL.

G. D. YOKOM,
PHYSICIAN AND SURGEON.
Office at Postoffice Drug Store.
RICO, - COLORADO.

J. P. LANDON,
PHYSICIAN AND SURGEON.
All Calls Promptly Attended, Day or Night.
RICO, - COLORADO.

JOHN KINCAID. W. A. WILLIAMS
KINCAID & WILLIAMS,
ATTORNEYS.
Office over First National Bank.
RICO, - COLO.

T. J. TARSNEY,
ATTORNEY-AT-LAW.
Room 5, Over First Nat. Bank.
RICO, - - COLORADO.

CHARLES A. M'LAUGHLIN,
JUSTICE OF THE PEACE
—AND—
POLICE MAGISTRATE.
Collections attended to and Prompt Remittances made. Office over First National Bank.
RICO, - - COLORADO.

W. B. HESS,
Real Estate
AND
Insurance Agent.
NOTARY PUBLIC.
RICO, - COLORADO.

JULIUS THOMPSON F. W. INGERSOLL
THOMPSON & INGERSOLL,
LAWYERS.
Durango, Silverton,
—AND—
RICO, - COLO.

RICHARD M'CLOUD,
Register of the U. S. Land Office at Durango, Colo, from September 1, 1886, to September 1, 1890.
ATTORNEY-AT-LAW
Durango. - Colo.

Land and Mining Cases before the Courts, U. S. Land Offices, General Land Office and Interior Department, a specialty.

Eighteen Years' Experience as a Lawyer.

Refers to present Register and Receiver of U. S. Land Office, Durango, Colo.; and calls attention to the following testimonials:

From Senator Teller, Ex-Secretary of the Interior.

UNITED STATES SENATE, WASHINGTON, D. C., Sept. 6, 1890.
To whom it may concern:
Mr. Richard McCloud has been for the past four years register of the Land Office at Durango, Colorado. He has, so far as I am informed, made an excellent record for promptness, efficiency, and integrity in that position. He is a graduate of the Columbia College Law School, of New York city, and I consider him a very competent man for the transaction of any legal business connected with public land matters. Respectfully, H. M. TELLER.

From Copp's Land Owner, Washington, D. C., September 15, 1890.
The land service has lost a valuable official in the retirement of Register Richard McCloud from the Durango, Colorado, office. He will, probably, practice law in Durango, and, doubtless, reap the reward due his strict attention to duty and the exhibition of industry and ability during his official term.

CHAUNCEY W. BLACKMER,
Attorney-at-Law.
Will Practice in all the Courts.
Cortez, - - Colorado

GEORGE N. LAW,
U. S. Deputy Mineral Surveyor
All Kinds of Surveying Promptly Attended to. Office on Commercial St.
Rico, Colorado.

RICO BATHING AND SHAVING PARLORS.
Next Door to Burley's.
Hot and Cold Baths.

THE FIRST
NATIONAL BANK
RICO, COLORADO.
CAPITAL, - - $50,000
J. E. McCLURE....................President
L. L. BAILEY....................Vice-President
D. E. MOREY....................Cashier
GEO. E. NOLTE....................Assistant Cashier

CLIFTON HOTEL.
Mrs. J. L. Berlin, Prop.
Board by the day or week. Everything First Class.
GLASGOW AVE. - RICO, COLORADO.

St. James Hotel.
ALFRED ARCHAMBAULT, Proprietor.
MANTZ AVENUE, - RICO, COLO.
Best Cook and Finest Bill of Fare in Southern Colorado. Board, $7.00 per Week. Meal Tickets, 21 Meals, $7.50. Three Meals for $1.00.
EVERYTHING FIRST CLASS.

PATENTS!
Caveats, Trade-Marks and Copyrights
Obtained, and all business of the U. S. Patent Office attended to for moderate fees.
Our office is opposite the U. S. Patent office, and we can obtain patents in less time than those remote from Washington.
Send model or drawing. We advise as to patentability free of charge, and we ask no attorneys' fee until patent has been allowed.
For book "How to Obtain Patents," with references to actual clients in your vicinity, write to
C. A. SNOW & CO.,
Opposite U. S. Patent office, Washington, D C

WHAT HAVE THEY DONE?

The American Congress has adjourned, and we can now look back and see what they have accomplished in the way of needed legislation. The term has been frittered away, and of the acts of that body we point to no single one that can be said to be a measure of relief.

They have negotiated a few treaties with some of the minor foreign powers looking to trade relations; and yet, out of these, we fail to see in a single instance where the American people have gained anything.

That which in our opinion, would have tended to restore confidence in this country, and have placed the business of the nation on a secure footing,—the enactment of financial legislation, thus determining the status of silver,—has been passed by; while whole weeks of the term have been spent in the attempt to fasten upon the country an election law, that had for its aim the perpetuating of a political party, by placing the machinery of federal elections in the hands of the United States marshals, backed by the military; an idea that is distasteful and repugnant to him who looks with any degree of solicitude to the perpetuation of free institutions.

The people looked to the Congress just closed for the enactment of just and salutary laws in the interest of the whole people, and looked for the passage of measures in the interest of the millions, from whose toil comes all the material wealth of the country. They had reason to hope that the day of class-legislation and the passage of sumptuary laws had passed.

We had hoped that the demands of the people for relief, as evidenced by the popular uprising of the masses at the recent election, would have its effect on those who are now about to retire—some of them we hope forever—from a legislative body in which they have by their course been false to every pledge made by themselves and by their party platforms. From those who will succeed them in the next congress, have we any hope of the fulfillment of *their* pledges?

We think we know the sentiment of the people on the proposition to give the country a free coinage silver bill,—to restore that metal to the place from which it ought never to have been degraded. In view of the fact that a large majority of the people favor such a bill, does it not strike one that there is something radically wrong in the machinery of a so-called popular government, when the express will of a great majority can be set at naught, through the workings of a party caucus? Illustrative of this, we site the case in the recent Congress, where two at least, both able and eloquent advocates of a free coinage bill, who spoke the sentiment of their constituents; yet whose voices were smothered and their vote cast against it in the caucus of their party. We speak of Payson of Illinois, and Butterworth of Ohio.

The American people are slow to wrath, yet they have never yet failed to visit their condemnation on any man or party who has proven recreant to a public trust. It is a hopeful sign when we see men of the rank and file, as well as the representatives of their party breaking away from the restraints of party rule, and thoughtfully and patriotically dealing with questions vital to the interests of the people in an independent and manly way; men for whom the party lash has no terror.

Honesty the Best Policy.

He (trying to play a trump card)—As I passed your house last evening I thought I heard an angel sing.

She (stiffly)—I was at the theatre last evening. Mrs. Mulhooly and her twins were at our house visiting the cook.—Good News.

In France steam races are not permitted. Instead, an official test of the power, speed and steadiness of locomotives on different lines of the republic has been made, and a prize awarded to the winner.

A few thousand years ago—in the geological yesterday, as we may say—the northern part of North America was generally covered by great fields of ice, which were a mile or more thick in their deeper parts, and extended south as far as central Pennsylvania and northern Kentucky.

Heavy floods prevail throughout the east. Johnstown, Pa., and other points are again inundated. The Ohio, from Wheeling to its mouth, has a big rise; while the Big Canawha and other of its tributaries are running bank full. Fears are entertained that great loss will be the result.

With the opening of spring, will come a season of activity in all kinds of business. With the building of the railroad both ways from Rico and the additions to the working force on all the mines, together with the fact that a dozen properties opened during the past year, will add greatly to the output of the camp. We may look for a season of prosperity such as the town has not witnessed since the first hardy prospector pitched his tent between the hills.

Governor Routt has an eye to the economic side of things. Before appropriating any larger sum to the Columbian exposition, the Governor is of the opinion that the duty of the legislature is to make some provision to meet the pressing obligations of the state. He says, however, that a reasonable appropriation might be made to get together an exhibit, and that two years hence another appropriation might be made, should the way seem clear to such action. The governor says to be just before we are generous.

It is not alone the Democrats that realize that in Senator Ingalls a brilliant man retires to the shades of private life. Senator Ingalls retires as does the meteor which, for a brief period, illuminates the sky without having had any perceptible influence on it. It is not true that he has earned a commendable reputation by putting himself on record against gag law. It is not true that Ingalls ably represented the west. Senator Ingalls was first and foremost of his party on every question presented for keeping alive the animosities engendered by the war; it was Senator Ingalls who from his seat in the Senate for years, on every occasion, fanned the smouldering embers of civil strife; it was Senator Ingalls who, as an appeal to the prejudice and not to the reason of his party friends, that played the role of a hyena and dragged from the grave the bones of the gallant Hancock, the Chivalier Bayard of American soldiery; hero of Gettysburg, and held them aloft to the gaze of his admiring partizan friends. No, it is with no regret, that we bid adieu to Messrs. Ingalls, Hampton and men of that stripe, whose stock in politics consists of ante-bellum thunder, can will be spared by the people. Such fossils must give way for men whose ideas are more in keeping with the progressive age in which we live. If Ingalls and Hampton had forgotten that the war was over, there were enough old soldiers in the Kansas and South Carolina legislatures to call their attention to the fact.

Canine Intelligence.

One of the intelligent dogs lives in Bar Harbor, Me. He was carrying a paper the other day when several canine companions began to bother him. He put the paper down on the ground, and when a dog attempted to touch it sprang on him and gave him a good shaking. These tactics he repeated several times, till at last he could not get any dog to touch the paper, and then he quickly picked it up and walked away.—Kennebec Journal.

A "Hello!" Raise.

One telephone was put in at a small town in Kansas, and the owner of a house to rent immediately raised the price $5 per month. Then he went over and called up a sawmill half a mile away, and burst a blood vessel trying to keep up a conversation over the wire.—Detroit Free Press.

Better Not Sigh.

A chemical analysis of the food cooked by our grandmothers, which men sigh for once in a while, would show it 35 per cent. more dangerous to the stomach than food prepared in the modern kitchen. As a matter of fact, American women knew nothing of cookery up to twenty-five years ago.—Detroit Free Press.

Stop and Think.

A New Yorker worth $25,000,000 says he took the most comfort when his wealth counted up about half a million. A man with a million can take all the comfort that one with $75,000,000 can buy, and he has only one-seventy-fifth of the cares and anxieties. Stop, young man—stop at a million.—Detroit Free Press.

If you wish to have your portrait taken on ivory, face an enraged elephant.—Texas Siftings.

DAVID SWICKHIMER, President. R. M. FREEMAN, Vice-President
W. W. PARSHALL, Cashier. LOUIS HABERMAN, Secretary

RICO STATE BANK
CAPITAL STOCK, - - $30,000.
UNDIVIDED PROFITS, - $11,000.
Does a General Banking Business. Gold, Silver, Bullion and Exchange Bought and Sold. Town and County Warrants Bought and Sold.
Exchange on all the Principal Cities of Europe.
CORRESPONDENTS:
FIRST NATIONAL BANK, Denver. COLORADO STATE BANK, Durango.
WESTERN NATIONAL BANK, New York.
BOARD OF DIRECTORS:
DAVID SWICKHIMER. L. D. SWICKHIMER. R. N. FREEMAN.
LOUIS HABERMAN. JOHN EDER.
THOMPSON & INGERSOLL, Attorneys.
Banking Hours from 9 a. m. to 3 p. m.
COLLECTIONS PROMPTLY ATTENDED TO

J. W. BURLEY & CO.
THE LEADERS IN
DRY GOODS,
Groceries, Boots, Shoes
—AND—
GENTS' FURNISHINGS.
We Carry the Finest Stock of the Above Goods Ever Brought to the City of Rico.
DOUBLE STORE!
GLASGOW AVENUE.
RICO, - - COLO.

CLARKE BROTHERS,
DRY GOODS,
BOOTS, SHOES
—AND—
Gents' Furnishings
The Oldest Firm, the Largest Stock, the Best Goods, the Lowest Prices. Call and see them. They Will Give You Good Bargains.
DRY GOODS AT COST FOR CASH!
GLASGOW AVENUE.
RICO, - - COLORADO.

J. H. GOLDSWORTHY,
Livery, Feed AND Sale
STABLES.
Hay, Grain and Coal.
Coal Delivered to the Mines at Reasonables Rates

STORIES FROM THE RICO DEMOCRAT

Published in Rico, Colorado, Dolores County, in 1891
Founded and Edited by the Hon. T. J. Tarsney

THE RICO DEMOCRAT was published weekly during all of the remarkable year of 1891. This year was considered to be a springboard to the future, and memorable as the beginning of an era of prosperity for Rico.

The respected *Pathfinder of the San Juans*, Otto Mears, was building his narrow-gauge Rio Grande Southern Railroad into this remote region of southwestern Colorado—through the rugged mountains and canyons south of Vance Junction, the Ophir Loop and Lizard Head Pass. Otto's remarkable Rio Grande Southern reached Rico late in the year, on October 15, 1891, and shortly thereafter, the rails connecting Ridgway with Durango were joined—between Rico and Dolores.

According to long-time Rico resident and newspaperman, Charles M. Engel, the editor of ***The Rico Democrat*** reported the news "...in comprehensive detail, without embellishment or window dressing. Important happenings that made history literally come alive from the pages printed with hand-set type."

Many of the stories Editor Tarsney objectively published on the front pages (and elsewhere) in ***The Rico Democrat*** are reproduced in this volume of ***The R • G • S Story.*** The articles are a factual condensed history of the early days in Rico and the surrounding Pioneer Mining District.

For 14 years Mr. Tarsney was a locomotive engineer. He had resided in La Plata and Montezuma counties, and for five years he was a practicing lawyer. When he moved to Rico, he established his law office in the First National Bank building, and he continued his practice from this location in what was then the county seat of Dolores County.

The Democratic Judicial Committee endorsed the Hon. Mr. Tarsney for the office of district attorney for Colorado's Sixth Judicial District during the autumn of 1891. It is presumed that it was in the interests of his political career that his publication of ***The Rico Democrat*** was terminated with the issue of October 30, 1891, shortly after the Rio Grande Southern reached Rico. The newspaper was continued as ***The Rico Sun,*** published by Rust & Hutt.

LARRY PLEASANT COLLECTION – COURTESY KEN HAZEN

THIS IS AN EXAMPLE of The Rico Democrat, the weekly newspaper printed at Rico prior to the arrival of the Rio Grande Southern Railroad. The front page of the newspaper carried advertisements for both the First National Bank (in Column 2) and the Rico State Bank. (David Swickhimer, the owner of the Enterprise Mine, was the president of the latter bank at this time.) Rico residents were anxiously waiting for the RGS to build into town, as stated in Column 4, Paragraph 2.

RGS Will Benefit Rico Mining

FEBRUARY 6, 1891—We are indebted to W. C. Brace of the Rico Sampling Works for a call at our office on Wednesday last. In a short talk with him he expressed the hope and prophesied the great benefit to mining interests of the early advent of the railroad into Rico. We are now too far distant from railroad transportation and telegraphic communication to warrant the sampler in buying particularly high-grade ore, on small margins, and the disadvantage of our position is now considerably magnified by the present uncertainty and fluctuation of silver prices.

The management of the Rio Grande Southern would confer a great favor, and one that would be gratefully appreciated, if they would issue an order to the effect that when the regular passenger train from Ridgway is late, that passengers from Rico can come to Telluride on the freight train [as a mixed train]. There are absolutely no comfortable arrangements at Ilium for the accommodation of people who come in on the Rico stagecoach, and the wait of three hours is bad enough, but when the regular passenger train is, as it was last Thursday night, three hours late, the discomforture grows into absolute suffering. Will President Mears make a note of this and issue the necessary order?

—from *The Telluride Journal*

FROM THE RICO DEMOCRAT
Published at Rico, Colorado, 1891

Tie-Choppers Busy on RGS

FEBRUARY 20, 1891—Tie-choppers are busy along the entire line of the Rio Grande Southern between Rico and Durango. Work of grading the road is going on at points where they have heavy cuts, or rock work. With the melting of the winter snow, work will be pushed rapidly, and before the season closes, connection will be made.

Rio Grande Southern News

APRIL 24, 1891—The Rio Grande Southern Railroad is an enterprise which is destined to assure the speedy development of southern Colorado, and also to be of large commercial advantage to this city. No Rio Grande Southern bonds have yet been sold. The road thus far has been built with Colorado money, raised in Denver, Colorado Springs and Pueblo. It has been completed as far as Telluride, to which point its trains have run regularly all winter. Work has been constantly pressed, and the road, it is expected, will be running by way of Rico to Durango by August 1. Its ultimate destination is Phoenix, Arizona, and surveys from Durango south and southwest have been in progress for many months. Although only opened for business late last fall, the completed section between Ridgway and Telluride has paid something more than interest on the investment and the fixed charges besides.

Statements have been made suggesting that the enterprise has been crippled for want of funds. All such mention is erroneous. It is being pushed ahead with vigor. It has had the earnest support of **THE NEWS** from its inception, as an enterprise which would not only develop a rich and productive section of the state, but largely increase the area of Denver's trade territory. The day on which it is completed to Phoenix, Arizona, will be a fortunate one in the commercial history of our city.

—from *The Rocky Mountain News* of Denver

The Work Force of the RGS Will Be Increased in May

MAY 8, 1891—The force of laborers on the Rio Grande Southern will be largely increased on May 15, and the road will be pushed to completion as rapidly as possible.

—from *The Montezuma Journal*

ARTICLES FROM THE RICO DEMOCRAT...

"Frail Sisters" Admonished

March 13, 1891—At a hearing before Justice McLaughlin last Saturday, his honor took occasion to admonish several of the "frail sisters" present. He said that "...should they continue the practice of visiting the saloons and making a meretricious [tawdry] display of themselves, that the provisions of the statute provided punishment [for prostitution] would be enforced."

HOTEL ENTERPRISE ENLARGED

March 20, 1891—Workmen are engaged in making alterations in the Hotel Enterprise. The dining room and office will be enlarged, and the saloon removed to the room known as the Hotel Hall. This house [hotel] is undergoing a thorough renovation, and on its opening in the near future, it will, we are assured, be first class in all its appointments.

April 7, 1891—Perry Hudson, proprietor of the Hotel Enterprise, left for Denver by Monday's stagecoach to secure additional help and furnishings, rendered necessary by the recent changes and additions to the house [hotel].

The floors of the mammoth bar and billiard rooms of the Hotel Enterprise have been covered with a neat and substantial carpeting this week.

May 8, 1891—The Hotel Enterprise was formally opened to the public on Sunday morning last. For weeks this house has been undergoing extensive alterations, which have produced a transformation from cellar to dome. A large and commodious office takes the place of the former circumscribed space, and the narrow dining hall has given way to a spacious room, well ventilated and lighted. A well-arranged washroom, blacking-room and a baggage department is among the additions. A bar and billiard room, fitted up artistically in the room known as Hotel Hall, on the ground floor, and electric lights in all parts of the house give to the entire place a new-and-pleasing appearance. The walls and ceilings, under the skillful application of the decorators' art, present a pleasing view.

The people of Rico feel justly proud of their principal hotel, and the **DEMOCRAT** bespeaks for it the liberal patronage which the energy and peculiar fitness of the management deserve. Mr. Hudson has surrounded himself with a corps of assistants, each of whom have the qualities essential to a satisfactory discharge of their respective duties.

May 8, 1891—H. J. Dey has removed his saloon and cigar store to the Carpenter building, next door to the post office, which has recently been repainted, papered and decorated. A new and unique sign ornaments the front. It is called the Bijou.

Business of RGS Increasing

MAY 8, 1891—The general officers of the Rio Grande Southern have assured the people of Durango that the road will be rapidly pushed to a connection with Rico at once. From the parties up from Dolores we learn that great activity prevails among the contractors, who have waited for suitable weather to resume the work of construction. Workmen are arriving daily, and all find work.

RGS Business Activities Have Steadily Grown

MAY 8, 1891—The business of the Rio Grande Southern has steadily increased since it completion to Telluride. With Rico added to its shipping points, its business will take an impetus that will largely increase its earnings.

Tracklaying Pushed on RGS

MAY 8, 1891—All doubt is dispelled in regard to the early resumption of work on the Rio Grande Southern. From a gentleman just up from Dolores, we learn that payment has been made to the contractors on that end of the line, and all of them are receiving accessions to their working force and are rushing the work with the same vigor that characterized their operations last season.

RGS Places Order For 100 Miles of Steel Rail

JUNE 5, 1891—The Rio Grande Southern has placed an order for 100 miles of steel rails with the Bessemer works [of the Colorado Coal & Iron Company] at Pueblo. While it is true that the work has not begun on the road as early as we had hoped, and the chances of its being completed this summer in consequence diminished, there is no doubt now that we will have rail connection with Durango, as well as with Ilium, before snow [falls]. From a gentleman up from Lost Cañon, we learn that not less 300 men are employed in that vicinity [37 miles south of Rico].

Otto Mears Visits Rico

JUNE 12, 1891—Otto Mears, president of the Rio Grande Southern, came in on Monday, and he has been is town since. Mr. Mears reports that work on his road is going on rapidly. Iron [rail] has been laid to Ames [at Milepost 41.31], at which point a bridge will soon be finished, when they will push on to the next. The bridges are built by a corps of carpenters, ready to put in place. The only delay is in putting them up. The work of construction, with this exception, being nearly finished. The work will be pushed with all the men that can be used to advantage, and its early advent to Rico is an assured fact. The portion of the road from Telluride to Ridgway is doing a heavy freight-and-passenger business, and with its completion to Rico, will secure an immense tonnage.

Men and Mules Arrive From Pueblo

JUNE 19, 1891—A shipment of a large force of men and 100 mule teams has been made from Pueblo to finish the railroad work between Ilium [Vance Junction] and Rico.

RGS Building Fast...

Otto Mears Returns From His Trip of Inspection

JUNE 26, 1891—President Otto Mears, of the Rio Grande Southern, returned [to Denver] from his trip of inspection over the line yesterday, and he appears to be well pleased with the work going on. He was seen by a ***News*** reporter, and when asked about his trip, he said:

"Everything is progressing smoothly, and we are laying track and building bridges as fast as money can do it. There is no hitch anywhere, and I feel well satisfied with the result of my trip."

"What about the reported sale of your Rainbow Route [Silverton Railroad]?" was asked.

The *Pathfinder* smiled grimly and said: "Whenever I leave the city, somebody reports a sale of my line. Now, I'll tell all there is to it. Negotiations have been pending for sometime, but there has been no sale yet. The English syndicate, who owns a number of mines around Silverton, want to buy the road and have made a proposition, which I have under consideration, but the deal has not yet been closed, and I can't say at present when it is likely to be."

It was learned from other sources that the syndicate has not yet made what Mr. Mears considers a fair proposition, and he will not sell unless his terms are complied with. It is understood that nothing less than $500,000 will purchase the Rainbow Route.

— *The Denver News*

Work Suspended on Mendota Spur

MAY 29, 1891—Work has been suspended on the branch from Telluride to the Mendota mill, until the road is finished to Rico, to the accomplishment of which a large force [of men] will at once be put on the line.

The time within which the railroad is to be finished to Rico or forfeit aid in securing right-of-way, has been extended to August 1, 1891.

Messrs. Pine & Townsend, with the contractors on the Rio Grande Southern, came over from the headquarters camp at Ilium, to arrange for work in the immediate vicinity of Rico. They report that tracklaying is progressing rapidly this way from Ilium [Vance Junction], and say that the road will be pushed to completion as fast as men and money can aid.

RGS Engineer Gibbs Visits Rico

JULY 19, 1891—Chief Engineer C. W. Gibbs, of the Rio Grande Southern, was in Rico this week. Mr. Gibbs informs us that work is progressing on the road in a satisfactory way, the rails having been laid to the crossing of the Lake Fork, on Wednesday last. A force of engineers is operating in the immediate vicinity of Rico.

Contractors at Work on RGS Bridges at Ames and Trout Lake

JULY 19, 1891—L. W. Carlile, of the contracting firm of Carlisle & Weitbrec, who is building the Rio Grande Southern, was in town Monday. Mr. Carlile informs us that the company has a large force of bridge-builders at work between Ames and Trout Lake, and he says we will get the road as far as Rico during the summer. At present, Carlisle's forces—a portion of them—are building a branch road from Telluride to the Sheridan Mendota mill [at Pandora].

Carlisle & Weitbrec Firm Files Liens Against the RGS Railroad

JULY 17, 1891—The Messrs. Carlisle & Weitbrec filed liens this week in the counties of San Miguel, Dolores, Montezuma and La Plata against the Rio Grande Southern for sums aggregating to $225,000, due them for construction work. These gentlemen were the principal contractors.

From [other] parties just in from Ophir, we have learned that bridge building and tracklaying is progressing rapidly in that vicinity.

FUNDS SECURED FOR THE RGS

JULY 17, 1891—The Rev. S. C. Dickson, while at Ophir a few days since, was informed by Mr. Charles Vaile, cashier of the First National Bank of Durango, that the Rio Grande Southern had secured funds to complete the road, and that work would be pushed until it was finished.

Mr. A. Levy, tie contractor on the Rio Grande Southern, registered at the Hotel Enterprise on Monday.

FROM THE RICO DEMOCRAT

Published at Rico, Colorado, 1891

Rio Grande Southern News

JULY 31, 1891—Wilson Sinclair, of Red Rock Ranch, was in town on Tuesday. Mr. Sinclair is engaged in clearing the right-of-way for the Rio Grande Southern, and he reports the grading of the road is practically finished to the West Fork [of the Dolores River].

RGS Representatives and Financial Backers Visit Rico

AUGUST 7, 1891—A party of representative railroad men and capitalists arrived in Rico on Saturday last, and whatever doubts might have been entertained as to the completion of the road at an early day were dispelled by the statements of the gentlemen composing the party, and who are in a position to know whereof they speak. President Mears of the Rio Grande Southern, headed the party, which was composed of the following gentlemen:

S. K. Hooper, general passenger-and-ticket agent; J. W. Gilluly, treasurer; R. M. Ridgway, general superintendent; N. W. Sample, superintendent of motive power of the Denver & Rio Grande; and G. W. Cramer, superintendent of the D&RG Express. Mr. William Barth and Mr. John L. McNeill were also of the party. The two last-named gentlemen are leading bankers and capitalists of Denver.

In conversation with these gentlemen, we learned that their visit was in connection with the building of the Rio Grande Southern, and that its early completion is no longer a matter of doubt.

Carlisle & Weitbrac Hired to Complete RGS

Arrangements have been made with the great contracting firm of Carlisle & Weitbrac, who will at once resume operations. The road [i.e., branch] from Telluride to Pandora having been completed by them, the large construction force engaged there has this week begun operations between Trout Lake and Rico.

Active operations are not confined to the road north of Rico, but the force has been and is being increased on the Durango end, and the entire route will be finished this season.

Now, let us get ready for the prosperous times that are to come to us with the completion of this road. Rico has mineral wealth, and only lack of transportation has kept her from the front rank of mineral-producing points in the state.

RGS Rapidly Building South

AUGUST 7, 1891—We are informed, upon authority, that there will be no new terminal established this side of Ilium [Vance Junction]. The Rio Grande Southern will be pushed into Rico at once.

Dr. McNeill made a trip last week to the Big Bend. The doctor reports the work of grading on the RGS is nearly finished to the West Fork.

C. T. Pease, assistant engineer of the Rio Grande Southern, is located at Rico, where he will remain until the road is finished. He is engaged at present in locating the road through the town, laying sidings and planning the depot grounds. The depot will be located at the foot of the bluff, between Mantz Avenue and Campbell Street.

Promising Prospects For RGS To Complete Line This Year

AUGUST 7, 1891—Well, how is the railroad coming on? Will they build this summer? How far is it completed? Do you think they can get here this summer? Has the company money with which to finish the work?

These questions and a thousand others of this kind have been heard on every hand for the past month. In order to be able to answer the above interrogatories, we have endeavored to find out the conditions surrounding this enterprise, that we might lay before our readers the situation as it is. To that end, we were groping in the dark until the officers of the Denver & Rio Grande Railway came in last Saturday, and we called upon them at their hotel. From the information given by these gentlemen, who are in a position to know, and who cheerfully gave information, we can in all confidence say to our readers that the following answers to the queries are correct:

The railroad work is progressing favorably, and it is now finished to a point one and a half miles below Trout Lake. The delay at that point is in building two bridges that span one branch of the San Miguel River near Ophir, but for which the road would have been finished to the lake before this time.

RGS to Be Finished Before Winter

The road will be finished to Rico as fast as men and money can accomplish it. No doubt now exists that it will be completed and the cars running into Rico by the middle of October.

As to the question of the financial ability of the company to build the road, the statement made by the officials [of the D&RG] that the road would be built, leaves no doubt of their ability to do the work. Complications that had arisen between the company and Carlisle & Weitbrec, the principal contractors, have been satisfactorily arranged, and the entire construction force, that has just finished the Telluride-to-Pandora branch has been transferred to the line between Trout Lake and Rico—two camps having been established on Monday—and the work began that will in a short time end in the finishing touches that are required to complete the work.

Lew Carlisle left for the Mancos on Tuesday morning, and the object of his visit to the lower end of the road will manifest itself in a few days in increased activity along the entire line.

Railroad to Be Completed By November 1, 1891

AUGUST 14, 1891—Everything at this time points to a confirmation of the view entertained by us two weeks ago, based on the statement of railroad officials, that the road would be finished to Rico by November 1st. A large force of men are employed on both ends, and accessions to their members are arriving at the rate of two carloads daily, four camps are now established on the 15 miles between Rico and the [Trout] Lake, and in all probability the coming week will see a camp in sight of town.

Contractor's Workers Receive Meat

AUGUST 21, 1891—James Lavender and Mont Butler were in town on Tuesday. Mr. Lavender and the other gentlemen are furnishing beef to the railroad contractors north of Rico and have a camp near the stage station on the meadows. The present force are consuming five beeves daily.

More Men Arrive to Work on RGS

AUGUST 21, 1891—During the week past large accessions have been made to the working force on the railroad north of Rico. Bridge building is being done as fast as men can do it. The bridges over Howards Fork, two in number, are completed, and there are yet four or five small ones to finish north of Trout Lake. Ties have been distributed and placed to receive the iron to a point nearly around the Lake. A very few days will elapse before the construction train will reach the Lake. From the divide [Lizard Head Pass] the Lake the grading is nearly finished to Rico, and there being very little bridge building to do to complete the road to this place, the probabilities are that it will be finished by the first of November—perhaps in a little less time. Another camp was established on Tuesday within sight of town and their presence so near has inspired confidence in the Doubting Thomases here.

An Advance on the Railroad Front

AUGUST 28, 1891—General Mears, with a strong force is approaching Rico. Couriers just in from the front report General Ryan with a division (of navies) throwing up earth-works within a mile of town. The advance here already seized the pass above the smelters, no resistance being offered. Large reinforcements are daily arriving by rail at Ilium [Vance Junction] and are immediately sent forward to participate in active operations. General Butler, (not B. F.) with a large force is pushing up the Dolores to reach Rico from the south.... It is reported at headquarters that the women, children and non-combatants here can leave by rail about October 1st.... There is renewed activity in the banking and business circles, in view of the close proximity of such a large force. We feel confident that should they enter the town they will be detained long enough to be pay-rolled.

Large Work Force in Sight of Rico

SEPTEMBER 4, 1891—The fine weather of the past two weeks has enabled the construction force on the Southern to prosecute the work continuously. Contractor Ryan has a large force in sight of the town, and at all points the work is being pushed with energy.

Lew Carlile was in town this week, giving his personal attention to the conduct of railroad operations on the Southern. Ryan's forces are throwing dirt at a lively rate within sight of the town on the north, while several large bodies of men have passed through town during the week to join the force building in from the south.

Silver Spike Being Crafted

OCTOBER 9, 1891—George L. Sheldon, at his jewelry established, is manufacturing a solid silver railroad spike, which will be driven at the depot grounds as part of the ceremony attending the formal opening of the road on the 15th.

With a continuation of the present beautiful weather for 40 days, the railroad will be finished, which will connect Durango, the great smelting point of the San Juan, with the greatest mineral-producing region in the state.

The activity that characterized the building of the railroad to Rico has not abated. Close upon the road came the corps engaged in building the telegraph line, and on Friday last the wire was strung along the avenue to the Hotel Enterprise, where an office has been established, and now our good people who have waited and watched for years for its coming, find themselves in touch with the world outside.

Palace Hotel Opens in Durango

OCTOBER 9, 1891—The Palace Hotel is a new enterprise just opened in Durango. This elegant hotel is near the Denver & Rio Grande depot, is first class in all its appointments, and was built during the present summer at a cost of $14,000 [a large sum in 1891]. We were shown through it last week by the genial proprietor, Mr. McNeeley. The hostelry is furnished with plumbing of first-class workmanship from top to bottom, well lighted and ventilated, and it has a ladies toilet room in connection. Mr. McNeeley informs us that rates are in keeping with the present condition of the money market.

MICHAEL KOCH COLLECTION

RIO GRANDE SOUTHERN No. 14 was the only 0-6-0T switcher owned by the railroad, and she was affectionately known as a "yard goat." This small saddle-tank engine was the motive power for the first train to arrive in Rico—hauling a construction train that arrived at 10:00 p.m. on September 30, 1891. Since the train came into the yard after dark, no photographs were taken of this momentous occasion. The little 0-6-0T formerly served on the Denver & Rio Grande as their No. 105, built by Baldwin Locomotive Works in 1881 as one of the D&RG's Class 48 engines. The RGS sold No. 14 to the Yellow Pine Lumber Company of Mobile, Alabama, in 1899. This photograph was taken while the engine was on the turntable in Ridgway.

Carloads of Freight Arrive in Rico

OCTOBER 9, 1891—Two carloads of merchandise for Rico's merchants was received Wednesday via the Rio Grande Southern.

A carload of beer arrived Wednesday over the Southern, consigned to Morgan & Simpson, of the Rico Bottling Works.

Those of our citizens who have spare apartments to place at the disposal of our visitors on the 15th, will leave word with W. W. Parshall, as a number beyond the capacity of our hotels are expected.

J. P. Nourse, local agent and [telegraph] operator for the Southern, is located in the Hotel Enterprise, where the click of the little joker is heard that reminds us that the great civilizer has caught up with us.

Should the weather remain favorable for outdoor operations, the Southern will be completed to Durango in 40 days. The management will bend all their energies to accomplish this much desired result.

Large Force of Men Move South

OCTOBER 9, 1891—The large force of men under Contractor Ryan have finished here, and yesterday broke camp and took up the line of march down the river a short distance, where Mr. Ryan has another piece of construction work.

Superintendent Ridgway was in Rico this week with his special [train]. While here, he personally superintended the laying of a wye in the Rico yard. When anything in the track line is to be done, Bob sees the things done right.

Locomotive Engineer Visits Editor of The Rico Democrat

OCTOBER 9, 1891—We were pleased to meet Mr. Alex Davidson, an engineer on the Rio Grande, who pulled Superintendent Ridgway's special into Rico this week. Alex is a good boy. Back in the 1880's, Alex put black diamonds in the firebox when the writer was sitting on the right side of a Baldwin.

Mr. Richard Plunkett, a San Juan Indian trader, was in town on Tuesday. While here arrangements were made by which Mr. Plunkett will bring to Rico, as an adjunct to the R.R. celebration on the 15th, a band of Navajo Indians, who will exhibit for the amusement of the large concourse of excursionists who will visit Rico on the formal opening of the railroad.

RGS WYE BEING LAID OUT IN RICO

OCTOBER 16, 1891—Engineer Pees has been busy during the past week laying out [a wye] in the Southern yards.

THE STORY OF RICO AND THE PIONEER MINING DISTRICT

A Chronology of Events

(Continued From Page 105)

1891 The first Rio Grande Southern train arrived in Rico during 1891. Ever since Rico's incorporation in 1879, growth and prosperity in the area had been held back by the high cost and unreliability of transportation to the outside world. During 1891 all the town's hope for ending its isolation depended on the construction of the RGS railroad. The RGS had a contract with Rico's city council for land and a right-of-way in town provided that the railroad should reach Rico by November 1, 1890. It was especially galling to Rico residents when the RGS completed its branch to Telluride in 1890, but failed to reach Rico during that year.

Perky little Otto Mears had achieved what most people—and some experts—considered impossible. He had built 170 miles of narrow-gauge railroad through country that was so rugged that only three miles of track was on level ground. Little locomotives struggled hour after hour on the 167 miles, climbing over four divides, each one of which led their trains to new watersheds. Otto Mears' political and financial skills also were truly remarkable.

(Continued on Page 207)

AN ARTICLE FROM THE RICO DEMOCRAT...

FIRE IN FIRST NATIONAL BANK

May 8, 1891—Monday evening, about 8:00 o'clock, an alarm of fire was given, which was promptly responded to by the [volunteer] fire department. The fire was located in the First National Bank building, over the suite of rooms occupied by Kincaid & Williams, lawyers. Investigation showed, however, that the burning out of a [stove] flue was the cause of the alarm. No damage was done to the building or contents further than that occasioned by a torrent of water turned into the flue from the top, which found an escape through a lateral flue-hole into the consultation room of the firm above named, which ruined the carpets and did some damage to a portion of their library. On the same floor is the offices of T. J. Tarsney, W. B. Hess, Charles McLaughlin and A. H. Mundee. The libraries, books and furniture was removed to places of safety, and but little damage was sustained except that above named. The manner in which the fire boys performed for the first time under Chief Gault, and the orderly manner in which they carried out his orders, reflects the greatest credit upon them, and is an assurance to our citizens of efficient service at their hands in the event of a fire.

FROM THE RICO DEMOCRAT
Rico, Colorado, July 17, 1891

DOUBLE TRAGEDY!

P. D. Hudson, Proprietor of the Hotel Enterprise, Shoots and Kills Rose Conway, and Then Ends His Own Life

The Hotel Annex the Scene of One of the Bloodiest Tragedies That Has Darkened Rico's History

AT ABOUT 10:30 O'CLOCK Sunday night the inmates [guests] of the Hotel Enterprise were startled by the report of a pistol coming from the annex of the hotel, followed in a short time by another.

As most people in town had retired for the night, although it became known in a few moments what had taken place, little excitement was occasioned. The only indication that anything unusual had occurred, was the grouping together in the hotel corridor and the little gatherings on the streets, where men with subdued voices discussed the thrilling episode. Many of the occupants of the hotel were unaware that anything at all had happened outside the ordinary routine of hotel life.

Hon. A. P. Adams, who occupies a suite of rooms in the main building, and Mr. George Mattern, an employee, were the first to reach the annex. Upon opening the door, they were horrified to see the lifeless form of Hudson prone upon the floor, and lying at right angles to him, drawing her last breath, the body of his victim, with a gaping wound in her right breast.

From Mr. James Houck, who occupies a room in the main building, we learn that a short time before the shooting, he heard an angry altercation between these two unfortunate people. The woman was in her room, and Hudson was demanding admittance, accompanied by the threat to kick in the door in case of refusal.

Houck heard the woman say she was going away in the morning, and soon thereafter said she would "go now," and leaving the room, passing Hudson in the hall leading to it, came out on the landing in her night dress, but soon returned to her room and began dressing, and had finished, excepting one foot, when the leaden messenger stopped her earthly career.

Mr. Houck further says that while knocking for admittance, Hudson, speaking from the outside, said: "You made a gun play on me today, now I have four guns with me, and you can have your choice," and going back to the door of a room occupied by other female help, exhibited four pistols; whereupon they cried out, "Rose, he has four guns!"

Immediately after exhibiting the guns to the other help, he returned to the woman's room, and going inside, found her as stated, nearly dressed.

Nothing is known of the cause for the commission of the terrible deed, further than that, as he stated, she had drawn a gun on him during the day, and the further fact that she had in her possession, belonging to him, a watch of unique design and of great value that he desired her to give to him, which she refused to do, and told him she intended to take it with her on her departure in the morning.

It was noticed by many that Hudson, late in the afternoon and evening, was somewhat under the influence of liquor, but that he contemplated the commission of such a fearful deed, none had the slightest intimation.

Hudson in 1888, was a cook for the contracting firm of LaCount & Grady, while engaged in the construction of the Farwell irrigating canal in the Montezuma Valley. He came to Rico about three years ago, and was employed as cook and afterwards became the proprietor of some of the large boarding houses at the mines, in which business he made considerable money.

He enjoyed fully the confidence of all, with whom he had dealings.

Sometime during the past winter, he had purchased the Hotel Enterprise, and this spring refurnished and made extensive alterations on it, and [he] has given his personal attention to its management.

At the time of his death, he was about 23 years of age. He leaves an aged father, who resides somewhere in Arkansas, to whom the news of his tragic end will bring sorrow.

His victim, Mrs. Rose Conway, is a sister to Mr. George Isgar, a former businessman of Rico, now of Durango, who was summoned and arrived on Tuesday morning.

The murdered woman was married two years ago. The marriage proving an unhappy one, she has been living apart from her husband for the past year, and we understand that divorce proceedings are now pending. At the time of her death she was only 19 years of age. Her parents reside at Rosita, Colorado, and the news of her fearful death will cast a dark shadow over their lives.

To Mr. Isgar and the stricken family, the sympathy of the community goes out, in this hour of sorrow.

The shot that killed the woman was from a 41-caliber pistol. The ball entered at the right nipple and ranged downward, coming out at the left side, passing in its course near the heart.

She lived about 20 minutes after the fatal shot was fired.

After firing the fatal shot, Hudson came out of the room and called to the other servants, who were listening to the war of words, and who had heard the shot, [exclaimed]... "Rose was shot!"

Instantly returning, he was heard to ask the woman, "Rose, are you dying?" Being answered in the affirmative, he said, "Then darling, I will die with you," and the last shot rang out.

The self-inflicted shot took effect in the center of the forehead, ranging slightly downward. Death must have been instantaneous. Those who assisted in preparing the remains for burial say the neck was broken.

Dr. J. P. Landon had been hastily summoned, but upon his arrival, they were beyond human aid.

Indications point to Hudson having torn open the dress of the woman after he shot her—whether to examine the extent of her injuries, or in search of the watch referred to, will never be known—as when Mr. Adams entered the room, he found both the dead man's hands smeared with blood and the imprint of a bloody hand on the jamb of the door.

At the time of shooting himself, he must have been lying on his back, or at least in a sitting posture, as those on the outside had plainly heard the conversation, but heard no fall.

Acting coroner, Justice Campbell, was summoned, but upon ascertaining the facts above stated, an inquest was deemed unnecessary.

The remains of these ill-starred people were prepared for burial by Undertaker Carey, who provided a neat casket for each, and as they lay side by side in the same room where their lives went out together, to the hundreds who viewed the remains on Monday, the features of neither presented traces of either fear or passion. Not the slightest evidence of a troubled life or violent death told those who viewed them that they had not gone to sleep in the hope of glorious resurrection.

The affair has cast a deep gloom over the town, where Mr. Hudson was well-known, and when through his well-known integrity and energy, he gave promise and success in the business world. The esteem in which he was held is best evidenced by the presence of the vast concourse of people who have poured in an incessant stream to gaze for the last time upon the marble features of one who sleeps the eternal sleep in the narrow confines of his last resting place.

Let us draw the veil of charity over his last earthly act. Let us hope that He who directs all things, will give both so suddenly called to Him rest in His keeping that was denied them in the stormy battle of life.

EARTH TO EARTH

THE FUNERALS of the victims of Sunday night's tragedy were attended on Tuesday by a large concourse of people. That of Mrs. Conway was held at the Congregational Church at 10:00 a.m., conducted by the Rev. Dickinson. Mr. and Mrs. Isgar, brother and sister-in-law of the deceased, were present, as was also the husband, Mr. James Conway. To the relatives upon whom has fallen this great load of grief, is extended the sympathy of Rico's people.

At 2:00 p.m. a solemn cortegé moved from the Hotel Enterprise, at the head of which went the hearse, bearing all that is mortal of its late proprietor, followed by the civic societies of the town, among them the A. P. Adams Hose Company, with the fire[-fighting] apparatus draped in mourning. At the Congregational Church, Rev. S. C. Dickinson rendered a solemn-and-impressive discourse, appropriate to the occasion. The choir sang "Along the River of Time" and "Thy Will Be Done" [The Lord's Prayer]. The solemn notes of the organ, mingled with the voices of the choir and the touching remarks of the pastor, drew tears from the eyes of many in the large throng. After the service, the procession formed and moved toward the cemetery. As they marched with measured tread, the bell tolled a death knell, which added solemnity to the occasion. Arriving at the grave, a short service was had, and the casket was lowered, the last solemn words of the man of God mingled with the sobs of strong men and sympathizing women and children, and all that remained of one so lately among them in the full enjoyment of manly vigor, was at rest to await the summons [from God Almighty] that will call the grave to give up its dead.

TUESDAY'S FIRE

September 15, 1891

ABOUT 3:00 O'CLOCK Tuesdày afternoon our citizens were startled by dense smoke coming from the rear of the Hotel Enterprise. The rapid ringing of the fire bell soon brought the company [the volunteer fire brigade] together, and the fire was found to be in the county jail.

It soon became known that there was a prisoner confined in one of the cells, and heroic efforts were made to save his life. The heat was so intense, however, that it was only with super-human effort that Marshal Heffernan opened the outside door.

After the water had reduced the heat and flames to some extent, the second door was thrown open, and a sheet of flame four feet thick met the firemen, and back of this awful furnace could be seen the parched-and-lifeless body of Daniel Sullivan.

The [fire-fighting] company soon got the fire under control. The damage to the jail amounting to about $200.

Sullivan, the man who lost his life, was a blacksmith in the employ of Maurice Ryan, of the... [Rio Grande] Southern, and is said to have been a good mechanic and hale fellow, when not under the influence of liquor.

Sullivan's breast and arms were a complete mass of tattoos; but with the exception of the "crucifix" over the heart could not be deciphered. His parents are supposed to reside in Pennsylvania. The remains were taken charge of and given burial by Contractor Maurice Ryan.

FROM THE RICO DEMOCRAT

Rico, Colorado, October 16, 1891

By Editor T. J. Tarsney

RIO GRANDE SOUTHERN RAILROAD FINISHED

Ball and Banquet in Honor of This Historic Event

A GREAT DAY FOR RICO

THE FORMAL OPENING of the Rio Grande Southern Railroad and celebration of the event was one that will long be remembered by the citizens of Rico [October 15, 1891]. The sun rose bright and clear over Dolores Mountain, exactly on the scheduled time, and as the beautiful orb of day peered into the Vale of the Dolores, it found the people busily preparing for the eventful day that had dawned upon them—gorgeous colored bunting draped the various business houses and numerous residences, while the many transplanted evergreens gave the town something of its pristine appearance before man had embellished Nature's handiwork with art.

And why all this demonstration?

Why this gala-day appearance of the town?

Simply because the Rio Grande Southern, the pride of the great Southwest, the crowning glory of Otto Mears' wonderful achievements was completed to the greatest mining camp in all southern Colorado—because the patient burro (quadruped) was no longer our only means of transporting the rich product of our mines, and can now find a well-earned rest from the years of tiresome toil to which they have been subjected in doing their humble part in bringing Rico's unparalleled resources to notice, and because the other class of burros (bipeds, i.e., humans) who have enfeebled their constitutions and wearied their jaws (and their neighbors) in declaring that the railroad would never be built. And if it was [built], it would make the cost of transportation higher instead of lower, and would bring the price of groceries and other necessities up, rather than down, and would put the price of coal up to $14 or $15 per ton, instead of leaving it at the customary price of $8 per ton, or bringing it below that, because this biped class can now give their jaws, the conglomeration they call their intellects, and their forbearing neighbors a rest. Is this not enough to make our people... or anybody else's people rejoice? We think it is, and we rejoice accordingly!

During the forenoon, it was announced that the reception committee, or at least a portion of them, were to meet the trains at Burnsville. A wagon was engaged, but only room for a few of the committee was thus procured, and more vehicles could not be obtained owing to everybody being so busy getting ready for the "blowout."

Among those who met the train at Burnsville was Julius Thompson, David Swickheimer, W. W. Parshall, Mayor C. F. Middaugh, J. W. Winkfield, Joe Hendershot, and others. The first trains to arrive reached Rico about 1:00 o'clock p.m., bringing the Telluride tourists and the band from that place.

The other trains brought a number of distinguished guests from different parts of the State, among whom were Dr. Rowan of Ouray and David Day of *The Solid Muldoon* [the Ouray newspaper], Mr. Botts of Colorado Springs, William Stapleton of *The Denver Republican*, Thomas Touge, who is prominent in journalistic circles and is the Denver correspondent of several British papers, Hon. Wolfe Londoner, ex-mayor of Denver, Mr. Jeffrey, the new president of the D&RG system, Governor John L. Routt, Otto [Mears], the "pathfinder and iron-trail builder," and several other noted gentlemen of various callings, most of whom honored Rico with their first visit. But if their expressions regarding our camp were sincere, we predict they will become frequent repeaters so far as visits to Rico are concerned.

On the arrival of the last train [the third one], a large concourse of people assembled to welcome the arrivals, and the customary formality of driving the silver spike was indulged in; but it is rumored that a little juggling was perpetrated, and that a plated spike was substituted for the solid silver one, which had been made of rich ore from the Enterprise Mine and properly inscribed to explain the occasion by Mr. Sterling, the expert jeweler and silversmith, who is in charge of that department in Mr. Shelden's jewelry store in Rico. After this proceeding, the people repaired to the Hotel Enterprise, where the numerous committeemen and other worthy citizens vied with each other in entertaining our guests with a varied assortment of fact and fiction, romance and reality, to the great delectation of the entertainers, if not the entertained.

It may not be inopportune to mention that the elements were in as happy a mood as were the people who were met to celebrate. A perfect autumn day deluged with a flood of golden sunlight, which lit up the landscape with a wealth of radiant beauty such as is known only in the Rocky Mountains, and must be seen to be appreciated or realized, [and] was the bounteous offering of Providence for the occasion. When evening's somber veil had mantled the landscape with that sweet, soft silence that engenders tender emotions in the breast of sentimental swains and dreaming maidens, the moon shone out with a refulgence that rendered the use of the electric street lights unnecessary, and made the beautiful-and-unique transparency at George Downs' business place appear as sickly as an Eastern youth the first time he feels the thrill of tender passion. The band filled the air with sweet dulcet strains of music, [and] bonfires shone like beacons of hope from every hillside surrounding the city. The detonations of giant powder reverberated over and anon from the

Historic Newspaper Stories...

RGS Rail Arrives in Rico

By T. J. Tarsney

FROM THE RICO DEMOCRAT

No. 5

neighboring mines and Nature wore her sweetest smile.

At an early hour, the merry-makers repaired to the spacious hall, which formerly served the people of Rico as a skating rink, and indulged in the fascinating pleasures of the terpsichorean art until the banquet was announced, when the glad throng repaired to the Hotel Enterprise and were seated in the spacious and appropriately decorated dining hall, to the grandest collation ever spread before a throng of San Juan epicures. After the corporal systems had been satiated with delectable viands, the party was treated to a rare intellectual feast in the way of speeches from some of the brainiest men in Colorado. Hon. Julius Thompson, in responding to the toast "Our Honored Guests," made many clear and forcible hits and showed wherein both the railroad and the people whom it is built to serve, can mutually benefit each other by working in harmony and fostering each other's interests, while by antagonizing each other, as is by far too often the case, ruin must be the inevitable result. Mr. Thompson's remarks were clear, plain and logical, and struck a responsive chord in every individual in the audience.

Judge A. E. Kennedy, in response to "Railroad Legislation," was brief, pithy, pungent and pertinent.

Governor Routt's remarks were versatile, interesting, amusing and instructive. Opinions seem to be divided as to the judgment of the Grand Junction lady, whom the governor told us about. The lady has quite a number of supporters among the fair sex of Rico, but if his excellency sees this, we have a satisfactory explanation to offer, so he need not let that deter him from his contemplated investment in Rico mines.

Mr. Jeffrey's remarks on the relations between corporations and the people was one of the most concise and comprehensive dissertations upon the subject we have ever had the pleasure of listening to, and we are led to believe from the gentleman's remarks that his plans for the management of the D&RG system will much improve the condition of affairs in southwestern Colorado, both for the road and its patrons, and we tickle ourselves with the hope that we have guessed right upon this subject, which is of great moment to us all.

The Honorable Wolfe Londoner's remarks were a happy combination of sarcasm and sense, fact and fun, and are, we believe, having a good effect on John Eder, as his *protégé*, Alec Goria, says John has in some way overcome his native modesty to the extent of removing the pantalets from the limbs of his household furniture, and he has quit trying to discourage mild

(Continued on Page 136)

W. H. JACKSON PHOTO –
DENVER PUBLIC LIBRARY, WESTERN HISTORY DEPARTMENT

NOT LONG AFTER the first construction train's arrival in Rico on September 30, 1891, William H. Jackson, well-known photographer from Denver, arrived on the scene to record the progress of the RGS. The noon hour had passed by the time W. H. Jackson had located a camera position on Newman Hill to photograph Rico. His location was above the southern end of town. Sandstone Mountain, a Rico landmark, rises near the center of the view, beyond the town. Glasgow

Avenue, the main street in this settlement, ran north and south through town. The town fathers found space for the Rio Grande Southern's station grounds alongside the east bank of the Dolores River, which is in view at the far left of this scene, taken during the fall of 1891. Railroad contractor's tents are visible at what would become the location of the first enginehouse. A string of boxcars was spotted on the siding, near the site of the proposed depot.

W. H. JACKSON PHOTO – RICHARD A. RONZIO COLLECTION

A CONTACT SOLAR PRINT was made of this picture from the original glass negative by Richard A. Ronzio. William H. Jackson shot this panoramic view from high on the east slope of Expectation Mountain, as he looked eastward toward Dolores Mountain during the fall of 1891—a few days prior to the October 15 train celebration. This print contains so much of historic value regarding the town of Rico that it was decided to enlarge it in two different stages. These enlargements are reproduced on the following pages. The Rio Grande Southern's Milepost 66 was later placed beside their track, directly above the handwritten "M" on the photograph (along the bottom edge).

(Continued from Page 133)
dissipation among the boys so far as beer and bologna is concerned, but he draws the line at Anheuser and Wheeling Stogas.

The remainder of the exercises were simple, harmless and enjoyable to the participants, and consisted chiefly of waltzes, quadrilles and sometimes lemon and sugar, with soda on the side.

The excursionists departed by early trains this morning, highly pleased with the appearance of our camp, and favorably impressed (we hope) with our people. Come again, gentlemen, and bring the ladies and children with you.

— FROM THE COLLECTION OF WILLIAM A. GRAVES

RICO

Milepost 66.2
Elevation 8,737 Feet

W. H. JACKSON PHOTO – RICHARD A. RONZIO COLLECTION

IN THIS GREATLY ENLARGED VIEW, taken from a corner of the photograph to follow, is a shot of the railroad yard in Rico, still under construction. An afternoon shadow from Expectation Mountain had already covered the scene, making it difficult to see all of the view. However, computer enhancement has brought out some interesting details. A small Class 60 (C-16) 2-8-0 was switching cars in use by the construction gang. Other construction cars and boxcars were scattered about the yard, beyond the temporary tents of the contractor's camp. Notice that a wye had been temporarily located adjacent to the tents in order to turn locomotives. Arthur Ridgway, the railroad's first superintendent, arrived ahead of tracklayers to design the trackage for Rico's station grounds. Apparently, this original wye at Rico was approved by him. In any case, the wye may have been partly destroyed by high water in the Dolores River during the following spring, forcing the railroad to place the wye farther south. Huge piles of ties, bridge timbers and building lumber had been stacked on the ground in preparation for constructing a depot, enginehouse and other support facilities for the Rio Grande Southern Railroad. The Dolores River flowed past Rico in the river bottomland west of town, adjacent to the new narrow-gauge trackage. Rico's street intersection at Glasgow Avenue and Campbell Street are at the upper left in this scene. The first "official" passenger trains into Rico for the town's formal celebration arrived on October 15, 1891.

W. H. JACKSON PHOTO – RICHARD A. RONZIO COLLECTION

THE W. H. JACKSON PRINT of October, 1891, is shown as a four-page spread, in order to graphically display the town of Rico. The original white-painted schoolhouse stands all alone in the upper left-hand portion of the view. Other structures in the picture can be located by studying the Sanborn Map of 1893. Block 23 contained residences along Silver Street, southwest of the schoolhouse, a few of which are still in use today (1996). The Congregational Church was built on Block 15, pictured on the right-hand page, where Silver Creek cut midway through the block. The original Dolores County Courthouse stands one lot in from the intersection of Commercial Street and Mantz Avenue. It stood on the northwest corner of the intersection of Silver Street and Mantz Avenue. Closer to the camera, near the middle

of the view, Silver Creek cut through the corner of Block 14, where a large two-door livery stable faced on Glasgow Avenue. Notice that the Hotel St. James had been relocated to the corner of Glasgow Avenue and Mantz Avenue. Its original site was now an empty hole near the livery stable. Across Glasgow Avenue, Block 13 contained the huge Hotel Enterprise, with its western (rear) exposure in sunlight. The east half of Block 13 and the west half of Block 12 contained houses that lined both sides of River Street between Mantz Avenue and Soda Street. Houses on the west side of River Street were located at the top of the embankment above the Rio Grande Southern's right-of-way and the Dolores River. The tents of railroad-construction workers are in view, along the bottom edge of the photograph.

W. H. JACKSON PHOTO – RICHARD A. RONZIO COLLECTION

THE SOUTHERN SECTION of Rico was brightly illuminated by the low afternoon sun when W. H. Jackson exposed his plate-glass negative. The fronts of all the commercial buildings lining the east side of Glasgow Avenue (in Block 6) are bathed in sunlight. A few of these buildings still exist in present-day Rico. On this side of Glasgow Avenue were more buildings that are located on the map in Block 1. The access road leading down to the RGS depot grounds made a sharp left turn from Mantz Avenue at the embankment edge. Farther south, along the embankment is Block 2, which contained the large Silver League Theatre. Four boarding houses for women were on lots behind the theater. Across the street from this theater building is Block 5, which contained an assay office in the tallest structure.

RICO
— 1890 —
SANBORN
MAPPING AND GEOGRAPHIC INFORMATION SERVICE

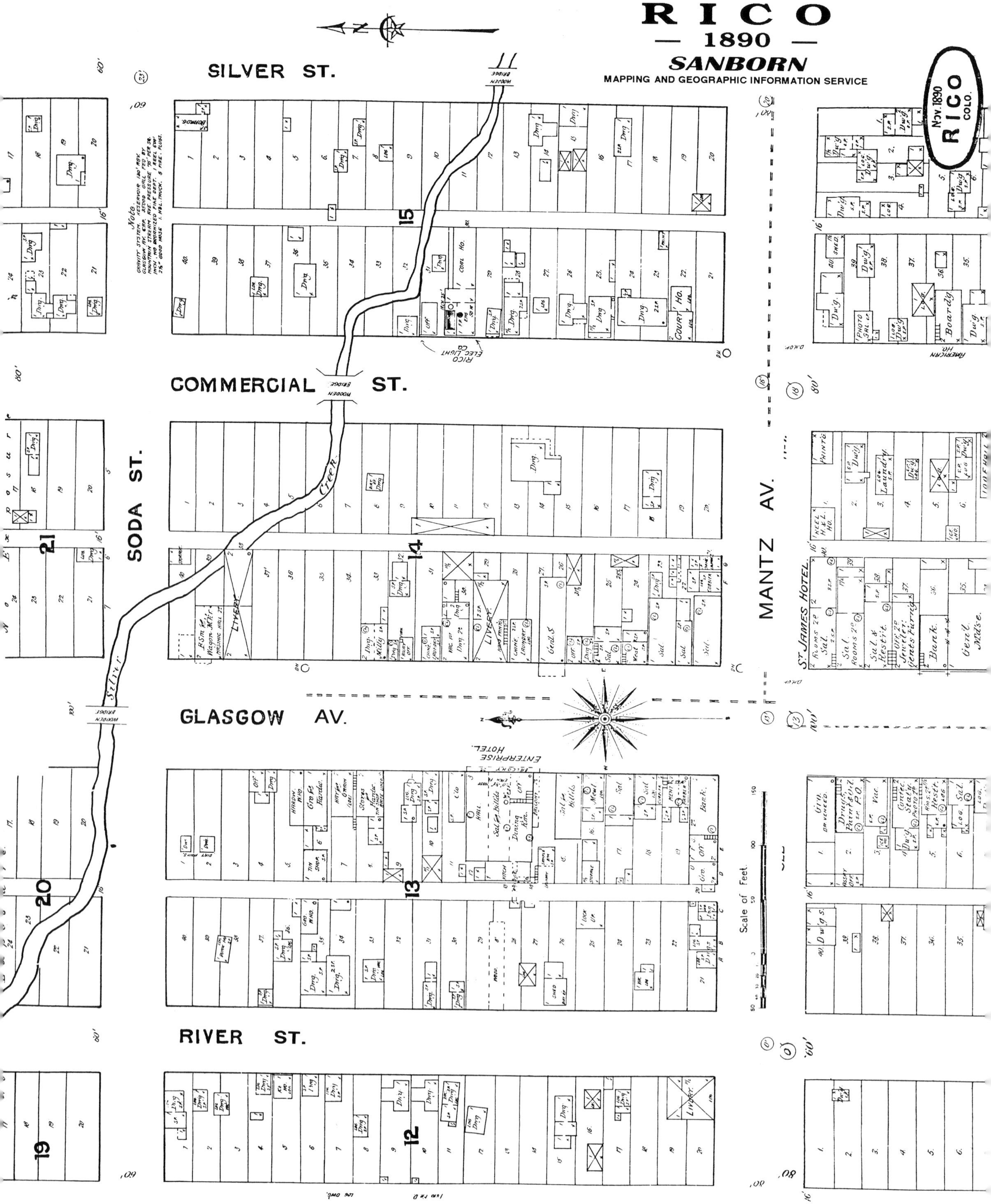

KURTZ PHOTO – WILLIAM A. GRAVES COLLECTION

J. W. BURLEY & COMPANY moved into a new building next door to the Rico State Bank on Glasgow Avenue in 1890, where his dry goods store operated. Read the newspaper account on Page 147. This building was later owned and utilized by Charles Engel, Sr.

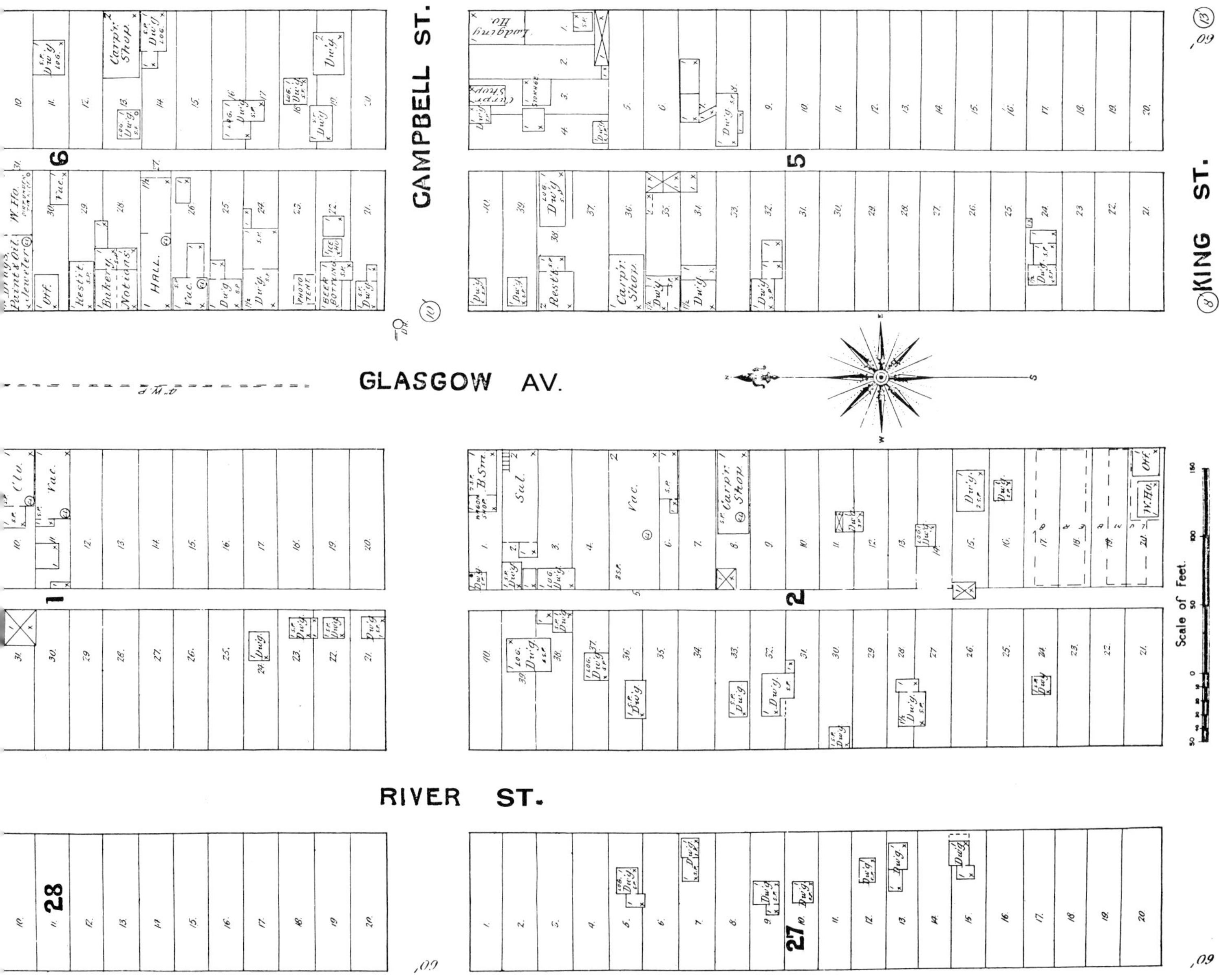

A FORECAST OF THE SAN JUAN

OCTOBER 23, 1891— ***THE REVIEW*** takes this opportunity to extend hearty congratulations to the patient long-suffering people of Rico, upon the auspicious completion of the Rio Grande Southern Railroad to their own town and superb mining camp. This event occurred on Wednesday, September 30, 1891, and occasioned profound rejoicing. The whole population turned out to welcome the messenger of salvation and redemption. But the road is not to stop there, but will be pushed on to its final terminus at Durango. Only 16 miles of grading remains to be done. The building of this railroad unquestionably gives the two companies, of which Otto Mears is president and the inspiring force, absolute control of the ore traffic of the entire San Juan Region. And it as unquestionably points to the concentration of the smelting interests at Durango. The man who cannot see this result at a glance, must be too stolid and stupid to recognize any potent conclusion. Everything in the situation now presented leads as inevitably to that result as that water and sunshine will produce harvests from the soil.

Otto Mear's "Rainbow Route" [the Silverton Railroad] from Silverton to [Red Mountain Town and Ironton, north of Sheridan Pass, later known as Red Mountain Pass], commands all the ores of San Juan County and of the Red Mountain district, one of the richest in the state. This road leads straight to Durango [via the D&RG's Silverton Branch]. The Rio Grande Southern strikes and penetrates and accommodates all the mines of Telluride and Marshall Basin by a branch, and those of Rico by the mainline, whose [southern] terminus is at Durango. It strikes not only the richer mines of the localities named, but [also] traverses the cattle ranges and farms and coal deposits of Dolores, Montezuma and La Plata counties. At Durango, there are inexhaustible quantities of splendid bituminous coal, which produces the finest smelting coke in the state. There is a magnificent stream of water at the feet of the only eligible sites for smelting works. The Valley of the Animas is very beautiful and filled with productive farms. It has every advantage that could be desired, and those in interest have full knowledge of the opportunities offered. The one smelter now running there is being enlarged, and a strong company has been organized to build another large plant. Undoubtedly, as the mines develop and the quantity of ore increases, other stacks will be put in.

Now, the trending of events leads to these conclusions—first, that in the course of a year, the Denver and Pueblo smelters will be deprived of large consignments from the San Juans, and secondly, that the Denver & Rio Grande Railroad will also lose a material part of its present ore traffic from that region. Instead, it will have bullion [to transport].

However, there is another element of considerable importance. It will go far forward enriching and developing to an unprecedented degree the southwestern [region]... of the state, all of which, in a commercial point of view, politically and otherwise is tributary to Denver. While we of Denver will lose a fraction of the output of ore, we shall gain more than its equivalent in the increased volume of trade. Instead of one or two or three struggling, poverty-stricken camps,there will be a multitude of brisk-and-thriving towns and cities. The Southwest is coming to the front with prodigious strides. It is an immense country, full of all manner of riches, and in our opinion, it will advance more rapidly henceforth than any other division of the commonwealth [of Colorado].

— *Mining and Scientific Review* [of Denver]

RGS TRAFFIC ASTONISHES RICO

OCTOBER 23, 1891—The traffic of the Rio Grande Southern to and from Rico will astonish the natives. Already, it is apparent that this important line will be placed on a paying footing at once. The full beneficial results of its construction will be realized when the connection is made with Durango, at which place the great bulk of our mining product will find a smelting point. The tireless energy of Mears, seconded by such lieutenants as Ridgway and others, will soon put the finishing touches on this, the most important line yet constructed in the state.

Dr. J. P. Wallace, manager of the Century mine near the head of Bear Creek, having struck a bonanza in that property, is engaged with a force of men in constructing a wagon road from the to intersect the Rio Grande Southern at Robinson's switch....

FROM THE RICO SUN

Published at Rico, Colorado, 1891

SAN JUAN SISTERS!

Linked Together by Everlasting Ties of Steel

ALL ADVANCING RAPIDLY

DECEMBER 12, 1891—Before another issue of **THE RICO SUN** reaches the public, the Rio Grande Southern Railroad will be completed, and trains running over the entire route. The work of laying rails will be finished in a few days, when through trains between Ridgway and Durango will be put on [i.e., scheduled].

The importance of this little line of railroad is not yet fully realized, nor will it be for some time

, & R. G. Form 477—4-92-2m.

BILL FOR VOUCHER.

Department No. 880 Auditor's Voucher No. 519

Chargeable to To The Bell Pressed Brick Co Dr.

81 $ Address Durango Colo

Pay

Address

92 Sept. Total, $ #929

Aug 17 For 4000 brick @ 7c 28 00 $28 00

Car No. 1035 sent to Rico.

The calculations herein are correct and no previous voucher has been passed for the items.

DISBURSEMENT CLERK.

Correct: Approved:

W. D. Lee Supt. C. M. H.

Place of delivery of all materials and Supplies must be stated in Bill for Voucher.

JOHN F. BELL, Prest. and Manager. E. H. HUDSON, Vice-Prest. HENRY J. ARNOLD, Secy. and Treas.

The Bell Pressed Brick Company,

CAPACITY, 50,000 PER DAY,

MANUFACTURERS OF

PRESSED AND ORNAMENTAL BRICK.

Plant, One Mile West of Durango, On the Rio Grande Southern Railroad.

DURANGO, COLORADO, SEP 13 1892

Rio Grande Southern R. R. Co

IN ACCOUNT WITH

THE BELL PRESSED BRICK COMPANY.

Aug 17 To 4000 brick at 7c $28.00

Car No. 1035, sent to Rico.

O.K. W. O. Beerbower 9/15

BILL FOR RGS VOUCHER No. 519 — Sept. 1892 — 4,000 Bricks From Durango

BILL AND SUE KNOUS COLLECTION

to come. It links together with inseparable ties some of the most prosperous and thriving villages in the West, and the country through which it passes is widely known for its fertility, varied resources and future promise. Cultivation has greatly advanced this portion of a once-savage wilderness, but there is yet much for the hand of man to accomplish ere our countless industries are forced to yield their vast riches and submit to man's usage.

Ridgway, Telluride, Rico and Durango, already prosperous towns, will derive numerous benefits from their present railroad facilities, and a mutual improvement will result from the various products of each. Durango, perhaps, will derive the greater aid from the union, as her advantageous location, superior facilities and immense riches will invite a trade from the mountain towns, which cannot but add to her already rapidly growing population and wealth. Telluride and Rico will now begin to mine in earnest, as the Durango ore market can purchase our products and leave the producers a much larger share of the profits than heretofore. The Dolores and Montezuma valleys will be taxed to supply the towns along the Rio Grande Southern with the full fruit of the soil, while Ridgway will thrive off her manufacturing and mechanical industries.

A new era has dawned upon this section, and the harvest will be great. Let us be happy.

ENTERPRISE MINE HAS NEW CITY OFFICE

DECEMBER 12, 1891—The Enterprise Mining Company's city office now adjoins **THE RICO SUN,** in the First National Bank building.

How to Boom Rico

DECEMBER 12, 1891—Lower your rents as an inducement for people to fill your houses.

Let those who have more claims than they can

work, give part of them to some one who will develop them.

Don't size every man up who comes here, and ask him as much for a six-foot hole in the ground as a live camp would for a mine.

The secret of success is more pay mines and less assessment work. One-hundred feet in one claim instead of ten feet on ten claims. If we do this next summer, we will have men from the outside who will take the other nine claims and do from 100 to 200 feet of work on them.

Don't for a moment think the advent of the railroad will sell your prospect for you. The days for selling staked claims for $500 and $1,000 have passed away.

Some of the old-timers may not like "How to Boom Rico," but this or something like it, will have to be done, railroad or no railroad.—Subscriber.

Although her population will not justify it, the number of Rico's suburbs is such as to lead the uninformed to believe we are a multitude of thousands. There is Piedmont, New Boston, Bellevue, Horse Gulch, Westport, Breezy Point and Mearsville. The coming year will doubtless give birth to a few more, and a city directory will be a need.

MORE RGS TRACK THAN ANY OTHER COUNTY

DECEMBER 12, 1891—Montezuma County claims 72 miles of the Rio Grande Southern Railroad, which is more miles of its track than is laid in any other county.

Enginehouse Framework In Place

DECEMBER 12, 1891—The framework of the roundhouse for the Rio Grande Southern at this point is in place, and the structure will be pushed to completion as early as possible. We have not yet been informed as to the number of engines the house will accommodate.

The Rio Grande Southern Railroad will run through trains from Ridgway to Durango after next week.

LIGHTS TO BE SET UP TO RGS DEPOT

December 5, 1891—The electric light company is now making arrangements to place lights on the boulevard leading from **The Rico Sun** office to the depot. Lights will also be put up in the railroad yard, near the S. J. Hood lumber and coal houses and on the streets leading to the RGS yard.

An electric [street] light has [already] been placed at the top of the grade leading to the depot, for the convenience of night travel.

WESTPORT VILLAGE STARTED

December 5, 1891—B. M. Picker has purchased the house and lots formerly owned by James Trimble, across the Dolores River, west of Rico [and the RGS yard trackage], and had the ground surveyed. These lots will soon be platted and placed on the market, and the little village will be known as "Westport." The location is very desirable as residence property, and no doubt next season will witness the erection of some neat little cottages on the new townsite.

GRAND VIEW SMELTER TO BE REACTIVATED

December 19, 1891—Considerable activity and enterprise has been displayed by W. C. Brace in his management of the business of the Rico Ore Market. He is about to renovate the small copper furnace at the Grand View Smelter, for the purpose of treating the slag in the Grand View Smelter's dump. Messrs. Klingender and Beffort worked this dump under lease two years ago, at a fair profit. The slag was broken by hand, screened and then shoveled into a sluice-box, whence it was conveyed into a plunge-gig. This mode of treatment being primitive in itself, a good part of the values were lost, which can undoubtedly be recovered by smelting. The slag itself is not high grade, but it is interspersed with matte and shot-bullion, which carry all the values. Mr. Brace has secured the services of a metallurgist and will blow-in as soon as practicable. No ores will be treated, as the nature of this experiment is not a smelting, but a melting-down process. The slag will be simply melted, allowing the lead to collect in the crucible, after which it will be run into bars of base bullion and shipped [on the Rio Grande Southern]. The result project will be watched with interest.

SANBORN
MAPPING AND GEOGRAPHIC INFORMATION SERVICE
— 1886 —

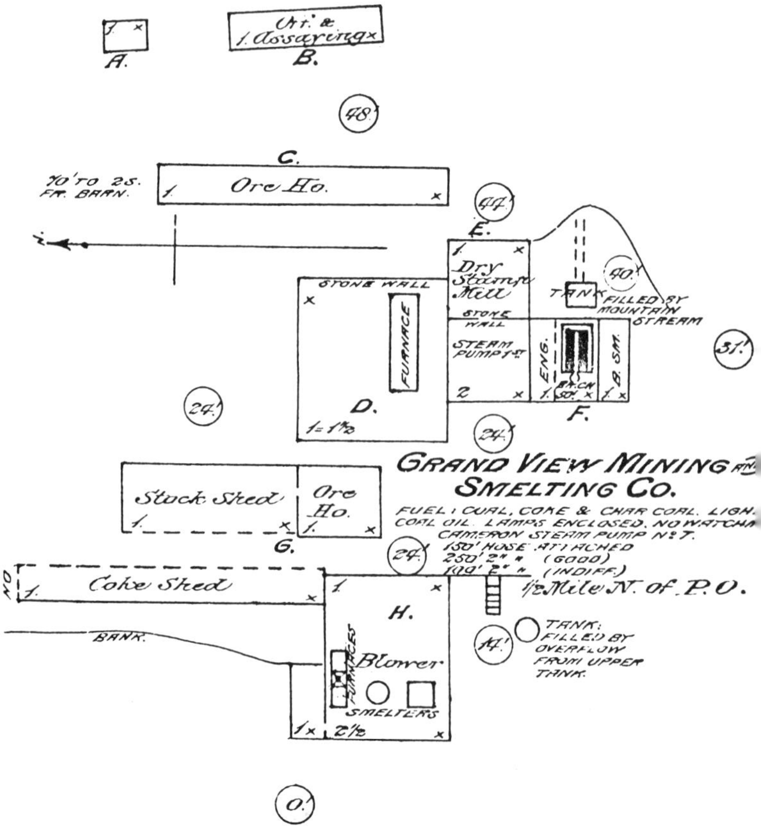

ORVAL L. JAHNKE COLLECTION

OVERLEAF: Judging by the absence of leaves on some of the aspen trees in the foreground, this view may have been taken during late October of 1891. The Rico depot of the RGS (which included a freight section) was newly completed, along with a wooden platform. Tents for the contractor's crews were still in position, beyond the new wooden 50,000-gallon water tank. Three Bridge & Building Department's work cars were in the yard (painted gray). The recently constructed two-stall enginehouse had been positioned very close to the Dolores River. Buildings located beside the RGS track (at the far left) were used for warehouses and coal storage. S. J. Hood, a lumberman, was making arrangements to build a large planing mill on the flats west of town at this time. According to a newspaper account, he was going to erect buildings large enough to carry a full stock of lumber, grain, hay and coal. Mr. Hood had made a proposition to the town council to buy or lease the land for a period of five years. Smoke was coming from the tall stack of the Grand View Mining & Smelting Company's plant at the base of Nigger Baby Hill. Two RGS passenger cars had been spotted in the yard, amidst long strings of boxcars that had brought in carloads of goods. Winter snow had not yet made its appearance, enabling construction to go ahead full speed.

MINER SUFFOCATED

By Carbonic Gas in the Early Sabbath Morning

THE SILVER GLANCE MINE

Scene of the Fatality and Barney Mundt is the Victim

DECEMBER 26, 1891—Barney Mundt was suffocated by carbonic-acid gas at the Silver Glance mine at 3:30 o'clock on Sunday morning, and his working partner, Billy Dunn, escaped death only by an accident.

Mundt and Dunn were working on the "graveyard" shift in the shaft, which is 300 feet deep. At 3:00 o'clock they were hoisted to the surface for lunch. Half an hour later, the men descended in the bucket, and when within 20 or 25 feet of the bottom of their lights went out. A body of gas had forced itself through the broken-up formation and filled the shaft to a depth of probably eight feet. The men realized this, and when the bucket struck ground, Dunn shouted to Mundt, who had stepped out, to ring the bell to ascend. He, however, was overcome by the poisoned air and fell in the bottom of the shaft. Dunn grasped the bell cord, rang three times, then fell into the bucket unconscious. Three bells means a man on the bucket, but as no signal was given to hoist, the engineer, Jas. Reveler, feared something serious had happened and pulled the bucket to the surface. By this time, foreman Berger, who was sleeping over the engine room, had been called to the scene, and in his night clothes stood looking down the shaft as the bucket was ascending. As soon as Dunn struck the air above the gas, he began to regain consciousness, and all the way up the shaft his wild shrieks could be heard by those above. He was lifted from the bucket, and Mr. Berger had himself lowered in the hopes of saving Mundt. Berger's light went out at a distance of 30 feet from the bottom and he began to grow faint. He found Mundt, however, who was quivering in his death struggle, but was unable to lift him into the bucket. To save himself, Berger rang to be hoisted, and when he reached the surface, he was

(Continued on Page 151)

THE J. W. BURLEY & COMPANY was constructing their third store when this photograph was taken (visible near the center of this picture). The structure was located on the Sanborn Map of 1893, in Block 1, Lots 5 and 6, on Glasgow Avenue. THE RICO DEMOCRAT carried the following story of their second new building, in what would become the Chas. Engel Merchantile store:

J. W. BURLEY & COMPANY NOW IN NEW QUARTERS

May 22, 1891—Standing at the head of the mercantile interests of Rico is the wholesale and retail house of J. W. Burley & Co. Their business was opened in Rico in the spring of 1888, and had so increased by the summer of 1890 as to necessitate their now commodious quarters to accommodate their fast-increasing patronage. They are located in a double building 50 x 85 feet, constructed during the past year entirely of stone, one side of which is at all times stocked with every article known to commerce, and in quantities that give the assurance that under no circumstance would there be a likelihood of a shortage in any of the articles of daily consumption. Here can be seen piled tier on tier of can goods of every kind and brand; immense piles of sack sugar from Speckles' refineries loom up beside the products of Colorado's flouring mills; and immense pyramids of Vermont maple sugar, the sweetness of which is an offset to the sour disposition of innumerable barrels of Wisconsin cranberries, both of which can be found in quantities sufficient to aid in preparing a wedding supper or in stocking one of our large mines with a full season's supply.

In the immense dry goods stock, with which the other half of this mammoth building is stocked, can be found any article in this line, from a paper of pins to a Persian shawl, and no house in the state carries a finer or more select stock of dry goods and gent's furnishings. This department is under the personal supervision of Mr. George Burley, who is a thorough man in this line, and his handiwork is seen on every side in this department in the neat-and-tasty display of the innumerable articles to be found there.

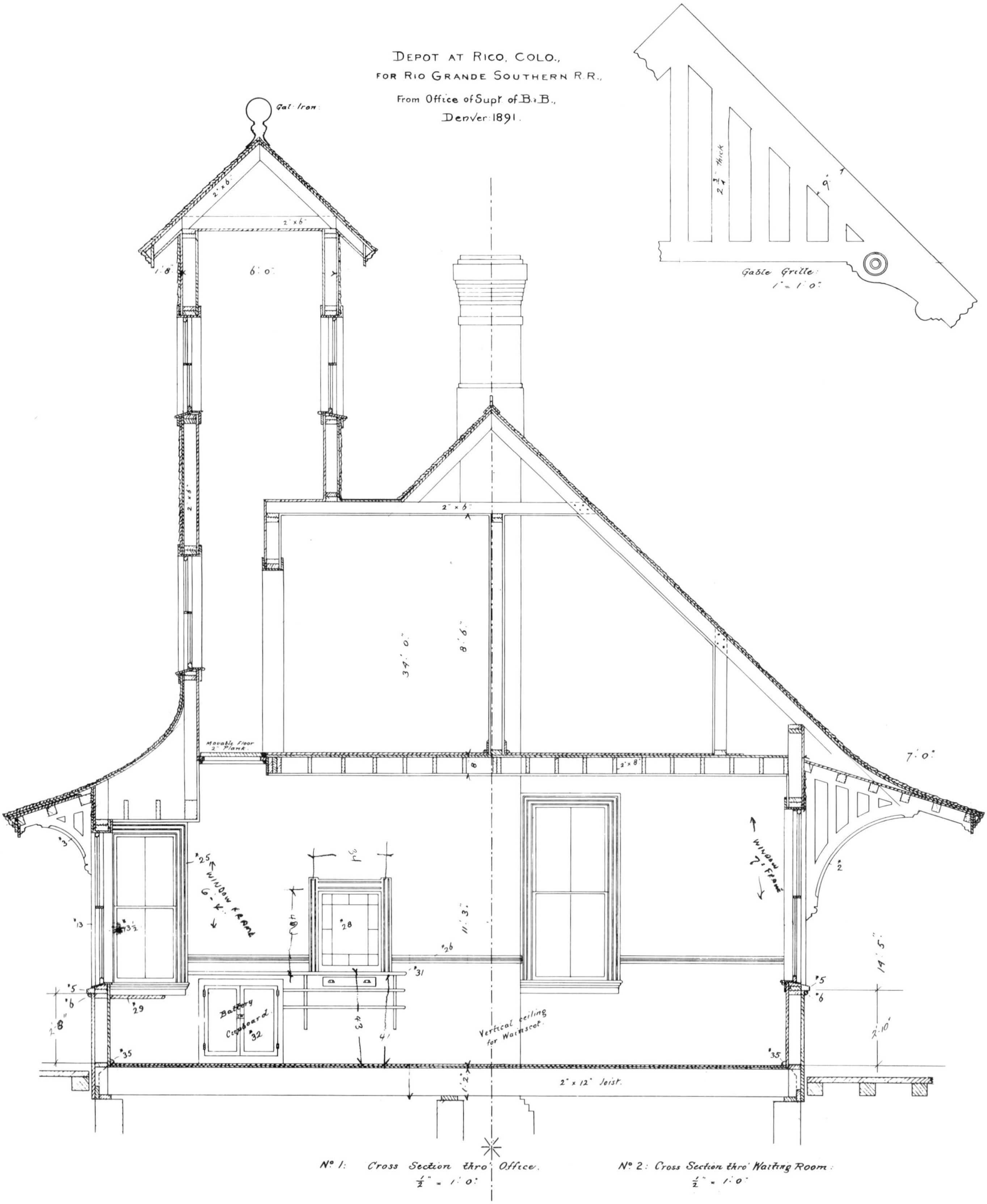

DEPOT AT RICO, COLO.,
FOR RIO GRANDE SOUTHERN R.R.,
From Office of Supt of B.&B.,
Denver 1891.
Gable Grille
1" = 1'0"
No 1: Cross Section thro' Office.
½" = 1'0"
No 2: Cross Section thro' Waiting Room.
½" = 1'0"
Battery Cupboard
Vertical ceiling for Wainscot
2" x 12" Joist
Movable Floor 2" Plank

DRAWINGS COURTESY ROBERT SHANK, SR.

DEPOT AT RICO, COLORADO

FOR THE RIO GRANDE SOUTHERN RAILROAD
FROM THE OFFICE OF THE SUPR. OF
BRIDGES AND BUILDINGS
DENVER, 1891

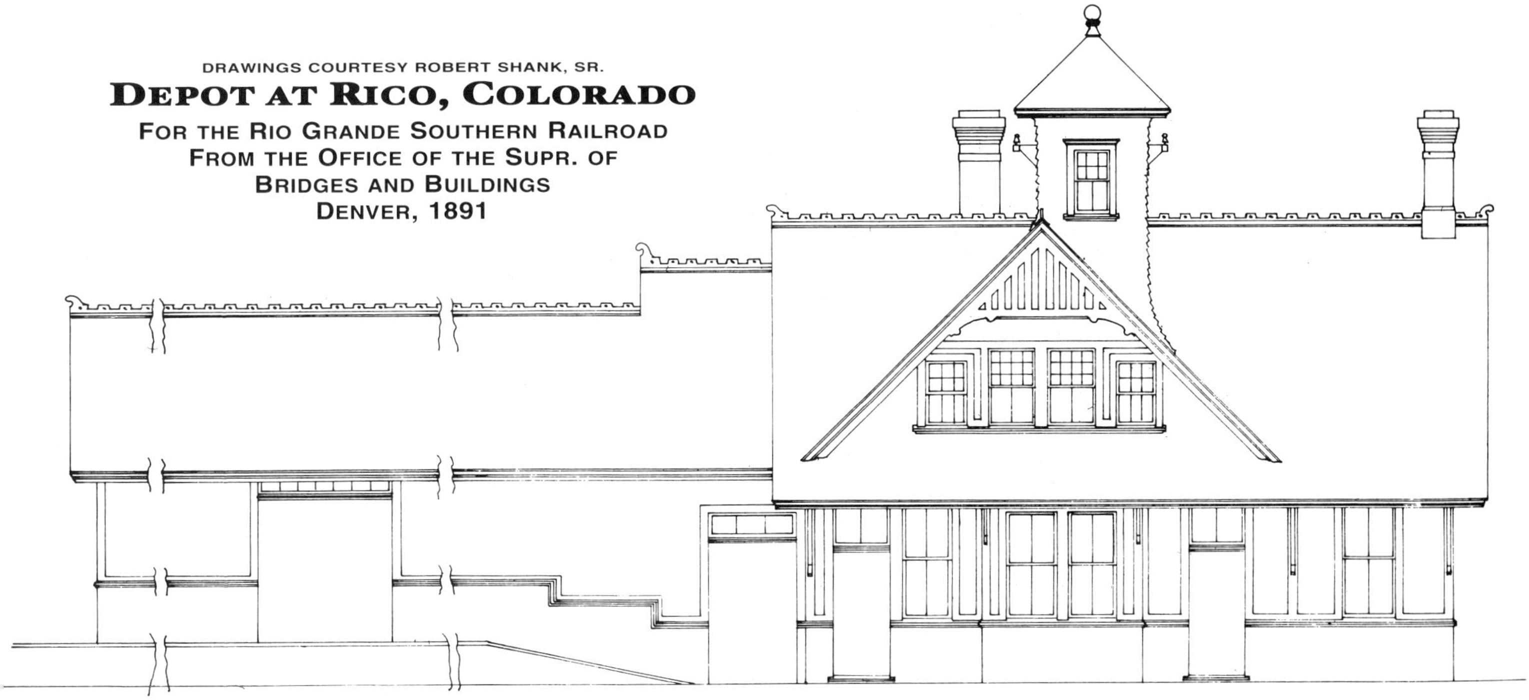

MAINLINE SIDE

TOWN SIDE

so overcome by the gas that his actions were like those of a madman. Recovering his senses, Mr. Berger again descended, taking a rope with him, but by this time, Mundt was dead. He was unable to remain below long enough to place the rope about Mundt's body, and gave up the recovery of the dead man until a hose had been brought from town, attached to the steam jet and the gas pumped out of the shaft.

No blame is laid to any one for Mundt's death, as it was an unavoidable accident. The presence of gas in the shaft had hitherto been unknown, and this time was not caused by blasting, as had been wired to the Denver papers. The men had been working with pick and shovel for three days, owing to the nature of the ground, which bore the semblance of a volcanic ash.

RGS MEN CELEBRATE COMPLETION OF ROAD

DECEMBER 26, 1891—Employees of the Rio Grande Southern, whose idleness was occasioned by the completion of the railroad on Sunday last, filled the town on the evening of the Sabbath and kept the businessmen busy cashing checks.

RIO GRANDE SOUTHERN DEPOT AN ATTRACTIVE BUILDING

JANUARY 2, 1892—Rico's depot, in point of architecture and finish, is excelled by no similar structure of its dimensions in the country. It combines economy, convenience and comfort.

Depot at Rico, Colorado

For the Rio Grande Southern Railroad
From the Office of the Supr. of
Bridges and Buildings
Denver, 1891

Cresting and Finial

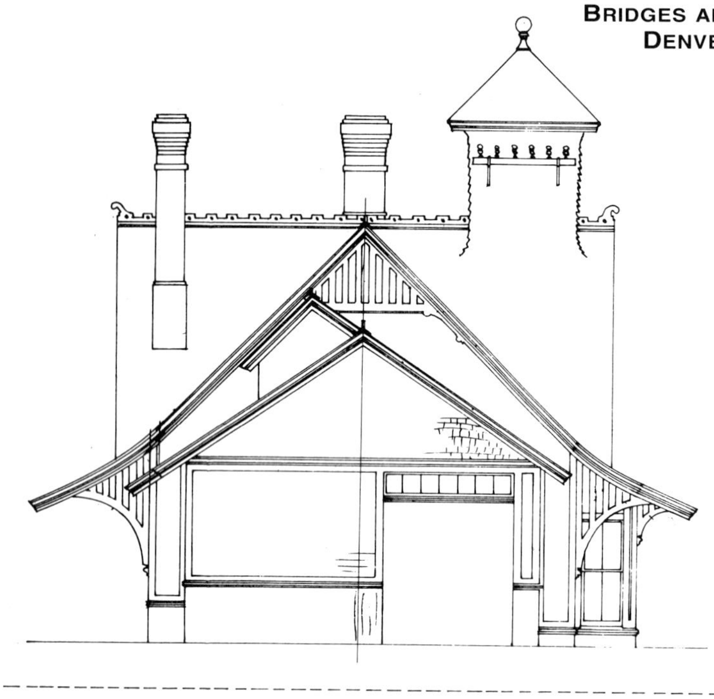

Freight Room End

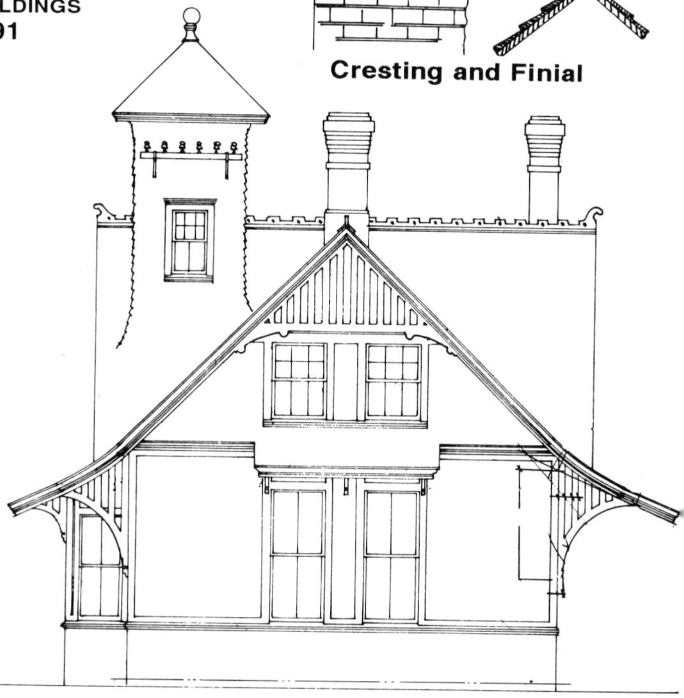

Waiting Room End

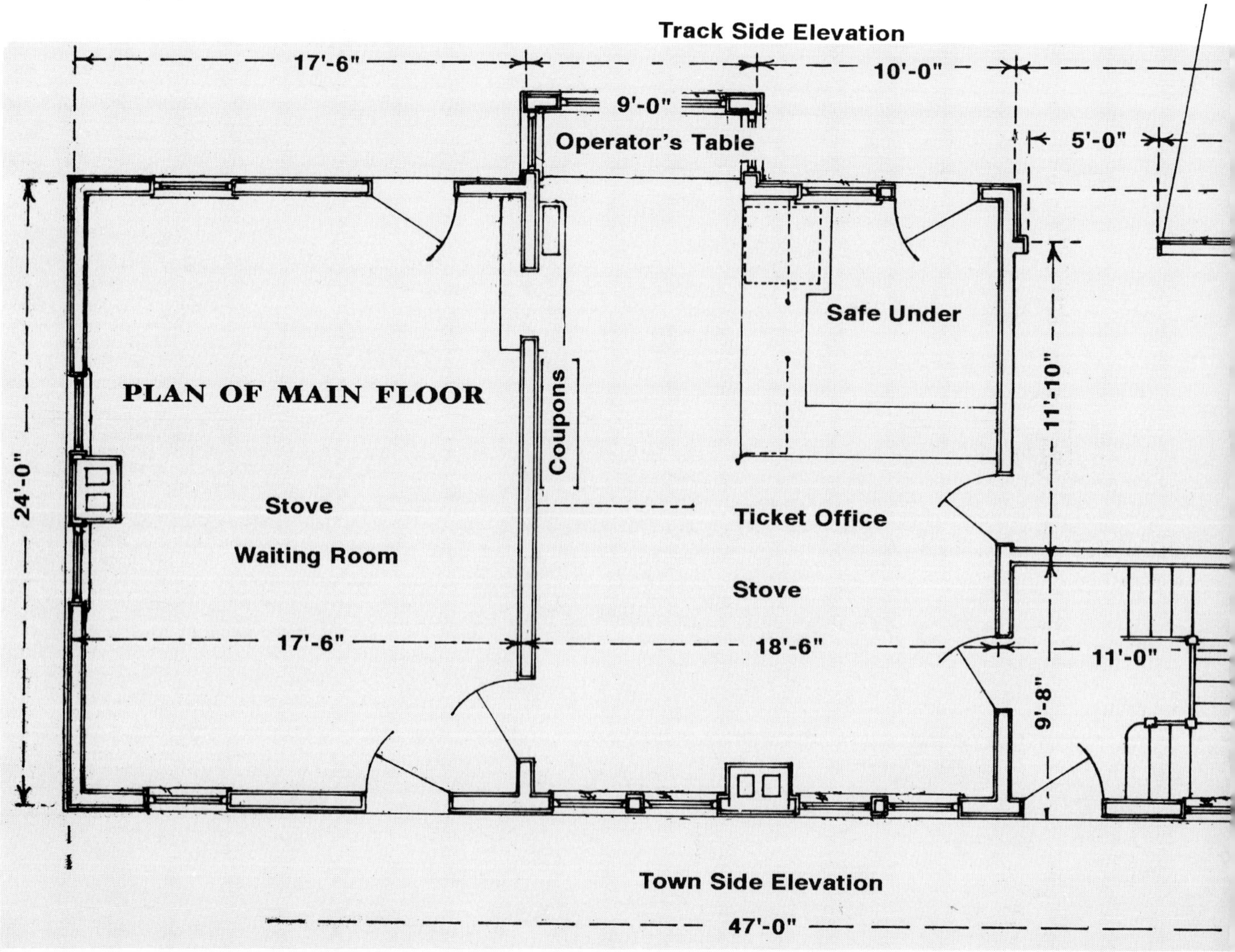

PLAN OF UPPER FLOOR

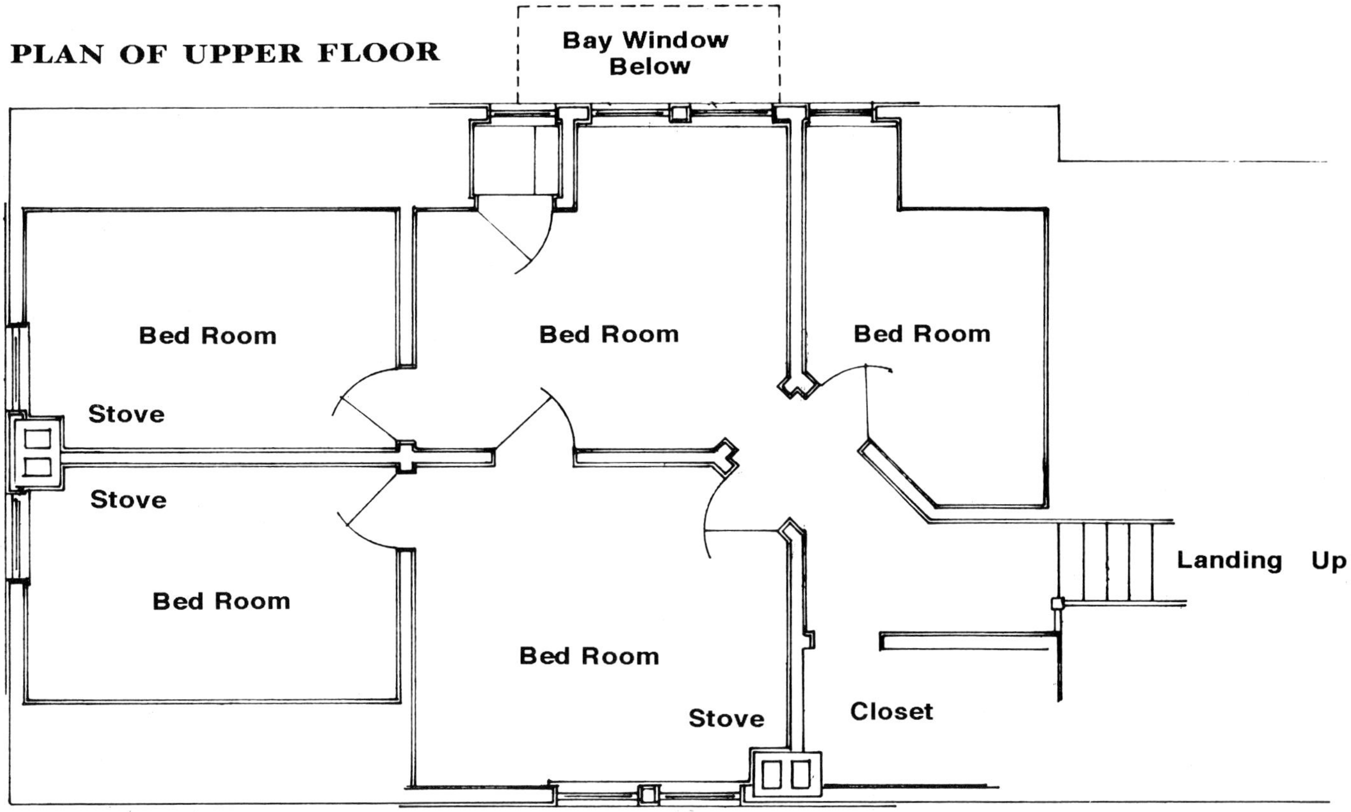

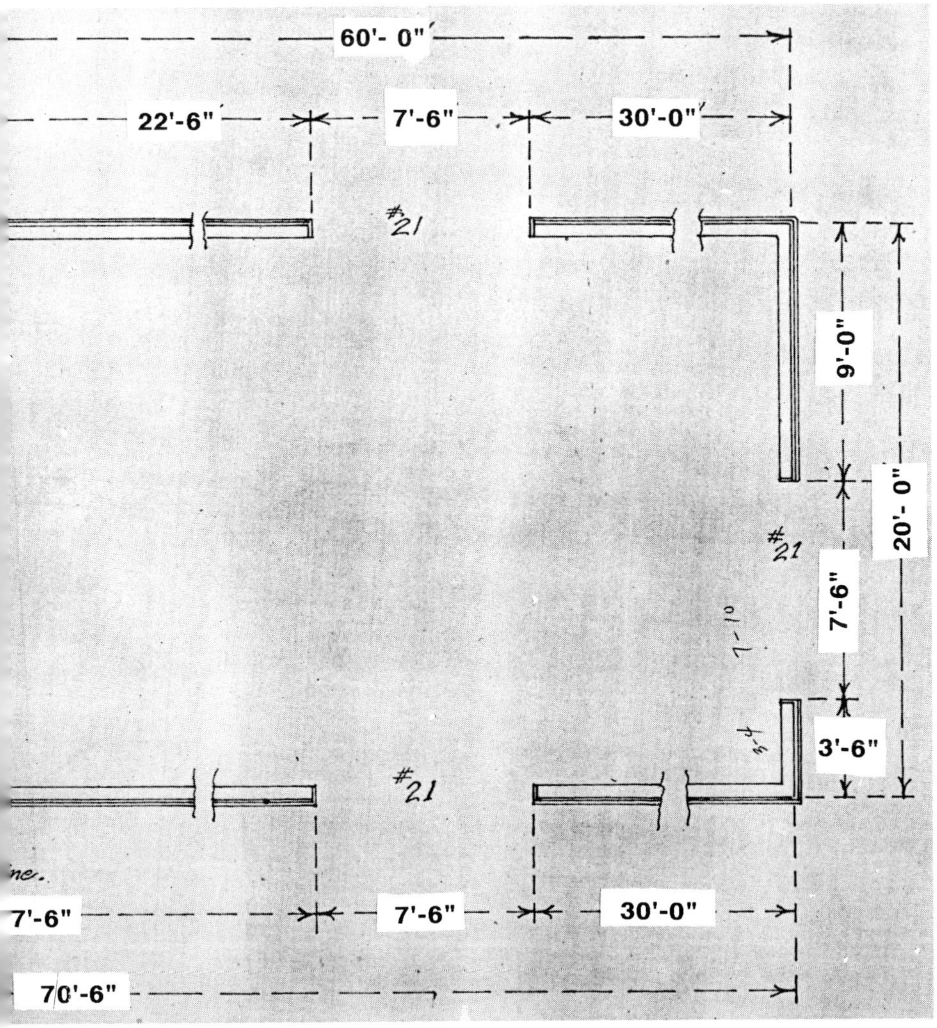

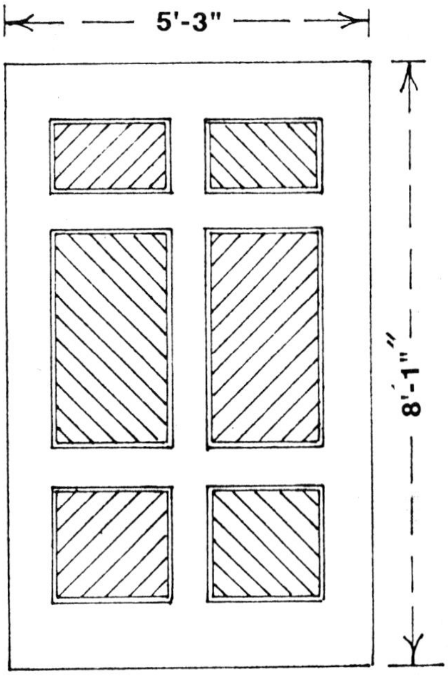

EXPRESS ROOM DOOR
Section & General
Finish Same As
Freight Room Door

First Rio Grande Southern Train From Durango to Rico Welcomed

DECEMBER 26, 1891—T. C. Graden, one of Durango's leading merchants, was a passenger on the first through train [from Durango] to Rico on Sunday. Mr. Graden is a partner of S. J. Hood, the coal-and-lumber merchant, and expresses himself as eminently pleased with the camp's future outlook.

Engineer Peas says he is not going to Denver at present—as he has some more side tracks to lay out. P.S.—This goes, if he don't get side-tracked. N.B.—He got side-tracked and left for Denver Thursday morning.

First RGS Train Arrives in Durango

DECEMBER 26, 1891—At 4:45 p.m., says ***The Durango Herald*** of December 22, the first train arrived over the Rio Grande Southern, making its initial round trip to Rico [and return]. The long-talked-of railroad connection has been made between that place and Durango. The second railroad route to Denver is ours, and we are delighted. Preparations are in the making for a grand "blow-out" over the affair, in which the sister cities will unite to sing praises and rejoice over the prospects of great future prosperity to both. No doubt that the completion of the new smelter in Durango will reduce all Rico ores, and thus the towns have a mutual interest. What helps one helps the other. No rivalry exists between them, but each has a warm interest in the advancement and welfare of the other.

Passengers on Train From Durango

DECEMBER 26, 1891—Mrs. Ruby Kelly, sister of Ed Rust of **The Rico Sun**, accompanied by her little son, McGinnis, was a passenger on Tuesday night's train from Durango, which was 14 hours on the road, having encountered more snow than was expected. Mrs. Kelly bears the distinction of being the first lady passenger over the new railroad from Durango to Rico.

The Durango train leaves Rico at 6:45 a.m., reaching its destination at 4:00 p.m. More time is required for the return trip, as it is almost a continuous up-hill pull, and an enormous lot of freight comes by way of Durango. A mixed train will be run during the winter, as the passenger traffic at present will justify but one train.

New RGS Rotar y Snowplow Sighted

JANUARY 1, 1892—The Rio Grande Southern's new rotary snowplow passed through Rico this morning on its way to the Artic regions near Telluride. Two engines furnished the power to assist the plow in its work.

Deadly Slide Above Rico

JANUARY 2, 1892—M. Q. Cretcher and Wm. Connel were killed by a snowslide on Nigger Baby Hill early on Friday morning while in the act of making a mining location.

They left the Rico Club Rooms saloon shortly before midnight and proceeded at once to climb the hill. It was understood among a few friends of the unfortunate men that they were to return in about an hour and a half. After a lapse of over three hours from the time of their departure, Fred Baxter expressed himself as feeling uneasy regarding their safety, knowing the character of the hill for slides and the increasing danger from recent heavy snowstorms. At Mr. Baxter's suggestion, a party composed of [six men]... started for the spot where Cretcher and Connel expected to make their [mine] location. They found a large slide at the base of the hill, and within a few feet of the trestlework of the railroad, which obstruction apparently somewhat broke the force of the mass of snow, and it was piled up in a confused heap of several feet.... A large crowd assembled at the scene of the slide, and after considerable exploration, uncovered the men....

The slide, which resulted in the death of the men, was about 400 feet long and probably 50 feet wide. It started at one of the branches of the Hope & Cross mine dump. The [mining] claim upon which Cretcher and Connel endeavored to place their stake was known as the Silver Drift and runs parallel to the Ella Mine, a patented mining claim.

RGS UNABLE TO SUPPLY DEMAND FOR COAL

JANUARY 2, 1892—The Rio Grande Southern is unable to supply the demand for Durango coal at this point. The scarcity of locomotive power is the cause. [No reason was given regarding the lack of engines along this portion of the line.]

NEW VEIN OF COAL FOUND AT RICO

JANUARY 2, 1892—J. H. Goldsworthy has just opened a new vein of coal in his [coal mine]... an extension of the old Pasadena Mine. The quality of this coal is said to be unequalled by any other fuel found in this locality, and it is being sold at prices which come within the reach of all. In lots of 10 tons, Mr. Goldsworthy has established a rate of $6.00 per ton. Heavy consumers would do well to note this rate and examine the quality of the coal before purchasing their winter's supply of fuel. Every pound of the coal will be found as represented. Call at Mr. Goldsworthy's office, on the north side of Silver Creek, and leave orders.

W. D. LEE APPOINTED TO NEW POST ON RGS

JANUARY 9, 1892—Howard Ridgway has resigned his position as assistant superintendent on the Rio Grande Southern, and W. D. Lee has been appointed in his stead.

(Continued on Page 156)

D. & R. G. Form 477—4-92-2m.

BILL FOR VOUCHER.

Department No. 810

Auditor's Voucher No. 628

Chargeable to

To Denver and Rio Grande R.R. Co. Dr.

(10 / 81) $ 1243

Address Denver Colo

Pay

Address

Total, $ 1243 #924

1892 Sept a/c

Aug	31	For time of Scale Repairer testing and regulating scales at following points, month of August 1892.			
		Rico. Scale Repairer 1 Day. @ 100.00 Mo	3 23		
		Telluride " " 2 " " 100.00 "	6 45	9 68	
		Ridgway " " 1 " " 100.00 "	3 23		
		R. G. So. propr at Ridgway 1/2		1 62	
		Supervision. 10 0/0 on 11 30		1 13	
		The calculations herein are correct and no previous voucher has been passed for the items.		$ 12 43	
		DISBURSEMENT CLERK.			

Correct: W. D. Lee Supt

Certified:

Approved: Otto Mears

Place of delivery of all materials and Supplies must be stated in Bill for Voucher.

BILL FOR VOUCHER
No. 628
— 1892 —
For Scale Repair at Rico
BILL AND SUE KNOUS COLLECTION

io Grande Southern
RAILROAD.

The New All-Rail Line to

elluride, Placerville, Ft. Lewis, Mancos and Rico.

W COMPLETED and OPEN for BUSINESS

rom RIDGWAY

the line of the Denver & Rio Grande R. R.

To TELLURIDE

And will be pushed to final completion to rango and Rico by midsummer, opening up e most Magnificent Scenery in the Rocky ountains, heretofore inaccessable, and passing rough the...

mous Montezuma and Shenandoah Valleys

The Great Agricultural Region of the

DOLORES RIVER.

is new line will bring the Tourist within an easy ride of the Wonderful

omes of the Cliff Dwellers

Immediate connections are made at Ridgay with trains of the Denver & Rio Grande ailroad to and from Denver, Pueblo and olorado Springs for Placerville and Telluride.

age Connections from Telluride to Rico.

As fast as road is completed, train service ill be extended, with stage connections from rminus to Rico.

OTTO MEARS,
General Manager

D. & R. G. Form 479—6-92-2m.

BILL COLLECTIBLE

In making settlement or Voucher please refer to Auditor's No. 1068

628

Month August 1892

R&P Dep't Div'n No. 631

Credit Account No. RRBtRm Pay Rolls 11.30 / 1.13

Memo. No.

Claim No.

Auth'd by

Date 189

Rio Grande Southern R. R. Co.

Denver Colo.

To Denver and Rio Grande Railroad Co., Dr.

REMITTANCES SHOULD BE MADE TO J. W. GILLULY, TREASURER, DENVER, COLO.

1892 Aug.	31	For time of Scale Repairer testing and regulating scales at following points, month of August 1892		
		Rico. Scale Repairer 1 Day @100.00 mo	3 23	
		Telluride " " 2 " 100.00 "	6 45	
		Ridgway " " 1 " 100.00 . 3.23 your propn 1/2	1 62	
		Supervision 10 0/0	1 13	$ 12 43

R. G. S. R. R. CO. Treasurer's Bill for Voucher. No. 924

Correct: C. S. Thompson

Approved: Wm H Rogers

Approved: Auditor:

Approved: E. T. Jeffery

Prest. and Gen'l Mgr.

[Note: The first through timetable between Ridgway and Durango became effective on January 31, 1892. A passenger train departed from each terminal each day, and each train stopped overnight at Rico. A passenger train operated over the Telluride Branch to meet each train at Vance Junction.

[Upon the resignation of R. M. Ridgway in March of 1892, W. D. Lee became superintendent, and four years later, Mr. Lee became the general superintendent of the road.]

The RGS shop at Ridgway will employ 10 men this winter.

TELEGRAPH LINE NOW OPEN FROM RICO TO DURANGO

JANUARY 9, 1892—The telegraph line has been stretched to Durango. One of the benefits will be to lessen the danger of collisions on the Rio Grande Southern. [See the story below.]

FIRST WRECK ON THE RGS

JANUARY 9, 1892—The first wreck on the Rio Grande Southern occurred Wednesday, about four miles below [south of] Rico. A doubleheader from Durango had reached the point where the accident occurred and "died" for want of coal and water. The southbound train from Rico was going down the grade at such a speed that it was impossible to stop, and ran into the dead train. No lives were lost, but the engines and a few cars were damaged to a considerable extent. The wreck was cleared away yesterday, and traffic again resumed.

[NOTE: The above wreck occurred at about 8:15 a.m. on January 6, 1892. The northbound train was a "work extra" and should have had a flagman out ahead of the stalled train; however, the conductor failed to send a man ahead for flagging. This accident happened only a day or two before the telegraph line was strung along the RGS into Rico, and the work extra was running without orders. Engines 17 and 7 were on the work train, while Engine 36 was on the mixed train, going down the 1.8-percent grade south of Rico. The mixed consisted of several gondolas loaded with coal, a flatcar with a lumber load, a combination baggage-mail car and a coach. Luckily, no one was injured; however, most of the cars were derailed, and Engines 17 and 36 received considerable damage to their pilots, lead trucks and headlights.]

MINING REPORTS VERY GOOD

MARCH 26, 1892—Official reports from mines are most satisfactory. The Enterprise declared a dividend of $50,000. Owing to the severity of the weather, the branch has not been built, and it is almost impossible to ship the ore being taken out of the mine.

— *The Mining & Engineering Journal*

Mile by Mile to...

The Rico Division Point

The End of the First District and the Beginning of the Second District

By Russ Collman

— *Text Continued From Page 41* —

Back on the mainline, at Milepost 65.5, a wicked snowslide ran down no less than three ravines at one time. And this combined bulk often roared down over the railroad. RGS men called this avalanche the Nigger Baby Slide, but it was normally not as bad as the Burns Slide.

RICO—Milepost 66.2, Elevation 8,737 feet—originally was called **CARBONATE CITY**, but the name was changed to "Rico" before the Rio Grande Southern arrived on the scene. Rico means "rich" in Spanish, and the name obviously was chosen because of the rich silver deposits in the local Pioneer Mining District.

Rico had an attractive wooden depot (of D&RG design), where the station agent and his family lived upstairs on the second floor (over the agent's office and waiting room). This colorful station was the division point office, separating the northern First District from the southern Second District.

The original intention was for the dispatcher's office of the RGS to be at Rico. However, this arrangement never worked out like it should have, and it would have been better if the division point had been moved to Dolores, 36 miles to the south—with the dispatcher working out of that station. Logistically, Dolores was a much more practical place for dispatching trains north and south, as well as for servicing and repairing engines and cars, but it seems that hindsight is always better than foresight. One newspaper reported that the division point, with a large freight yard, might be built at Red Rock, south of Rico. In any case, the dispatcher's office was moved to Ridgway after the D&RG took over the RGS late in 1893.

The freight yard at Rico consisted of five tracks, with one just east of the depot (the house track) and the other four just west of it. A total of 144 freight cars could be held on these tracks. A standard 50,000-gallon water tank was just north of the depot, while a Burnett & Clifton coaling facility was located at the extreme southern end of the yard, south of the wye.

Rico had a total of ***three*** wyes for turning engines and cars—although there was only one in place at any given time. The first wye was built during the construction period, and it was located where the first enginehouse was constructed after the yard was laid out. This wye was removed to a location that spanned the Dolores River with two short pile trestles, **BRIDGES 67-AS** and **67-ASS**. Both bridges were on 20-degree curves and accommodated the approach tracks to the tail of the wye. Bridge 67-AS was 96 feet long, while Bridge 67-ASS was 80 feet long, but it may have been lengthened to 96 feet after the bad flood of 1909. Following the extremely disastrous flood of 1911, which wiped out a number of the yard tracks, the wye was moved to the opposite side of the main track. This third and final wye remained at this site until the abandonment of the RGS in 1951.

Besides the buildings already mentioned, there were many other structures for various uses. The section house at Rico was the largest building north of the wooden enginehouse. The RGS used standard D&RG plans for this house, erected for the use of the section foreman and his family. During later years, it was painted a buff color, with brown trim. This followed the practice of the

STATION MAP OF RICO
1893 – 1900
DRAWN BY JIM KEY

3. SAMPLING WORKS
4. RICO P&L WAREHOUSE
5. S. J. HOOD LUMBER YARD
7. COAL SHED
16. RGS WATER TANK
17. RGS DEPOT
21. COAL GONDOLA
22. SAND HOUSE
23. COAL DOCK (FOR ENGINES)
24. FIRST ENGINE HOUSE (WOODEN)
25. SECTION HOUSE
26. BUNKHOUSE
27. COAL SHED
28. WOODEN BOARDWALK
34. RPO BODY FOR ROADMASTER
37. DEPOT OUTHOUSE
39. TRACK SCALE
40. SECOND WYE
41. RGS BRIDGE 67-AS
42. RGS BRIDGE 67-ASS
45. LOCATION OF FIRST WYE
46. RICO SMELTING CO. (ADAMS)
47. PASADENA REDUCTION CO. (TRACT)

D&RGW. (Incidentally, the buff faded to a straw yellow with the passage of time.) Just to the south of this building was the bunkhouse, used by section men and other workers and was painted like the section house.

The Rico depot's outhouse and coal shed shared a common wall. It was located to the south of the depot, just beyond the fenced yard and garden. Between the depot and outhouse there was a firehose shed. (At least one photograph shows firehose laid out on the ground to dry.)

All coal, lumber, foodstuffs and other supplies destined for booming Rico had to come from the north during the period between September 30 and December 19, 1891. As a result, the RGS rails over Lizard Head Pass no doubt were the busiest ever during these weeks.

The financial condition of Otto Mears' Rio Grande Southern had been stretched very thin during 1891, and the railroad had to defend itself against lawsuits for unpaid bills. It was imperative that President Mears complete railroad construction into Rico as rapidly as possible, in order to gain the revenue that service to that growing mining town would provide. To speed things up, Mr. Mears took men off construction of the Pandora spur to work on the mainline over Lizard Head Pass. With additional workers, the RGS resumed construction southward from Vance Junction to Rico during the spring of 1891, and construction crews were pressed hard to work night and day. Construction of the line was slow and extremely difficult at the Ophir Loop because of all the trestles, but Mr. Mears demanded that the work be speeded up. Needless to say, the mood at Rico was one of anxious anticipation.

At long last, a construction train, consisting of RGS engine No. 14, a small Class 48 0-6-0T saddle-tank switcher, pulling one small work car, approached Rico. Some railroaders called this switcher the "Goat," but others said the 0-6-0T looked more like a "mechanical burro." (See a picture on Page 128.) Reportedly, this short consist entered town late at night on September 30, 1891.

Eureka! The narrow-gauge tracks of the Rio Grande Southern Railroad had carried an engine across the town limits into Rico! At long last, the railroad had arrived! And virtually overnight, the assessed valuation of Rico soared from what was $480,000 during 1890, to nearly $1 million after the RGS arrived in town.

Almost everyone in the Rico area came to meet the RGS construction "train" as it arrived in the new freight yard. Thereupon, an unplanned celebration erupted like a volcano, and the unabashed revelry continued for at least two or three days. Of course, the men who came in on the construction train were treated to nearly anything they desired to eat and drink—and needless to say, the men made full use of their moment in the limelight!

As soon as the residents recovered from their inebriating celebration—which must have resulted in some mighty horrific hangovers—the townspeople began to plan a formal celebration. The plan was for a very traditional, elegant observance. Rico was in the mood for a "silver-spike ceremony," even though construction of the RGS was far from complete. Over 50 miles of track south of Rico had not yet been built. In any case, the little mining town was cleaned up, and many yards of red, white and blue bunting, as well as American flags, were strung up, providing an attractive backdrop for the gala occasion!

Luckily, some of the original train orders were saved, thanks to a Rico businessman, along with the first RGS train register. This register is now on display at the Gunnison County Museum.

Thanks to this surviving train register, existing photographs of the first passenger trains into Rico can be identified. The first train of October 15, 1891, was powered with RGS engine No. 12, a former D&RG Class 60 (C-16) 2-8-0 (No. 251), which hauled a combination baggage-coach and four coaches. This first train brought Telluride's well-known "Silver Cornet Band" to Rico for the celebration. (See this colorful marching band en route to Rico on Page 188 of Volume IV in this series.) The second train, pulled by D&RG No. 168, a Class 47 (T-12) 4-6-0 (which is now in the railroad park in Colorado Springs, near the former D&RG depot), consisted of a combine and two Pullman cars, also arriving from Telluride. The second and third trains are shown on Page 45 of Volume III. The third train came through from Denver with No. 174, also a D&RG Class 47 (T-12) 4-6-0, a coach, Otto Mears' private car (the "San Juan") and two Pullman cars. This last special train carried a number of notable people for the event. This group included Colorado's Governor John Routt and E. T. Jeffery, president of the Denver & Rio Grande. The Rio Grande Southern was represented by President Otto Mears, Superintendent R. M. Ridgway, Chief Engineer Charles W. Gibbs and other officials.

It was entirely appropriate that the silver spike for the occasion was made of silver from Rico's leading mine, the Enterprise. However, the spike actually used for the occasion was only silver-plated, while the all-silver engraved spike was reserved for display purposes only! In any case, shortly after 2:00 p.m., Governor Routt, Otto Mears, E. T. Jeffrey and other guests hammered down symbolic blows at the silver-plated spike, and Rico was officially connected with the outside world by railroad, albeit a narrow-gauge line! The crowd happily applauded and cheered with gusto, and both the Rico and Telluride marching bands provided lively music. Following the silver-spike ceremony, a considerable number of alcoholic drinks were consumed. After that, an elegant banquet was held at the impressive Hotel Enterprise, the town's largest hostelry.

Of course, Otto Mears was the guest of honor, and he sat at the head table. Numerous toasts and speeches were delivered, all of which predicted brilliant futures for the town, the mining district and the railroad. Following dinner, the guests danced the night away, with grand ballroom waltzes and other dances of the period—until after the proverbial 3:00 o'clock in the morning. After an all-too-short night for many of the guests, the trains departed northbound, beginning at the ungodly hour of 5:15 a.m., followed by the second one at 8:40, while the third train departed at a more respectable 9:20 a.m.

Rico had its railroad! However, the RGS had not yet been completed. As mentioned above, there was a gap in the track between Rico and Dolores. So, it was very important for the construction crews to connect the two ends of the RGS as soon as possible—so thatthrough trains could be operated all the way from Ridgway to Durango. The work force labored both north from Dolores and south from Rico. And the final connection was made less than 20 miles south of Rico, on December 19, 1891.

No formal celebration was held, and the construction crews had to be happy with turkeys and beer, delivered by the first train on the following day. After delivering the L-C-L freight for the workers, the first train from Dolores continued on to Rico, where 20 cars of coal were delivered.

The Rio Grande Southern had a very profitable year in 1892. Although the RGS had been operating trains as trackage was constructed, the Rio Grande Southern Railroad Company did not officially accept ownership of the railroad from the Rio Grande Southern Construction Company until February 1, 1892. In his report to the stockholders for the first five months of operation by the RGS Railroad Company, between February 1 to June 30, 1891, President Mears reported a handsome profit.

The figures were as follows:

Gross Earnings	**$244,894.25**
Expenses	**118,764.56**
Net Earnings	**$126,129.69**

RGS LOCOMOTIVE NOTES

By Robert W. Richardson

OTTO MEARS originally intended to have 36 locomotives on the Rio Grande Southern. This was based on expected traffic and his experience with the Silverton Railroad, as well as observing operations on the Denver & Rio Grande.

The D&RG had a surplus of narrow-gauge engines and cars after the 1890 opening of the new standard-gauge line to Utah. They sold Mears 19 Class 60 (later, C-16) engines, a 2-6-0, an 0-6-0T switcher and a 4-4-0. From the Rio Grande Western (in Utah) he bought 12 engines, including nine Class 56 2-8-0's (slightly lighter than a Class

COLORADO RAILROAD MUSEUM COLLECTION –
COURTESY OF MALLORY HOPE FERRELL

RGS FIREMAN SAM DAVIS and Engineer Tom McKinney were proudly posing beside the tender of handsome Engine 21 during the summer of 1892. The railroad had reached Rico on September 30, 1891, to the relief of everyone in the little mining town. Although No. 21 was built several years previously by the Baldwin Locomotive Works for the Denver & Rio Grande Railway—in 1882—her crew was proud of their diamond-stacked engine. Operated on the D&RG as their No. 261 until sold to the RGS in 1891, she served the RGS well. It is quite possible that this engine was used for helper service from Rico to Lizard Head Pass. Notice the link-and-pin couplers on this Consolidation, in use until 1903. No. 21 was dismantled in 1926.

60 2-8-0) and three 4-6-0's. The Ten-wheelers, from various original sources, were expected to haul the RGS passenger trains.

The Rio Grande Western engines needed repairs and overhauling, and they went directly to the D&RG's Burnham shops. Before all of them were sent to the RGS, the financial crash of July, 1893, took place. The overhauling work of Engines 29 and 35 was halted, and these locomotives sat at Burnham for years. No. 29 never got to the RGS, and No. 35 was sold to the Boston Coal & Fuel Company.

When construction began on the RGS at Ridgway during August of 1890, D&RG Third Division Class 60 engines and Third Division engine crews were used. Gradually, the newly purchased engines arrived at Ridgway and Durango. The Shay from the

THE SHAY GEARED ENGINE probably arrived when construction of the Enterprise Branch of the RGS was begun. This small Shay was built in 1890 for Otto Mears' Silverton Railroad (No. 269, the "Guston") by Lima Locomotive Works. She was supposed to be "the perfect answer" for switching cars up the 5.0-percent grade of the branch. The RGS gave the Shay the number "34" and put her to work on the switchbacks. Although Otto Mears felt that this geared-drive engine was better suited to the Silverton Railroad's curves and grades than a conventional 2-8-0, he traded her to his Rio Grande Southern for another 2-8-0, RGS Baldwin 2-8-0 No. 34. The Shay was "borrowed" for use on RGS construction trains, as well as to haul coal cars, on the southern end of the RGS until early 1892, when she went back to Silverton for a brief period of time before being sent to Rico for use on the Enterprise Branch. W. H. Jackson shot this picture in November of 1892, while the Shay was at the Silverton depot. After the Silver Panic of 1893, Shay No. 34 sat in the Ridgway yard until sold in 1899.

W. H. JACKSON PHOTO – E. J. HALEY COLLECTION

Silverton Railroad was put into service in work trains originating at Durango.

The nine Class 56 2-8-0's had short careers, and tally sheets show as little as three months of service for one. All were sold in the 1890's by the D&RG, except for No. 34, which was traded to the Silverton Railroad, and No. 35, which was still at Burnham in 1901. One RGW locomotive that was supposed to become Engine 26 never arrived from Utah.

Periodically, over the years, before the final receivership in 1930, numerous C-16 "200's" took over operations on the RGS. There were several reasons for this. Major shopping was generally done at Salida and Alamosa. If those places were busy with D&RG engines, the D&RG would supply engines from its Third and Fourth divisions to temporarily replace the out-of-service RGS C-16's. Sometimes this led to shortages of motive power on the D&RG, and sometimes a drop in revenues would lead to a reduction in shop forces. A third—and not so obvious—reason was in bookkeeping showing D&RG rentals as income, and reducing RGS net income by increasing the cost of operations on the RGS.

Generally, the D&RG kept its larger 2-8-0 classes on its own lines; however, around 1907, a Class 70 (C-19) engine would occasionally appear on the First District of the RGS. These engines performed well in Ridgway–Telluride service, and the RGS was well pleased with them.

With the D&RG's acquisition of the Class 70 (or 72) Florence & Cripple Creek 2-8-0's and the three Crystal River outside-frame 2-8-0's in 1916 and 1917, and their assignment to the Third Division, the RGS began to receive more of the Class 70–71–72 engines. And in 1916, the RGS bought three ex-F&CC 4-6-0's. Class 70 engine No. 420 was favored for the Ridgway–Telluride passenger train, which actually was a mixed train, usually with one freight car in the consist. Before the arrival of No. 420 (which was renumbered "42" in 1918), Engine 6 was the favored engine for this run.

The three F&CC Ten-wheelers acquired in 1916 initially were restricted to the Durango–Rico district, as they were classed as "85" (T-19) engines, and they caused some damage to some sections of the First District trackage. With the increased water capacity of their tenders, they handled the mixed runs

out of Durango. They were tested on the D&RG's passenger trains out of Salida and Alamosa, and a trade for a Class 125 (K-27) "Mudhen" 2-8-2 was nearly consummated. RGS officials preferred the Mudhen.

During 1910–1914, the original wrought-iron boilers on the D&RG system were replaced with steel boilers, and at the same time, extension front ends were installed, and new straight stacks replaced the old diamond stacks. Engines deemed not worth the change were scrapped.

In the 1920's, RGS track and bridge improvements enabled the road primarily to use C-19 (Class 70) engines, and its C-16's gradually were set aside. However, by 1927, most freight operations were conducted by the D&RGW's K-27 Mudhens. This situation ended with the severe "Crash of 1929" and the final RGS receivership. Receiver Victor Miller sent back all leased engines to the D&RGW, and the RGS endeavored to handle the reduced traffic with their own engines—Nos. 40, 41, 42, 20, 22 and 25. As the D&RGW had many idle engines, the RGS occasionally could rent a K-27. However, because the RGS was slow in paying its bills, the D&RGW eventually would only rent an engine on authority of high officials in Denver, delaying delivery of a K-27.

The C-16 locomotives gradually were set aside during the early 1920's, and by 1926, none were to be used again. All except Engine 6 sat idle at Durango until they were scrapped there during the 1930's. No. 6 had gone to Alamosa for shopping, but when partly dismantled the work was stopped. Unable to return C-16 No. 6 to Durango, the D&RGW resorted to buying the engine, then filed the necessary papers for "D&RGW No. 6" to be scrapped—and the shop proceeded to cut up the 2-8-0!

Gradually, over the early years, nine of the C-16's went to scrap. No. 1 was demolished in a head-on collision near Mancos in November of 1901. Another C-16, No. 4, survived for a few years longer, while Engines 2, 5, 7, 8, 9, 12 and the original No. 20 went to scrap when needing major overhauling, such as firebox or boiler repairs, or when not converted to steel boilers.

During 1916, new ICC regulations went into effect, and the original No. 20 4-6-0 did not survive that edict. The 2-6-0 and 0-6-0T engines were sold in the 1890's, while the sole 4-4-0, No. 36, was heavily damaged in January of 1892, in a head-on collision near Montelores, and she was set aside and sold. The Shay was set aside after a brief period of use on the Enterprise Branch, and she sat in the Ridgway yard until sold to a Wisconsin lumber line in 1899.

In 1916, the RGS had ample motive power for most of the year. Ten Class 60's, three Class 70's and three Class 85's made up the roster. These locomotives were Nos. 3, 6, 10, 13, 15, 16, 17, 18, 19, 20, 21, 22, 25, 40, 41 and 42. The Class 85 4-6-0's handled the Durango–Telluride passenger train (a mixed train in practice), while No. 42 hauled the Ridgway–Telluride passenger train.

The engine terminals were at Ridgway, Telluride, Rico, Dolores and Durango. Dolores had only an ashpit, and coal had to be hand-shoveled into the tenders. Despite the numbers of freight engines laying over there, its facilities were never improved. Rico's enginehouse sheltered the rotary plow. Telluride's small enginehouse and water tank were supplemented by a coaling facility at Vance Junction, but the junction did not have a much needed water tank! Durango was jointly used with the D&RGW until the 1930's; however, the RGS frequently accused the D&RGW roundhouse force of appropriating parts from RGS engines! Receiver Miller had a small enginehouse built at the edge of Durango, and for years RGS engines and Galloping Geese were serviced at that point, using only the D&RGW turntable and passenger depot. Interline billing at times was acrimonious.

Among the oddities was the second engine No. 1, which was the Boston Coal & Fuel Company's "Perins Peak," No. 1. When the D&RG's Calumet Fuel Company took over the BC&F Company in 1906, this engine was re-lettered for the RGS. She was mostly used out of Durango on coal runs to Perins Peak and Hesperus. However, since No. 1 was a Class 56 2-8-0 and not deemed worth a new boiler, she was worked over the line to Ridgway for one final run, and she sat there until around 1926. The RGS was not able to scrap their second No. 1 because the engine was the property of the Calumet Fuel Company. By 1928, old No. 1 was finally cut up; however, the engine had been stripped of parts before then.

Second No. 1 was built for the D&RG in 1880 as their No. 80, the "Chico." She went to the original D&RGW in Utah as their No. 80 in 1886 (later, Rio Grande Western No. 80). Then, the 2-8-0 became RGS No. 35 in 1891, only to become the Boston Coal & Fuel Company's No. 1, and then she became the second RGS No. 1.

For many years, the Ridgway roundhouse serviced the D&RG's Third Division engines. The D&RG never equipped Montrose with a repair shop, despite the need there. So, RGS forces provided overnight hostling, even applying the new numbers to Class 70 engines during their overnight stay. During 1909, the Third Division was using various F&CC engines, and these would stay overnight at Ridgway, too. During late 1916, the ex-Crystal River engines were tried out on the Ouray Branch, but they derailed on the light rail and worn frogs at switches, as they had rather wide tires. A couple were tried on the RGS, and a rumor was spread that they were to become RGS engines. [See the story elsewhere in this volume about No. 361 working across the line in regular service.]

One of these handsome ex-CR engines was sent out of Durango to work on the lumber company's line at McPhee. However, the 2-8-0 was refused because of excessive weight for their lightweight rail.

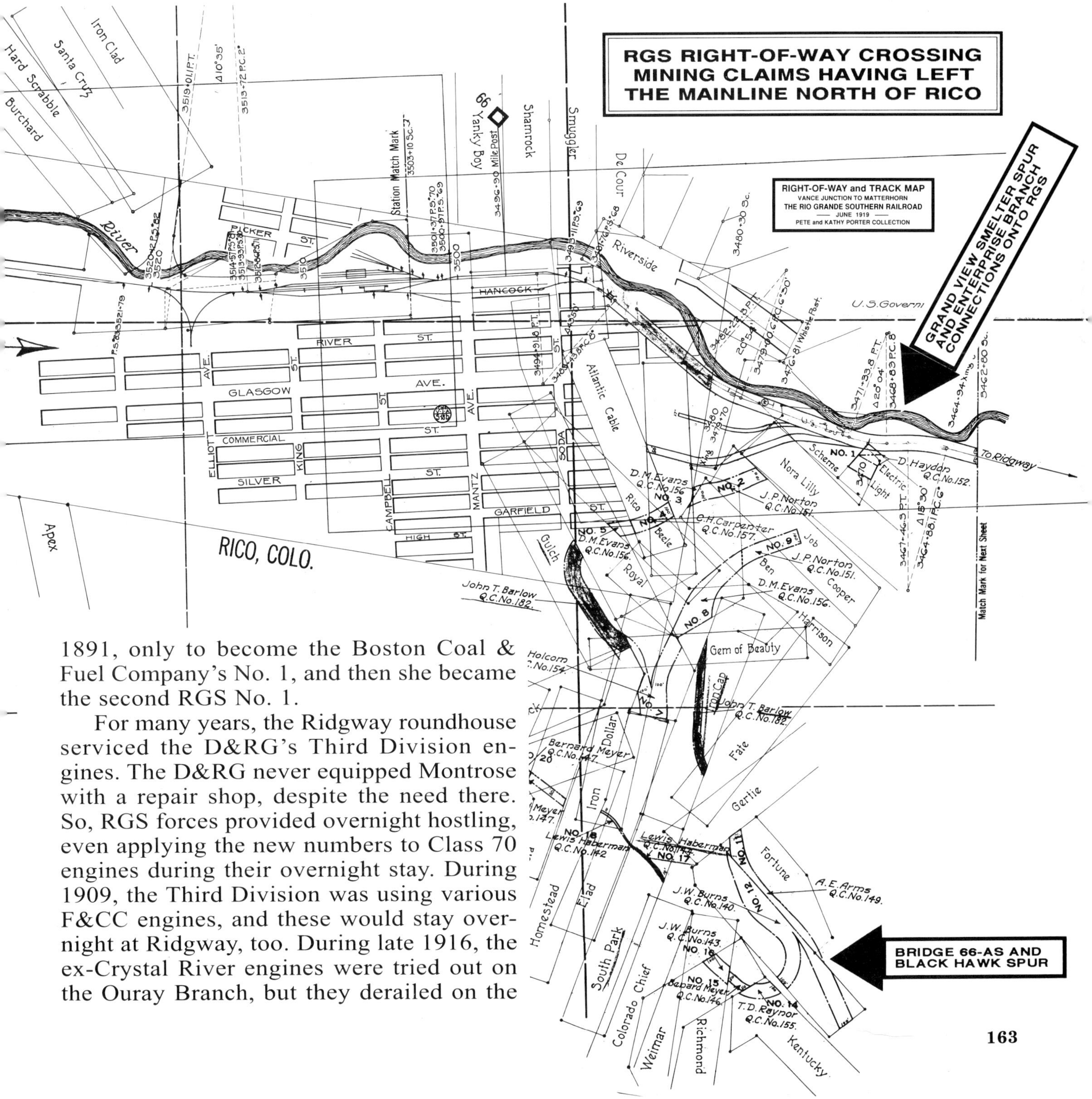

"RICO" ON TWO ENGINES

By Robert W. Richardson

IN PROMOTING the Denver, South Park & Pacific Railroad, the former territorial governor of Colorado, John Evans, could not resist "needling" the rival Denver & Rio Grande. His fourth locomotive, one of a number of Mason Bogies, was named the "San Juan." Others he gave names to, representing places in territory the D&RG people considered theirs, were the "Granite," "Crested Butte," "Lake City," "Ouray," "Silverton" and finally, No. 26, a 2-8-6T turned out by Mason in June of 1880, which Evans named the "Rico." This meant that in the early 1880's there were two locomotives named for Rico, and both of them could have been in Denver's Union Station at the same time, as both roads used that terminal.

By the late 1880's, both companies had given up on naming engines, largely for economic reasons. The DSP&P engine, of course, never got anywhere near Rico; however the D&RG's No. 244, which became the Rio Grande Southern's No. 4 long after her name had been removed and forgotten, did operate through the little mining town.

The motive-power people of the Union Pacific did not like the Mason Bogies, and the U.P. gradually set them aside after taking over the South Park line. By the early 1890's, all except one had been sold or scrapped. But the U.P. inherited and kept alive Evans' grand dreams, and as late as the 1890's, the U.P. was still eyeing surveys through the Gunnison Country to points beyond, hoping to add the San Juan Region to their narrow-gauge empire.

DAVID C. BUTLER COLLECTION

AFTER A SNOWSTORM had dumped a moderate amount of snow in the San Juan Region of southwestern Colorado during 1931, the RGS sent out Plow-Flanger 02 to clear the line north of Rico. The crew was ordered to work on the grade up to Lizard Head Pass and down the other side of the hill. This view was taken at the northern edge of Rico, and it appears as though some snow melt had already started, since some of the rail was visible in the foreground. The intense rays of the sun at this high altitude will rapidly shrink the depth of the snow.

Mile by Mile to...

The Rico Division Point

The End of the First District and the Beginning of the Second District

By Russ Collman

THE ENTERPRISE BRANCH OF THE RIO GRANDE SOUTHERN

No. 6

THE ENTERPRISE BRANCH—Milepost 65.4—connected with the main track of the RGS less than a mile north of Rico. This branch was primarily built to serve the Black Hawk Mine, situated on a part of Telescope Mountain called Nigger Baby Hill, and the Enterprise Mine, on the western slope of Dolores Mountain, east of Rico, on what was known as Newman Hill. Although supplies were delivered to the mines, the main purpose of this branch was to ship out ore from the Black Hawk and Enterprise workings.

As of September 3, 1892, the ore-loading facility and spur to the Black Hawk Mine were ready for use. Then, on September 5, 1892, the trestle over Silver Creek (Bridge 66-BS) was finished, and it was expected that the line to the Enterprise Mine would be completed by September 15. After the Enterprise Branch was placed in service in 1892, the line remained in use until about 1907. Reportedly, the rail was pulled up by 1910; however, Robert W. Richardson states that some of the rail was removed by the Rico roadmaster before that date.

This branch started from the RGS mainline at a point eight-tenths of a mile north of Rico, and four switchbacks and a maximum 5.0-percent grade were needed to get up and around the lower slopes of Nigger Baby Hill to cross Silver Creek. From Bridge 66-AS, the branch went on around the part of Dolores Mountain called Newman Hill on a lower grade.

At a point about one and one-half miles up from the mainline, this branch divided into two spurs. The shorter-and-higher spur continued upgrade to the Black Hawk ore bins—at Milepost 67.2—on the north bank of Silver Creek. The Black Hawk Mine was 400 feet above the ore bins, and ore from the mine was carried down to the loading station on the branch by a 65-bucket aerial tramway, about 1,600 feet long. Some reports credit the Black Hawk Mine with shipping as many as 150 carloads of ore per

CHARLES MILLER ENGEL COLLECTION – COURTESY OF RACHEL A. HARDWICK

THE ROADMASTER OF THE RGS had called the Rico dispatcher to clear out the line on this particular day, in and around Rico. The snowfighting train was made up of RGS Plow-Flanger 02, two locomotives and a caboose. The train had backed north, out of Rico, and pulled to a stop near the site of the Enterprise Branch switch at Milepost 65.4. Footprints in the snow mark the grade of the branch line. The snow lay undisturbed elsewhere along the track, except for the photographer's footprints, made when he left the locomotive to record this scene. Dolores Mountain is beyond the ridge.

month in 1892 and 1893; however, it may not have shipped this much ore. In any case, the ore proved to be of rather poor quality.

There were four switchbacks on the Enterprise Branch, and the line was on a steep 5.0-percent grade as far as the Black Hawk switch. From this switch the line ascended on an easier grade along Silver Creek for about 1,100 feet in a northeasterly direction to a point where it reversed itself by means of a sharp loop, crossing Silver Creek on a curving trestle, **BRIDGE 66-AS**. This pile trestle was a curved 26-degree span over Silver Creek, 112 feet in length and 22 feet in height. The Black Hawk ore bins were near this trestle, as was the Rico-Argentine mill. From Bridge 66-AS, the Enterprise Branch continued around the slope of Newman Hill in a southwesterly direction, on what was a grade of about 1.5 percent, to the workings of the Enterprise Mine. There was another bridge on this branch—**BRIDGE 66-BS**—but little is known about it. Mike Blazek states that this trestle was 48 feet in length and 15 feet in height. Presumably, it crossed a rocky area where there was no stream and was filled-in. The distance of the line from the Black Hawk switch to the end-of-track at the Enterprise ore bins was 1.35 miles, and this point was listed as Milepost 68.5 by the RGS.

The Enterprise Branch has been shown on some maps as having a spur to the Rico-Aspen Mine, south of the Enterprise Mine. However, there is no indication that a grade ever went beyond the Enterprise workings, at an elevation of 9,236 feet. However, a passing siding was located north of the Enterprise Mine, listed as being 674 feet in length. Cars for one or more local mines may have been loaded with ore at this point; however, no information is known about the purpose for this siding.

This unusual RGS branch started up the hill from the mainline, near the base of Nigger Baby Hill, to the east of the Grand View Mining & Smelting Company's structure. If you study the U.S. Geological Survey map of 1899, printed with 50-foot-contour intervals, it becomes clear how the switchbacks of the branch climbed the base of the moun-

tain on the steep grade, before crossing Silver Creek. The grade leveled out somewhat after crossing the creek en route to the Enterprise Mining Company's works at the end of the branch.

The Silver Panic of 1893 caused the Enterprise Branch to shut down for a time, since the mines had closed. However, operations were resumed at least in part because photographs taken during 1896 and shortly thereafter show boxcars on the line. Otto Mears' Shay geared engine, No. 34, had been sent to Rico to work on this branch. However, after the financial crash caused the Black Hawk and Enterprise mines to close for a period of time, this little engine was sent to Ridgway, reportedly because repairs were needed, but the Shay never returned to Rico. While No. 34 was away, the RGS went into its first receivership, and Otto Mears was forced to step down from his position as president.

CHARLES MILLER ENGEL COLLECTION –
COURTESY OF RACHEL A. HARDWICK

ONLY THE HEADLIGHT was visible behind flying snow, thrown in the air by Plow-Flanger 02. It was being shoved upgrade by locomotives attempting to open the first leg of the Enterprise Branch, near the backside of the Grand View Smelter, located at the northern edge of Rico.

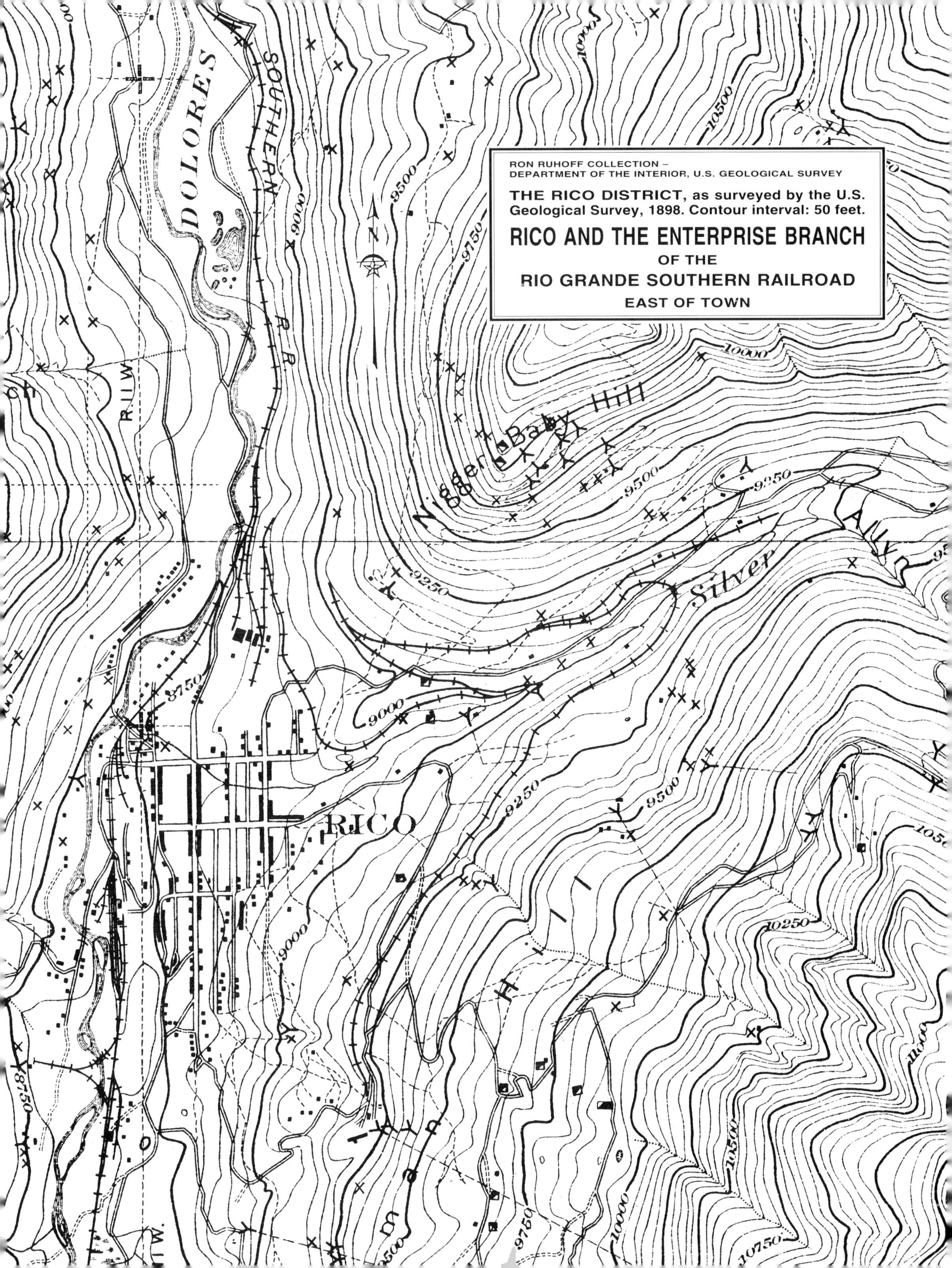

RON RUHOFF COLLECTION – DEPARTMENT OF THE INTERIOR, U.S. GEOLOGICAL SURVEY

THE RICO DISTRICT, as surveyed by the U.S. Geological Survey, 1898. Contour interval: 50 feet.

RICO AND THE ENTERPRISE BRANCH OF THE RIO GRANDE SOUTHERN RAILROAD EAST OF TOWN

DAVID AND FAYE WOLFORD COLLECTION

ON THIS OCCASION, the rotary snowplow of the RGS was working up the steep 5.0-percent grade of the Enterprise Branch. In all likelihood, the plow was opening the track near the Grand View Smelter. The men had poked around in the snowslide and shoveled tree trunks and branches out of the pathway of the plow. They were standing above the grade as the rotary plow roared through the snow, throwing it below the line—slowly advancing up the branch. Hopefully, the snow thrown out by the plow cleared the RGS mainline below, where a gondola stood near the Dolores River. Previously, the rotary plow probably had plowed open the line from the Rico yard, to a point beyond the Enterprise Branch switch. It then had to be pulled back to Rico, turned around on the wye and backed up the line toward Burns, far enough to clear the turnout leading off the main track, up the severe grade of the branch. Sandstone Mountain formed a landmark to the north of Rico, in view to the left. It is interesting to note that a reading of 16 degrees below zero (F.) was reported in THE RICO DEMOCRAT for February 13, 1891.

DAVID AND FAYE WOLFORD COLLECTION

THE DEPTH OF THE SNOW from the so-called "Nigger Baby Slide" was above the chute of rotary snowplow No. 2, as it continued to chew its way through the accumulated mass of snow. One of the crewmen was kneeling on the roof of the rotary snowplow, behind the whistle, giving hand signals to the second crewman protruding from a hatch through the roof. He is barely visible in back of the smokestack. In turn, he relayed signals to the pilot inside the hard-working machine. Hand signals were necessary because of the noise from the plow's machinery and the constant roar from the smokestack, making verbal instructions impossible. Notice that a small portion of the Dolores River appears at the left. Also notice all the men standing atop the slide.

RICO – ROTARY HEADQUARTERS
By Robert W. Richardson

ALMOST ALL THE "NEEDS" for a rotary snowplow were on the stretch of RGS track between Rico and Vance Junction. As there were no facilities at Vance Junction for storing the big machine, and Telluride was a cramped place, Rico was selected as the place for the rotary plow to be stationed. Rico had car men and others working there, so maintenance could be accomplished. Moreover, there were always engines at Rico —or so it seemed.

Otto Mears overestimated the need of rotary snowplows, and he had bought two of them. As it turned out, most of the RGS trackage rarely needed the services of a rotary plow, and this was particularly true for the portion of the road between Rico and Durango. So, one of the big machines was sold to a Canadian road. Actually, the rotary was the least used piece of equipment. Often, an entire winter would go by without the machine making even one trip. Engines with

COLORADO HISTORICAL SOCIETY COLLECTION

THE MINING TOWN OF RICO experienced a building boom shortly after the arrival of the Rio Grande Southern. At the time this view was photographed, passenger-train tickets were sold at the Hotel Enterprise, which was the largest building in town, standing at the left. This may be the first photograph to show the railroad grade of the Enterprise Branch of the RGS, which was completed on Newman Hill (east of Rico) on September 24, 1892. It is believed that this picture was taken during the summer of 1892, shortly before the branch line was finished—primarily because electric power poles had already been placed at intersections throughout the settlement. (See the accompanying newspaper article, dated December 5, 1892, regarding the extension of electric lighting down to the RGS depot.)

From The Rico Democrat...

HOTEL ENTERPRISE DOING LAND-OFFICE BUSINESS

September 4, 1891—The Hotel Enterprise, under the management of Mr. C. F. Middaugh, ably assisted by his chief of staff, Tom Lewis, is doing a land-office business, and winning golden opinions from its patrons. Verily, the Hotel Enterprise is a superior inn.

pilot plows, or engines pushing a wedge plow usually could keep the track open.

When needed, the rotary plow usually made a round trip to Vance Junction, and at times, there would be a side trip to Telluride. In the days of small locomotives, C-16's and C-19's, three of them would be just right to push the rotary most of the time. Even with the 16-hour law, a round trip often could be made. Because winds scoured the mountainsides, much of the track would have little or no snow on it, so the machine mainly had to tackle places where cuts filled-in, large snowdrifts and avalanches. Water was a major need, and the lack of a tank between Coke Ovens and Trout Lake made some trips extra long, due to the necessity for backing trains to either of those tanks.

The Enterprise Branch above Rico was a real problem. The rotary plow would need a full day there, due to the switchbacks. On the worst occasions of very-deep snow, the crew was forced to go back to Rico after plowing one leg of the switchbacks. There, the plow was turned on the wye, and then went back to tackle the next leg of a switchback. This contributed to management's view that this branch was not a profitable part of the RGS. This was particularly true when they considered that the Black Hawk Mine's ore was low grade. As a result, the RGS received low revenues from this mine.

ORE SHIPMENTS FROM RICO

OTTO MEARS was an investor in the

Black Hawk mining firm. In those days, secret—and not-so-secret—rebates were made to favored shippers, and the Black Hawk was so favored. The normal rate on rock ore going to the smelter at Durango was based on the net value per ton of the ore, as determined by the smelter. The Black Hawk management people had their great expectations dashed when the returns from the smelter showed that most of their shipments were valued at only $10.00 per ton, or less. After Otto Mears saw to it that they got a rebate as well, the RGS no doubt hauled their ore at a definite loss. As the heaviest coal cars (gondolas) only had a capacity of 15 tons, the value of their loads was under $150.00. Worse yet, because of conditions on the mostly unballasted track, such cars had to be carefully loaded, and just as carefully handled. Mears' grandson reported that Otto had invested in many mining operations; yet, never did any of them turn into a profitable investment.

The Rico-Argentine Mining Company

(Continued on Page 182)

RICO MINERS UNION
Nº 36.
W. F.

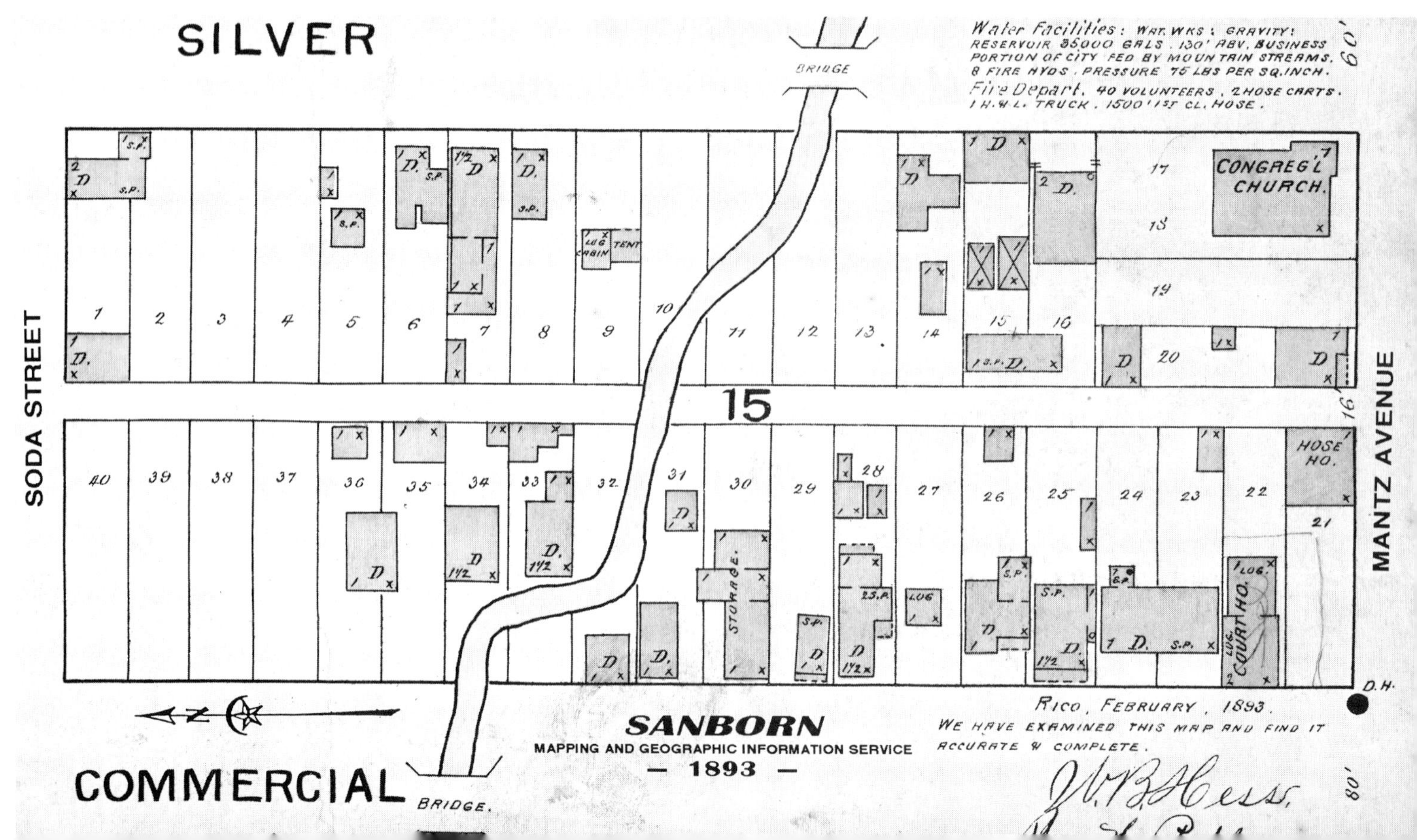

SILVER
BRIDGE
Water Facilities: WAT. WKS: GRAVITY: RESERVOIR 35000 GALS. 130' ABV. BUSINESS PORTION OF CITY FED BY MOUNTAIN STREAMS. 8 FIRE HYDS. PRESSURE 75 LBS PER SQ. INCH.
Fire Depart. 40 VOLUNTEERS, 2 HOSE CARTS. 1 H. & L. TRUCK. 1500' 1ST CL. HOSE.
60'
CONGREG'L CHURCH.
LOG CABIN
TENT
STORAGE
LOG
HOSE HO.
LOG COURT HO.
SODA STREET
MANTZ AVENUE
15
16'
D.H.
Rico, February, 1893.
WE HAVE EXAMINED THIS MAP AND FIND IT ACCURATE & COMPLETE.
SANBORN
MAPPING AND GEOGRAPHIC INFORMATION SERVICE
— 1893 —
COMMERCIAL
BRIDGE.
80'

MARGARET MATZICK COLLECTION

THE RICO MINERS' UNION, No. 36, affiliated with the Western Federation of Miners, had this representation down at the Rico depot for an unknown special event. The staircase in the background provided access to Campbell Street, above the embankment. Members of the Rico Cornet Band were at the left, while wives and children of the miners had gathered on the hillside and staircase.

FROM – "THE RICO ITEM," NOVEMBER 26, 1909
MARGARET MATZICK COLLECTION

Rico Miners' Union, No. 36, W. F. of M.
Meets first and third Saturday nights of every month at Odd Fellows Hall, at 7:30 o'clock. Prest, F. D. Roam; Financial Secretary, Chris Wold.
Box 470, - - - Rico, Colorado

COLORADO RAILROAD MUSEUM – COURTESY OF GORDON O'NEIL

THIS VIEW OF THE TOWN of Rico is an enlarged portion of the picture printed on the following page. It was taken during May of 1893. It was a sunny afternoon, before the lengthening shadow from Expectation Mountain spread across the valley. The RGS station grounds continued to change. Two short cabooses, one having three windows on its side (the original setup), were standing on yard trackage, with their respective trains. Notice the long string of flatcars to the left of the three-window caboose. Two RGS engines, with diamond stacks, were in view on the lead tracks going to the wooden enginehouse. Carpenters had now completed the RGS section house (at the left) and bunkhouse, and they had been painted oxide red, the same color as the water tank, north of the depot. An outhouse had been placed to the right of the road leading down to the depot, also painted oxide red. A lumber yard—complete with an enclosed warehouse and a covered lumber shed and office—fronted on the access road, beyond the depot. A wooden staircase was built from Campbell Street to the depot access road, and a boardwalk continued down to the south end of the two-story station building.

OVERLEAF: This is a full-frame view of the above photograph, and you can see that much of the snow had melted on the mountains above Rico. The picture was taken during May of 1893, shortly before aspen leaves had made their appearance (which usually takes place during the first half of June at this elevation). This allows you to see the original enginehouse before the aspens had leafed out. The Enterprise Branch of the RGS was still active at this time, although no activity is apparent on the grade back of Rico. Notice how Campbell Street came to a sudden halt at the edge of the embankment above the RGS depot. Mantz Avenue suddenly ended in the same way; however, an unnamed depot-access road made a sharp turn to the south at the embankment's edge and then traversed the face of the escarpment as it descended to the depot grounds. A small "red-light district"—with girls for hire—was located along River Street, shown at the lower left, on the edge of the terrace above the railroad yard. The magnificent Hotel Enterprise stands out as the largest building in town, the site of the celebration dinner and ball after the completion of the RGS into Rico.

COLORADO RAILROAD MUSEUM – COURTESY OF GORDON O'NEIL

Rico, Colo, Looking N.E. May 1893.

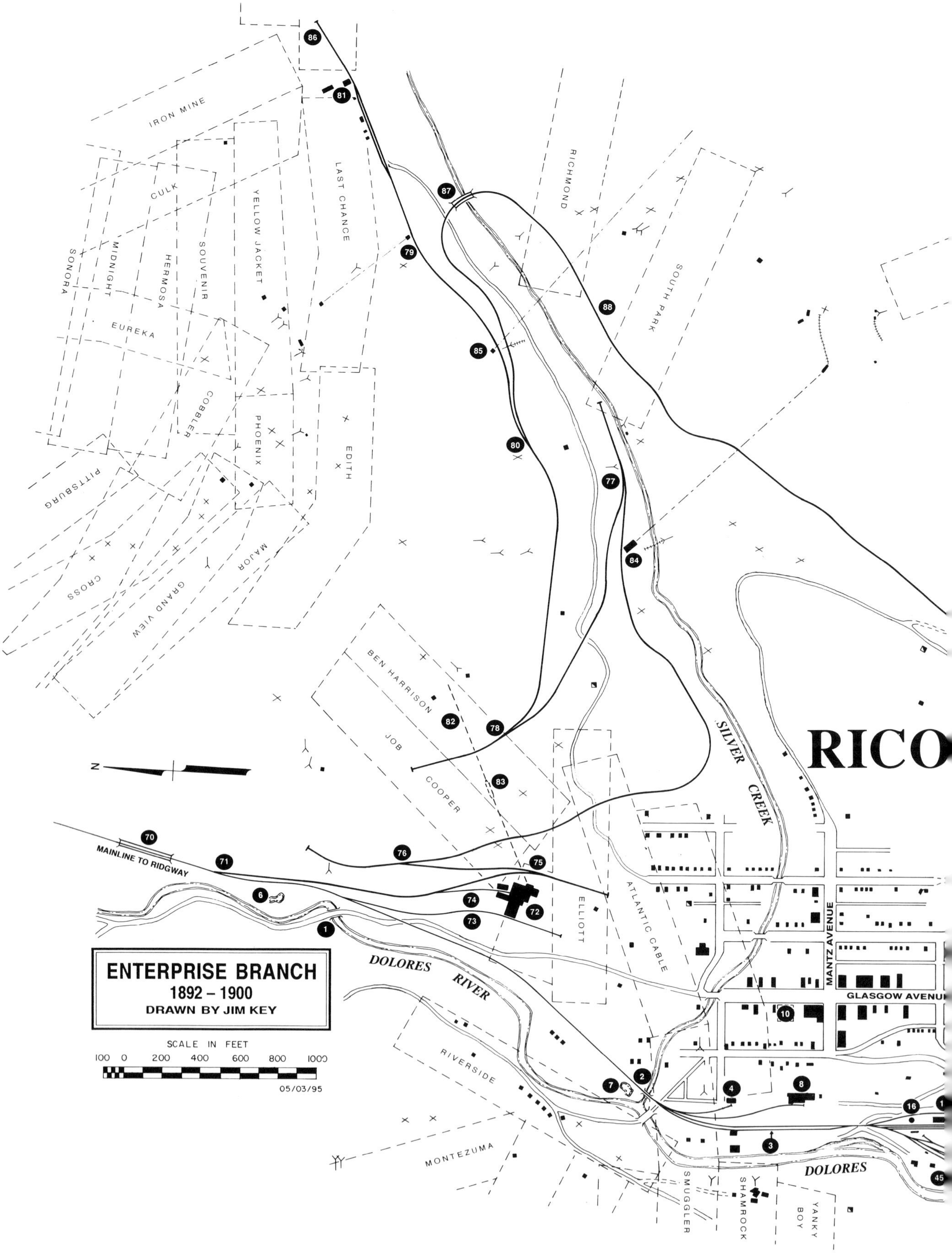

ENTERPRISE BRANCH
1892 – 1900
DRAWN BY JIM KEY
SCALE IN FEET
100 0 200 400 600 800 1000
05/03/95
RICO
MAINLINE TO RIDGWAY
DOLORES RIVER
SILVER CREEK
DOLORES
MANTZ AVENUE
GLASGOW AVENUE
IRON MINE
CULK
LAST CHANCE
YELLOW JACKET
SOUVENIR
HERMOSA
MIDNIGHT
SONORA
EUREKA
COBBLER
PHOENIX
EDITH
PITTSBURG
MAJOR
CROSS
GRAND VIEW
RICHMOND
SOUTH PARK
BEN HARRISON
JOB
COOPER
ELLIOTT
ATLANTIC CABLE
RIVERSIDE
MONTEZUMA
SMUGGLER
SHAMROCK
YANKY BOY
N

ENTERPRISE BRANCH OF THE RGS AT RICO

1892 – 1900

DRAWN BY JIM KEY

1. HIGHWAY 145 BRIDGE
2. RGS BRIDGE 66-C
3. MILEPOST 66.4
4. SAMPLING WORKS
6. HOT SPRINGS
7. SODA SPRING
8. RICO P&L POWER PLANT
10. HOTEL ENTERPRISE
16. RGS WATER TANK
17. RGS DEPOT
24. FIRST ENGINE HOUSE (WOODEN)
40. SECOND WYE
41. RGS BRIDGE 67-AS
42. RGS BRIDGE 67-ASS
45. LOCATION OF FIRST WYE
46. RICO SMELTING CO. (ADAMS)
47. PASADENA REDUCTION CO. (TRACT)
70. RGS BRIDGE 66-B
71. ENTERPRISE BRANCH SWITCH – M.P. 65.4
72. GRAND VIEW MINING & SMELTING CO.
73. LOADING SPUR
74. LOADING SPUR
75. SWITCHBACK No. 1
76. SWITCHBACK No. 2
77. SWITCHBACK No. 3
78. SWITCHBACK No. 4
79. YELLOW JACKET TRAMWAY
80. BLACK HAWK SPUR SWITCH
81. BLACK HAWK TRAMWAY TERMINAL
82. FUNICULAR RAILWAY (ABANDONED)
83. FUNICULAR PASSING TRACK
84. NEWMAN MILL
85. STANDING ORE BINS (1995) UNION CARBONATE
86. END-OF-TRACK – M.P. 67.2
87. RGS BRIDGE 66-AS
89. PASSING SIDING
90. ENTERPRISE MINE
91. LOADING BINS
92. MINE DUMP
93. END-OF-TRACK – M.P. 68.5
94. COLORADO MILLING & CONCENTRATING COMPANY

MERCHANT'S EXCHANGE.

Reading Room in Connection. Open Day and Night.

GEO. C. DOWNS, Proprietor.

Imported Wines, Liquors and Cigars.

ALE AND PORTER

ON DRAUGHT.

Bottled Goods a Specialty.

GLASGOW AVENUE,

RICO, - - COLO.

MINER'S MEAT MARKET

JOHN GAULT, Proprietor.

Butter, Eggs, Fish, and Oysters,

Game, Poultry and Vegetables.

FRESH and SALT MEATS

GOODS DELIVERED TO ANY PART OF THE CITY.

West Side Glasgow Avenue, Rico, Colorado.

MORGAN & SIMPSON

Manufacturers Of

SODA, BIRCH, GINGER ALE,

And Bottlers of the Celebrated

Anheuser Beer.

RICO. - - COLO.

St. James Hotel.

ALFRED ARCHAMBAULT, Proprietor.

MANTZ AVENUE, - RICO, COLO.

Best Cook and Finest Bill of Fare in Southern Colorado. Board, $7.00 per Week. Meal Tickets, 21 Meals, $7.50. Three Meals for $1 25.

EVERYTHING FIRST CLASS.

W. H. JACKSON PHOTO – RICHARD A. RONZIO COLLECTION

ON SEPTEMBER 24, 1897, William H. Jackson, the famous Denver-based photographer, shot the glass-plate negative that allowed him to make this solar print of Rico. You can plainly see the Enterprise Branch in this view, behind (and above) the town. This RGS branch started up the hill from the mainline, near the base of Nigger Baby Hill, to the left of the Grand View Mining & Smelting Company's structure, Rico's first smelter (built in 1882), near the center of this scene. W. H. Jackson set up his tripod on the slope of Expectation Mountain, to record this view portraying the progress taking place at Rico. By studying the USGS map of 1899, with 50-foot-contour intervals, it becomes clear how the switchbacks of the branch climbed the base of the mountain on a 5.0-percent grade, crossing Silver Creek, where the grade slacked off en route to the Enterprise Mining Company's works at the end of the branch. The RGS station grounds at Rico show no activity at this time, and the depot can be seen between the two conifer trees, near the center of the view. Snowcapped peaks near Lizard Head Pass are in view (at the far left).

DETROIT PHOTOGRAPHIC CO.

H. H. BUCKWALTER PHOTO – RICHARD A. RONZIO COLLECTION

AT FIRST GLANCE, this photograph looks as though it is nearly identical with the one printed on the previous pages. However, it happened that both William H. Jackson and Harry H. Buckwalter toured the San Juan Region south of Lizard Head Pass on the same day, during the autumn of 1897 (along with another Colorado photographer by the name of Reynolds). Buckwalter's glass negative is on file at the Colorado Historical Society in Denver. You can see that the trees in the foreground were virtually in the same position as in Jackson's picture.

continued to ship ore concentrates from Rico until the RGS was abandoned. During November of 1951, the last two revenue trains from Rico had four or five boxcars of about 25 tons each, consisting of frozen concentrates. As a result, there was no problem with the loads shifting en route! At Montrose, such loads were transferred to standard-gauge cars. However, the crew had to use a paving breaker to remove the concentrates! These were the last revenue loads on the northern district of the RGS.

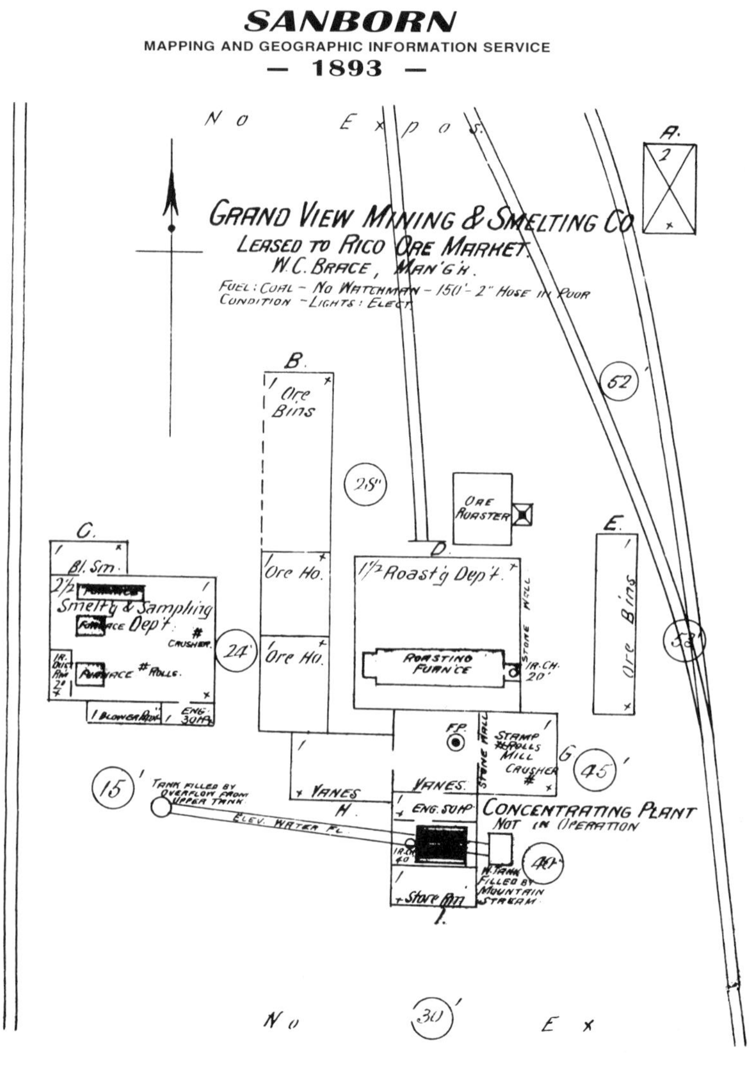

H. H. BUCKWALTER PHOTO – RICHARD A. RONZIO COLLECTION

OVERLEAF: This photograph has been enlarged to 420 percent of the original contact print by computerized scanner equipment, allowing the picture to be enhanced to a degree that was impossible only a few years ago. The results are reproduced on the following four pages. The smelter of the Grand View Mining & Smelting Company is visible near the center of the next two pages. It is interesting to find that a funicular railway had been constructed behind the smelter, ascending the slope of Nigger Baby Hill to the Grand View Mine. This funicular cable line probably was in use for only a year or so because the grade of the narrow-gauge Enterprise Branch cut through the cable railway at two elevations. This enlargement clearly shows a string of boxcars that had been spotted on the loading spur into the Grand View smelting-and-concentrating plant (No. 73 on Jim Key's map). This spur came off the RGS mainline south of the turnout to the Enterprise Branch. The RGS mainline can be seen curving below the string of boxcars.

THE ENTERPRISE BRANCH SWITCH off the mainline cannot be seen; however, it was just north of the hillside sheared off for the branch track. It climbed steeply above the boxcars and passed behind the Grand View buildings, with the switchback tail track ending just behind the two-story schoolhouse. The switch to the second level (labeled as switchback No. 1 on Key's map) was just behind the small dark building located midway between the Grand View plant and the school. The second level of track was behind and above the roof gable of the Grand View building with switchback No. 2's tail track ending at the row of trees. The third level climbed steeply along the slope of Nigger Baby Hill, cutting the cable track and passing behind Rico's original small white schoolhouse. It then curved sharply through a deep cut and headed east along Silver Creek. Just above a group of buildings at the Newman mill, it curved again and passed out of sight behind the hillside that also hid the switch for switchback No. 3. The fourth level of track came back out of the valley and passed behind the knoll—through which the deep cut on track level No. 3 runs. The big fill that carried the branch can be seen just to the left of the rounded knoll through which the lower level of track cut.

THE LINE CONTINUED northward across the slope of Nigger Baby Hill, cutting through the cable railway again to reach the switch for switchback No. 4. The tail track ended in the grove of pine and aspen trees. The fifth level of track came back southeastward across Nigger Baby Hill, cutting the cable track again and curving to the east, where it climbed along the side of the valley of Silver Creek, heading for curved Bridge 66-AS, out of sight in the trees. When the grade comes into view again, the branch was heading to the southwest, along the slope of Newman Hill (at the base of Dolores Mountain), just above the big mine dump near the top of the view. The switch for the Black Hawk Mine spur was directly above and a little to the left of the Newman mill buildings, and both levels of track can be seen beyond that point, with the Black Hawk spur disappearing among the trees and buildings. Presumably, the large dump was at the Iron Mine. The highest road resembled a railroad spur climbing from the the Enterprise Branch to the Grand View Mine buildings, at the left. The RGS cut the Grand View's cable railway three times when they built the Enterprise Branch. Notice the new brick schoolhouse, at the intersection of Commercial and Soda streets. Rico's Catholic Church probably had moved into the former public school building by this time.

THIS IS THE FIRST overall photograph showing the replacement (second) building constructed for the Dolores County Courthouse. This brick building still stands on Mantz Avenue.

PASSENGER TRAFFIC TO RICO

By Robert W. Richardson

BEING ISOLATED IN RICO, the local inhabitants were good customers for any excursion offers the RGS might make. There were always baseball games between teams of the various mining towns, and such specials were frequent. Horse races at Mancos and Telluride were other events meriting at least extra coaches on the regular trains.

However, the big event each year during the 1890's was the Festival of Mountain and Plain in Denver. For this, the D&RG apportioned its fleet of narrow-gauge Pullman sleeping cars to the most important points on its lines, and Rico was designated as one. Reservations could be made, and one or two Pullmans would be assigned to the regular trains for days at a time. Extra movements were scheduled for other events, such as lodge conventions. And, of course, conventions of the Grand Army of the Republic (GAR) and political conventions were a matter of course. But this use of Pullmans ended by 1902, when the D&RG tore up the third rail between Denver, Colorado Springs and Pueblo. RGS patrons then had to go by coach to such places as Alamosa, Montrose or Grand Junction to board standard-gauge sleepers to Denver.

Unlike Telluride, Rico never had parlor-car service. The Telluride parlor car came through from Salida for many years. The

(Continued on Page 190)

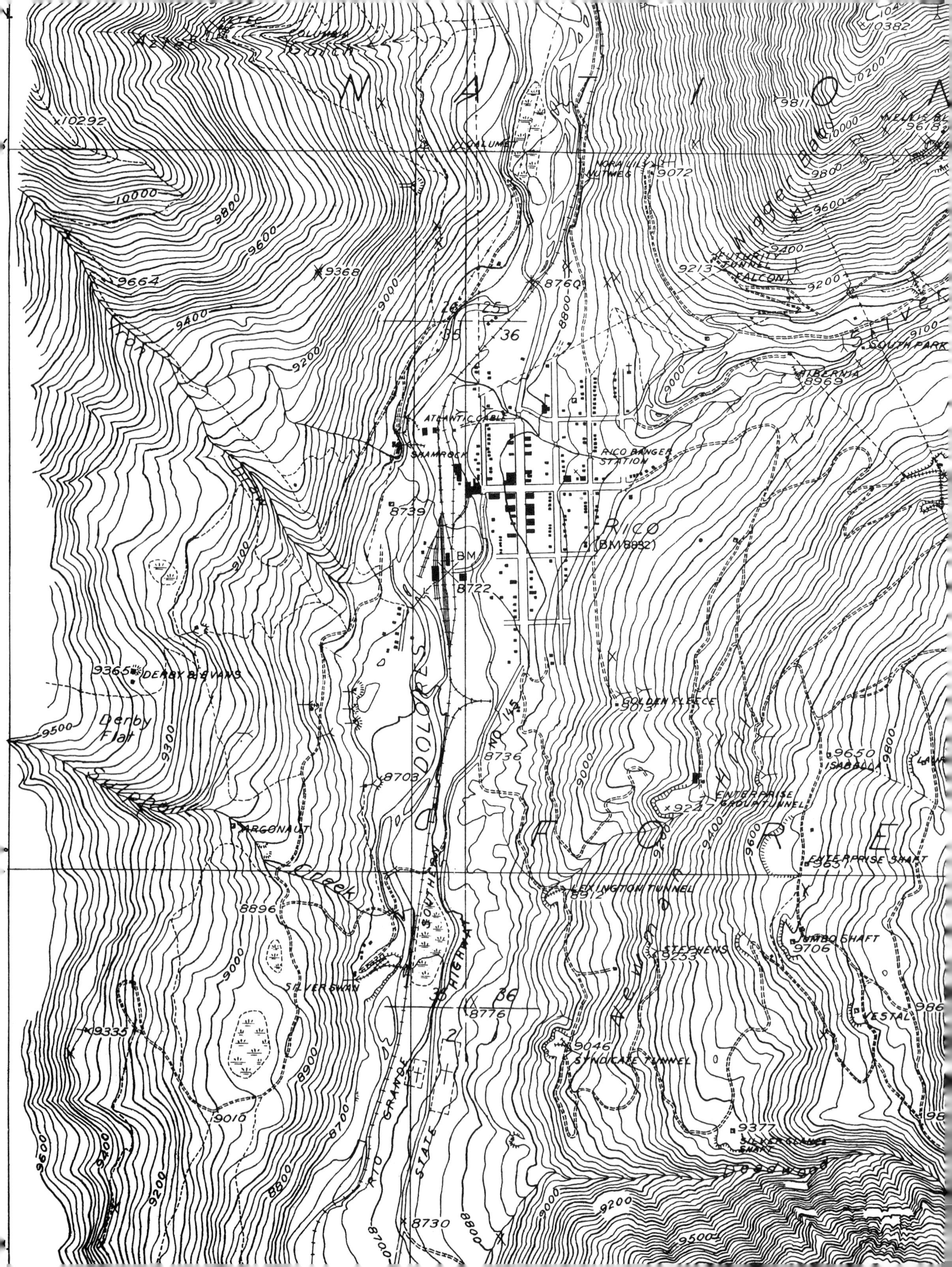
AZTEC
COLUMBIA
Dutch
x10382
x10292
10000
9800
9600
9664
9400
9200
9000
x9368
CALUMET
NORA LILY
NUTMEG
9072
9811
10200
9618
9800
9600
9400
FUTURITY TUNNEL
FALCON
9213
9200
9100
8760
8800
25
36
36
SOUTH PARK
HIBERNIA
8969
9000
ATLANTIC CABLE
SHAMROCK
RICO RANGER STATION
8739
Rico
(BM 8832)
BM
8722
9100
9365
DERBY & EVANS
Derby Flat
9500
9300
GOLDEN FLEECE
9013
NO. 145
8736
DOLORES
8703
9000
9650
ISABELLA
9800
ENTERPRISE GROUP TUNNEL
x9227
ARGONAUT
9400
9600
Creek
ENTERPRISE SHAFT
LEXINGTON TUNNEL
8896
SOUTH
HIGHWAY
STEPHENS
9233
JUMBO SHAFT
9706
SILVER SWAN
9000
35
36
8776
VESTAL
x9335
2
9046
SYNDICATE TUNNEL
8900
9010
8700
RIO GRANDE
STATE
9377
SILVER GLANCE SHAFT
Deadwood
9600
9400
9200
8800
8700
x8730
8800
9000
9200
9500

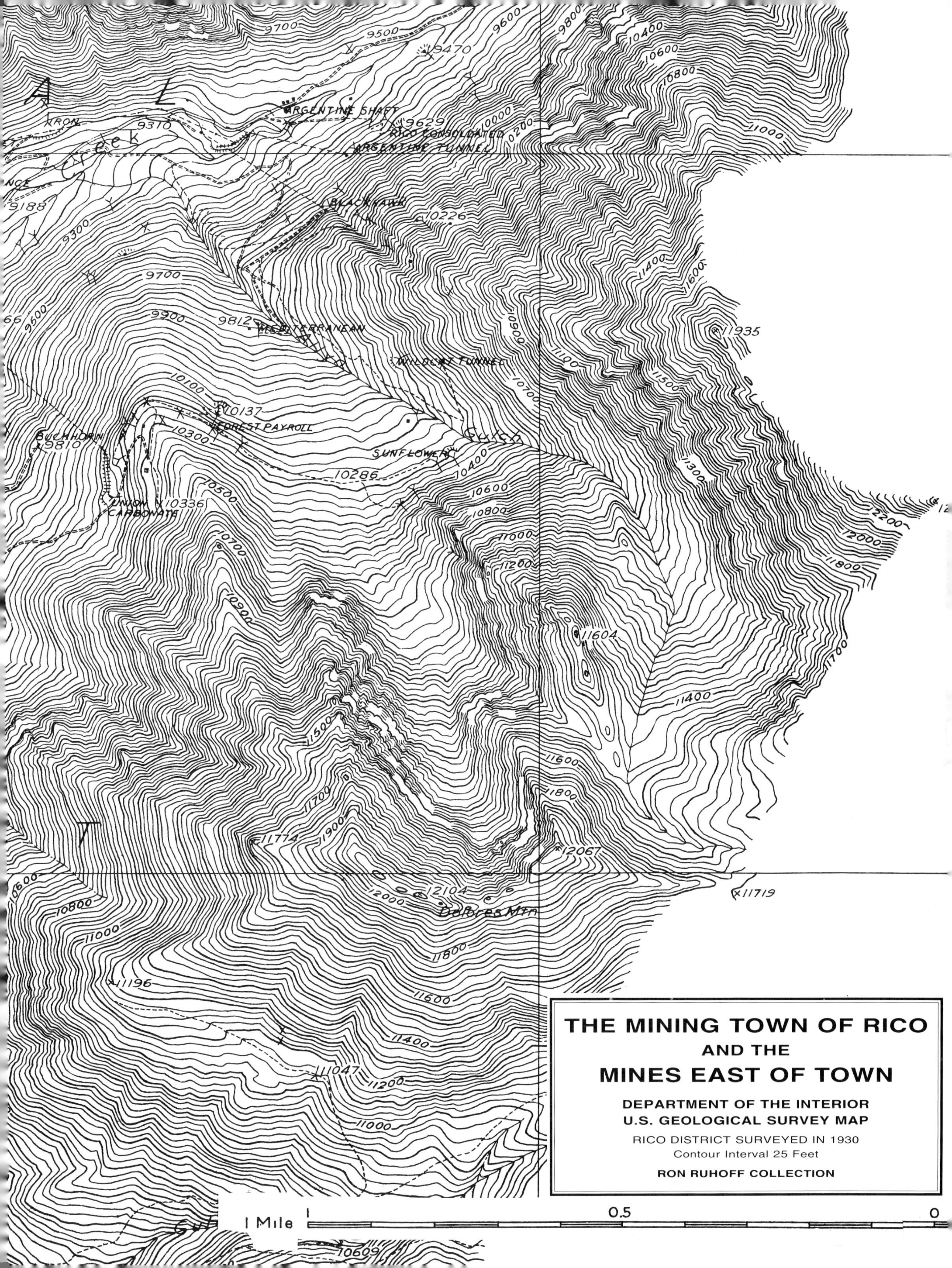

THE MINING TOWN OF RICO
AND THE
MINES EAST OF TOWN

DEPARTMENT OF THE INTERIOR
U.S. GEOLOGICAL SURVEY MAP

RICO DISTRICT SURVEYED IN 1930
Contour Interval 25 Feet

RON RUHOFF COLLECTION

general movement of passengers by train from Rico was northward, as the connections to distant points were much better in this direction than via Dolores and Durango. Much of the time, this latter service was a mixed train, with consequent erratic schedules, entailing late trains and long delays en route, due to switching.

TRACKAGE IN THE RICO AREA

OTTO MEARS used large quantities of surplus Denver & Rio Grande material after the D&RG standard-gauged its new mainline across Colorado by way of Tennessee Pass and Glenwood Springs. This released a great amount of light rail and fittings. There was a large amount of 30- and 35-pound rail, and this was used by the RGS for laying its line over the Ophir Loop, Lizard Head Pass and on down to Rico. Even though the price was a bargain—or so it seemed—in the end, it was a costly error to use so much light rail on all those 3.0- and 4.0-percent grades and sharp curves. And the fact that much of the rail was well worn did not help. The hard-working engines made it difficult for the track crews to maintain the line.

By the late summer of 1892, Otto Mears ordered enough 57-pound rail from the Colorado Coal & Iron Company in Pueblo to re-lay the entire trackage from Vance Junction to Rico, and much of the Rico yard as well. Impressed by the patented Greer ("Jeffery") double-headed spike, this type was ordered for use with the new 57-pound rail. Special tie plates for the spikes were also ordered, while angle bars came from a Milwaukee plant, and bolts and nuts came from Kansas City. The Greer spikes were manufactured in Chicago for the vendors, the Q&C Company.

After the track re-laying work was completed in 1893, the RGS had large amounts of the lighter rail on hand—which the railroad had in storage for years. Some of it was sold to mines and logging companies. However, when the RGS was abandoned in 1951, thousands of pairs of splice bars for 30-pound rail were still inside the freight room at the RGS depot in Rico.

New switchstands also were purchased at this time, a peculiar one produced by the Buda Foundry & Manufacturing Company, which required that a lever be raised before it could be turned to a different position. Split switch points also were installed, except in the Rico yard (for some unknown reason). There, stub switches prevailed, and the throat of the yard at both ends featured three-way stub switches, which then fanned out to additional switches. These stub switches lasted until the end of operations. However, the old "harp" switchstand was left on some of the yard tracks, as well as at derail switches. The new rotary switchstands had a good feature, in that unlike the harp stands, a switch lamp could be used, a much-needed safety feature.

(Continued on Page 195)

ORVAL L. JAHNKE COLLECTION

THE PHOENIX MINE, at the left, and the Yellow Jacket Mine, are on the southern slope of Nigger Baby Hill. The Yellow Jacket's cable-tramway line carried ore in buckets down to the Black Hawk spur of the RGS and emptied into the chute (at the right). A new trackside loading chute, still standing today (1995), was under construction, at the lower left, to connect with the Union Carbonate Mine by an aerial tramway. The photographer stood on the edge of the tailings dump at the Pro Patria Mine on Dolores Mountain and pointed his camera toward the north for this view.

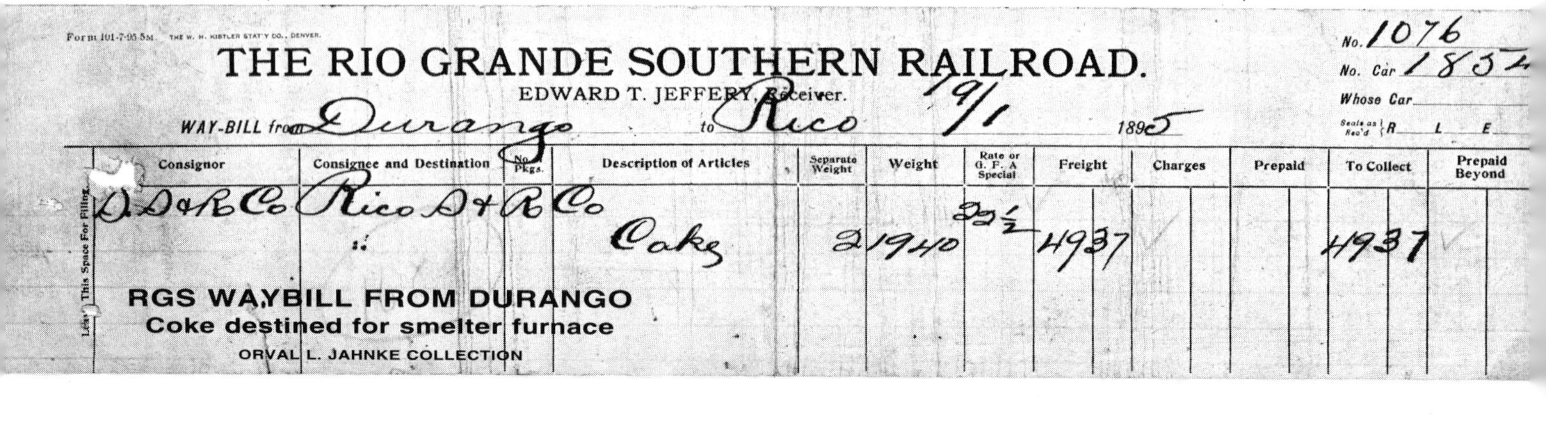

Form 101-7-95 5M. THE W. H. KISTLER STAT'Y CO., DENVER.

THE RIO GRANDE SOUTHERN RAILROAD.

EDWARD T. JEFFERY, Receiver.

No. 1076
No. Car 1852
Whose Car
Seals as Rec'd R L E

WAY-BILL from Durango to Rico 9/1 1895

Consignor	Consignee and Destination	No. Pkgs.	Description of Articles	Separate Weight	Weight	Rate or G. F. A Special	Freight	Charges	Prepaid	To Collect	Prepaid Beyond
D&RG Co	Rico D&RG Co		Coke		21940	22½	4937			4937	

Leave This Space For Filing.

RGS WAYBILL FROM DURANGO
Coke destined for smelter furnace
ORVAL L. JAHNKE COLLECTION

DELL A. McCOY PHOTOS

THE UNION CARBONATE ore bins were photographed during the fall of 1966, long after the Enterprise Branch was abandoned. This view shows the south and east sides of the structure. Also notice the old aerial-tramway cables. The grade of the Black Hawk spur curved toward Rico, in the valley (out of sight) at the left.

THESE ORE BINS are still standing today, a structure originally located on the Black Hawk spur of the Rio Grande Southern's Enterprise Branch. This tipple was connected with the Union Carbonate Mine by a cable tramway. The Black Hawk spur climbed along Silver Creek in an easterly direction to the end-of-track at Milepost 67.2.

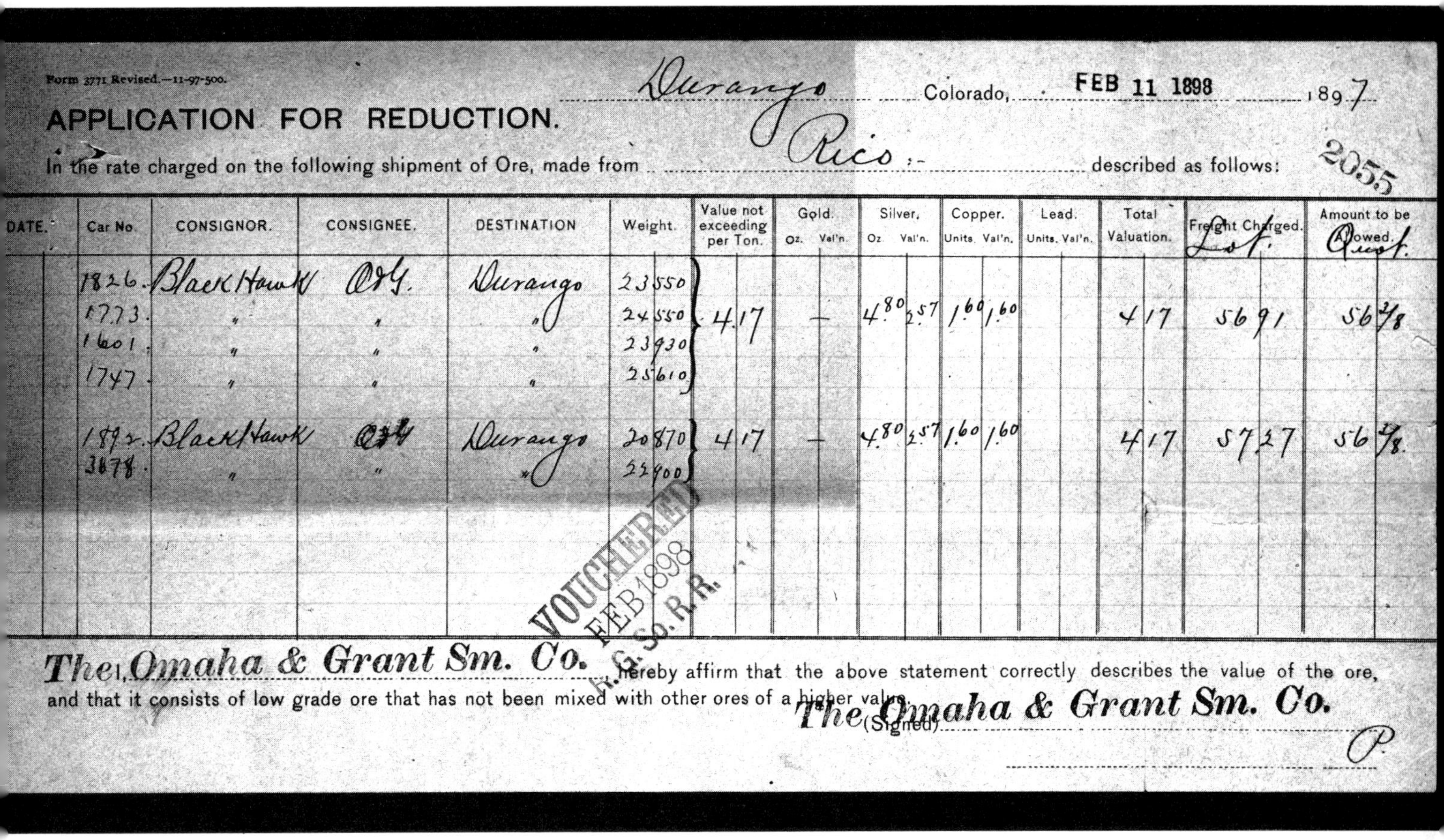

Form 3771 Revised.—11-97-500.

Durango Colorado, FEB 11 1898 1897

APPLICATION FOR REDUCTION.

In the rate charged on the following shipment of Ore, made from Rico :- described as follows: 2055

DATE.	Car No.	CONSIGNOR.	CONSIGNEE.	DESTINATION.	Weight.	Value not exceeding per Ton.	Gold. Oz.	Gold. Val'n.	Silver. Oz.	Silver. Val'n.	Copper. Units.	Copper. Val'n.	Lead. Units.	Lead. Val'n.	Total Valuation.	Freight Charged. Lot.	Amount to be Allowed. Quot.
	1826.	Black Hawk	O&G.	Durango	23550												
	1773.	"	"	"	24550	4.17		—	4.80	2.57	1.60	1.60			4.17	56.91	56 3/8
	1601.	"	"	"	23930												
	1747.	"	"	"	25610												
	1892.	Black Hawk	O&G	Durango	20870	4.17		—	4.80	2.57	1.60	1.60			4.17	57.27	56 3/8
	3678.	"	"	"	22900												

VOUCHERED FEB 1898 R.G.So.R.R.

The Omaha & Grant Sm. Co. hereby affirm that the above statement correctly describes the value of the ore, and that it consists of low grade ore that has not been mixed with other ores of a higher value.

(Signed) The Omaha & Grant Sm. Co.

E. J. HALEY COLLECTION

THIS APPLICATION FOR REDUCTION was for ore shipped from the Black Hawk Mine to the Omaha & Grant Smelter Company in Durango. The rubber-stamped notice read: "Vouchered, Feb. 1898, R.G.So. R.R." Notice that this shipment was for low-grade ore, not to be mixed with other ore.

There were quite a few kegs of double-headed spikes leftover after all these track changes. The Rico–Argentine Mining Company at Rico still had at least two kegs full of them at the time the Rio Grande Southern was abandoned. In the meantime, the section crews—disliking them—discarded thousands of them. They dumped them down mountainsides, and in one place, near Coke Ovens, they dumped a large quantity into a backwater of the Dolores River. However, by 1951, most of the spikes on the mainline into Rico were of the old Greer double-headed variety.

DELL A. McCOY PHOTO

THE BLAINE MINE was photographed during June of 1966. This mine was located a short distance east of Rico, beside Silver Creek. The lower slope of Dolores Mountain is at the left, and Expectation Mountain is in the distance, at the right (west of Rico). The Blaine Mine had cut into the base of Nigger Baby Hill through two mine portals, one at the left, and the second at the right, in the foreground. The mine tunnel at the left went into the base of Blackhawk Mountain, while the one on the opposite side of the creek went into Nigger Baby Hill.

MONTE BALLOUGH PHOTO – COURTESY MARGARET BALLOUGH PALMER

OVERLEAF: Looking southwest—down the valley of Silver Creek—toward Rico in the Dolores River Valley, this view was visible from above the dump of the Iron Mine. Bridge 66-AS carried the Enterprise Branch from the slopes of Nigger Baby Hill, over Silver Creek onto the slopes of Dolores Mountain, where the track headed southwest en route to the Enterprise Mine. A flume that carried water from Silver Creek was just down the hillside and parallel to the railroad grade. Two men and two horses were standing on the Iron Mine's dump, which was between the mine tunnel and the upper tramhouse. The lower tramhouse and ore bins stood beside the siding on the Black Hawk spur. The whole operation apparently was abandoned. The Pro Patria Mine was high on Dolores Mountain, at the upper left side of this view. The switch to the Black Hawk spur was out of sight, around the bend in the track.

3.
2.
1. IRON MINE
TRAM TERMINA
2. BLACKHAWK
3. PRO-PATRIA
4. RICO.

WILLIAM A. GRAVES PHOTO

AN ENLARGED VIEW of Bridge 66-AS shows it to be a combination pile-and-frame trestle, which spanned Silver Creek on a 26-degree curve. It was 112 feet in length and 22 feet in height above the creek, and the track was level on the trestle. A wooden flume accessed Silver Creek to take water down and around Newman Hill, so it could be used to generate electricity at mines. An early-day map artist mistakingly drew this flume as if it was a spur on the Enterprise Branch. This map was done for the American Institute of Mining Engineers during September of 1896.

MONTE BALLOUGH PHOTO – COURTESY MARGARET BALLOUGH PALMER

THE IRON MINE'S rock dump came from two openings on the slope of Nigger Baby Hill, at the right. The Yellow Jacket Mine—farther to the left—was at a higher elevation, while the Nellie Bly (Phoenix) was the highest opening. The Black Hawk spur ended near the open meadow.

Rio Grande Southern Railroad Company

C. W. PURINGTON PHOTO –
U.S. GEOLOGICAL SURVEY COLLECTION

IN THIS SCENE you are looking toward the northwest, across the valley of Silver Creek. Several empty boxcars had been spotted at the upper end of the Black Hawk spur on the Enterprise Branch in 1896. They are in view at the lower center of this scene, below the cable (wire rope) coming off Blackhawk Mountain, where the Black Hawk Mine's adit was located. The Iron Mine's tailing dump and mine structures appear above the end of the spur. Trails zigzag on the southern slope of Nigger Baby Hill, above the Yellow Jacket and Phoenix mines. Expectation Mountain rises to the west (at the left), while Darling Ridge crosses the scene at the center.

COLOR VIEWS: WILLIAM A. GRAVES PHOTOS

FROM THE SLOPE of Nigger Baby Hill, Bill Graves was standing on the Iron Mine's rock dump as he aimed his camera toward the east. In this 1995 view, the Rico–Argentine mill was in sight beside Silver Creek. It was near the original opening of the Argentine Mine. The base of Blackhawk Mountain is at the right.

THE RICO–ARGENTINE tunnel opening originally was named the "Argentine," and it is shown in the color view on the following page. Silver Creek flows beneath the bridge that led out of the mine. Mine rails carried cars to the waste dump and the nearby mill.

THE ARGENTINE MINE above Rico had this appearance ca. 1900. It was located on the bank of Silver Creek, about two miles up from the town (toward the east).

MONTE BALLOUGH PHOTO – COURTESY MARGARET BALLOUGH PALMER

THE ARGENTINE
BALLOUGH PHOTO.

MONTE BALLOUGH PHOTO – COURTESY MARGARET BALLOUGH PALMER

BLACKHAWK MOUNTAIN (originally called Blackhawk Peak) is the highest formation on the skyline of this view (at the left), at 12,677 feet in elevation. This set of two pictures was shot from the 12,210-foot summit of Telescope Mountain to produce this panoramic effect. You are looking toward the south, and Rico is out of sight to the right (west), far below in the Dolores River Val-

THE RGS ENTERPRISE BRANCH

By Robert W. Richardson

FOR SEVERAL YEARS after the Enterprise and Black Hawk mines closed as a result of the severe Silver Panic of 1893, the Enterprise Branch was left to rust. The mining people always hoped to reopen their mines, and they wanted the RGS to keep the trackage of the Enterprise Branch intact. However, the RGS roadmaster was not among the optimists, and when the railroad needed a few lengths of rail, he had a section crew go up and remove them from the Black Hawk spur. His thinking was that if the mines returned to operation sometime, he could always replace the rails.

[Photographs taken in 1896, 1900, etc., show boxcars spotted on the trackage of the Enterprise Branch, so the presumption is that rails were put back in place on this branch when the Black Hawk and Enterprise operations resumed. Perhaps it was after the Panic of 1907 that the branch closed permanently.]

At the time the RGS was being built, the price of silver was relatively low, fetching only about $1.00 per ounce. Yet, it cost nearly double that to produce the metal. However, the Rico mines kept on producing sil-

ley. Dolores Mountain, at 12,112 feet (at the right) shows a long snowslide chute, running from the top to the bottom. Whitecap Mountain is the higher ridge of rock to the left of Dolores Mountain. Dolores Mountain is due south from Telescope Mountain. The La Plata Range of mountains rises to heights of over 12,000 feet, barely visible in the far distance (at the right).

ver, as the mining industry believed that the price of silver would rise and become profitable to produce. This contributed to the severity of the financial panic of 1893 in Colorado. The mining firms had used a large part of their capital in stockpiling silver ore.

The Rio Grande Southern's sole Shay geared locomotive was obtained in 1892 by trading one of the Class 56 2-8-0's, No. 34, to the Silverton Railroad for the Shay. This Shay, which Otto Mears had purchased from the Lima Locomotive Works in April of 1890, was No. 269, the "Guston," on the Silverton Railroad. Otto Mears controlled both railroads at this time, and engines of the RGS were often loaned to the Silverton line. The SRR Shay was used on construction trains working out of Durango on the RGS. And when the line was completed in March of 1892, the engine was sent to Rico to be employed on the steep grades and switchbacking track of the Enterprise Branch. This line was better suited for a Shay than the 2-8-0 types of the RGS, even though its top speed was only about 10 mph on the branch.

However, with the Crash of 1893, that summer saw mining virtually close down in the Rico mining district, and the Shay (in need of repairs) was sent on to Ridgway. There, it remained until sold to a Wisconsin lumber road in 1899 (the Siskiwit & Iron River Railway).

F. L. RANSOME PHOTOS – U.S. GEOLOGICAL SURVEY COLLECTION

TWO MINE TRAMHOUSES were clinging precariously to the precipitous slope of Allyn Gulch. The upper building was at the portal of the Little Maggie Mine, and it is connected to the lower structure by an aerial tramway. The lower building was at the adit of the Black Hawk Mine, and was the upper tramhouse for the aerial tramway to the ore bins located near the end of the Black Hawk spur. The Little Maggie Mine also was the property of the Black Hawk mining company. The material sliding down the gulch was tailings from both mines. (Enlargement above from Page 206.)

DURING THE YEAR 1900, F. L. Ransome furnished the photograph printed at the left to the U.S. Geological Survey of the Interior Department. It shows Allyn Gulch, the location of the Black Hawk Mine. Mr. Ransome shot his picture from the ridge of Nigger Baby Hill, northeast of Rico.

THE STORY OF RICO AND THE PIONEER MINING DISTRICT

By William A. Graves

(Continued From Page 129)

1892 The Rio Grande Southern's division point at Rico was a busy operation in 1892, and this helped to make the town a bustling place. The RGS was organized into two divisions, with Rico as the division point. That meant that many railroaders lived in Rico. An extensive rail-

ORVAL L. JAHNKE COLLECTION

ONE OPENING of the Argentine Mine and tailings dump was located above Silver Creek, and it was photographed at the time when mules were used to pull the mine's tramway cars. These little four-wheeled dump cars had been coupled together, ready to be hauled into the mine. The thoughtful miners had attached a carbide lamp on the front of the harness, aimed at the ground so the mule could see where he was stepping, in order to avoid fallen rocks.

road yard and engine-servicing terminal was constructed to handle the railroad traffic and maintain the equipment. The RGS contributed a good-sized payroll to the town's economy.

After the RGS arrived in Rico, the life style of the town changed dramatically. No longer did the residents have to depend on freight wagons struggling over snowy or muddy roads to get necessities during the winter. The days of shortages and $50.00 sacks of flour or $3.00-a-dozen eggs were a thing of the past. The RGS would have more than its share of troubles, but to Rico residents, it was their lifeline.

The benefit of lower transportation costs was immediate. Freight rates that had been $50.00 per ton in 1880 were only $8.00 or $9.00 per ton in 1892 on the RGS. Production of gold and silver doubled in one year. Lead production tripled. There was no official census in 1892; however, unofficial estimates re-

(Continued on Page 218)

RUSS COLLMAN PHOTO

THE RICO–ARGENTINE'S processing mill was not operating when this photograph was taken in June of 1966. This flotation mill was built in 1938, and it was equipped to handle 135 tons of ore per day. It was located approximately 1.5 miles east of Rico, beside Silver Creek. The Argentine Tunnel's mine dumps were above the mill. Flotation waste (tailings) from the mill extended down along Silver Creek for some distance, with the unsuitable ore being dumped in piles along one edge of the settling ponds. J. A. Hogle of Salt Lake City, Utah, was in financial control of the Rico-Argentine Mining Company at this time, and he appointed C. T. Van Winkle as the manager. The Blaine Tunnel on Silver Creek produced a large amount of the ore processed by this mill.

RUSS COLLMAN PHOTOS

A TRAVELING COMPANION from Canada, Eric W. Johnson—a railfan photographer—was standing near a storage shed looking at the remains of the Rico–Argentine mill property in 1966. This is the southeast side of the mill building. An assay office is in the foreground (at the right), with a sawmill in the shadow area. Mine timbers are at the right, while a timber car is on the track.

SUPPORTING STRUCTURES of the Rico–Argentine mill were photographed from within the mill grounds—as the camera was pointed toward the southeast.

RUSS COLLMAN PHOTO

A CONVEYOR BELT inside this covered structure carried ore up into the top of the Rico–Argentine mill shown in this view.

MILL TAILINGS from the Rico–Argentine operation were photographed during 1978. Blackhawk Mountain is off to the right in this autumn scene

DELL A. McCOY PHOTO

FORT LEWIS COLLEGE, CENTER OF SOUTHWEST STUDIES COLLECTION
COURTESY WILLIAM A. GRAVES

ONLY A SHORT TIME had passed since rail had been laid on the Enterprise Branch when this photograph was taken during 1892 or 1893. It looks as though RGS Shay locomotive No. 34 was moving a string of cars on the new track. A number of flatcars had been cut in ahead of the engine, as it switched five boxcars at the Enterprise Mine's ore bins. Notice that a grade has been prepared for a second track at this facility. As mentioned, the Enterprise Branch ran a distance of 4.79 miles from the mainline connection at Milepost 65.4.

TOP RIGHT: The Enterprise Group buildings at the main adit into the Enterprise Mine were photographed during the winter of 1892. The large ore-loading bins and chutes are at the left, and the roof of the office building is in view above the snowshed. The snowshed had been built to cover the mine's tramway track, located on top of the mine dump. Although the Sanborn map incorrectly lists the railroad serving the mine as the D&RG, it actually was the Enterprise Branch of the RGS. The mine employed over 300 men and used 35 mule trammers (according to newspaper accounts about the mine).

CHARLES MILLER ENGEL COLLECTION –
COURTESY OF RACHEL A. HARDWICK

CHARLES MILLER ENGEL COLLECTION – COURTESY OF KEN HAZEN

DAVID SWICKHIMER and his pet hunting dog were the subject of this photograph, taken below the end-of-track at what was called the "Enterprise Group Tunnel." During the fall of 1891, he had given up control of this extremely rich mine, the Enterprise. Boxcars can be seen on the loading track, and they were being loaded with ore at the left-hand part of this scene, as this picture was being taken. David Swickhimer was visiting the Enterprise Group Tunnel in 1892, after selling out for a reported $2 million.

FROM THE RICO SUN...

DAVID SWICKHIMER VISITS THE ENTERPRISE GROUP

January 2, 1892—David Swickhimer returned to Rico on Sunday, after a two-month's absence. While away, he visited his big ranch at Beeville, Texas, and reports investments in the Lone Star State as doing well. Mr. Swickhimer left for Montrose yesterday, where he was joined by his wife. They will spend the next 30 days in California.

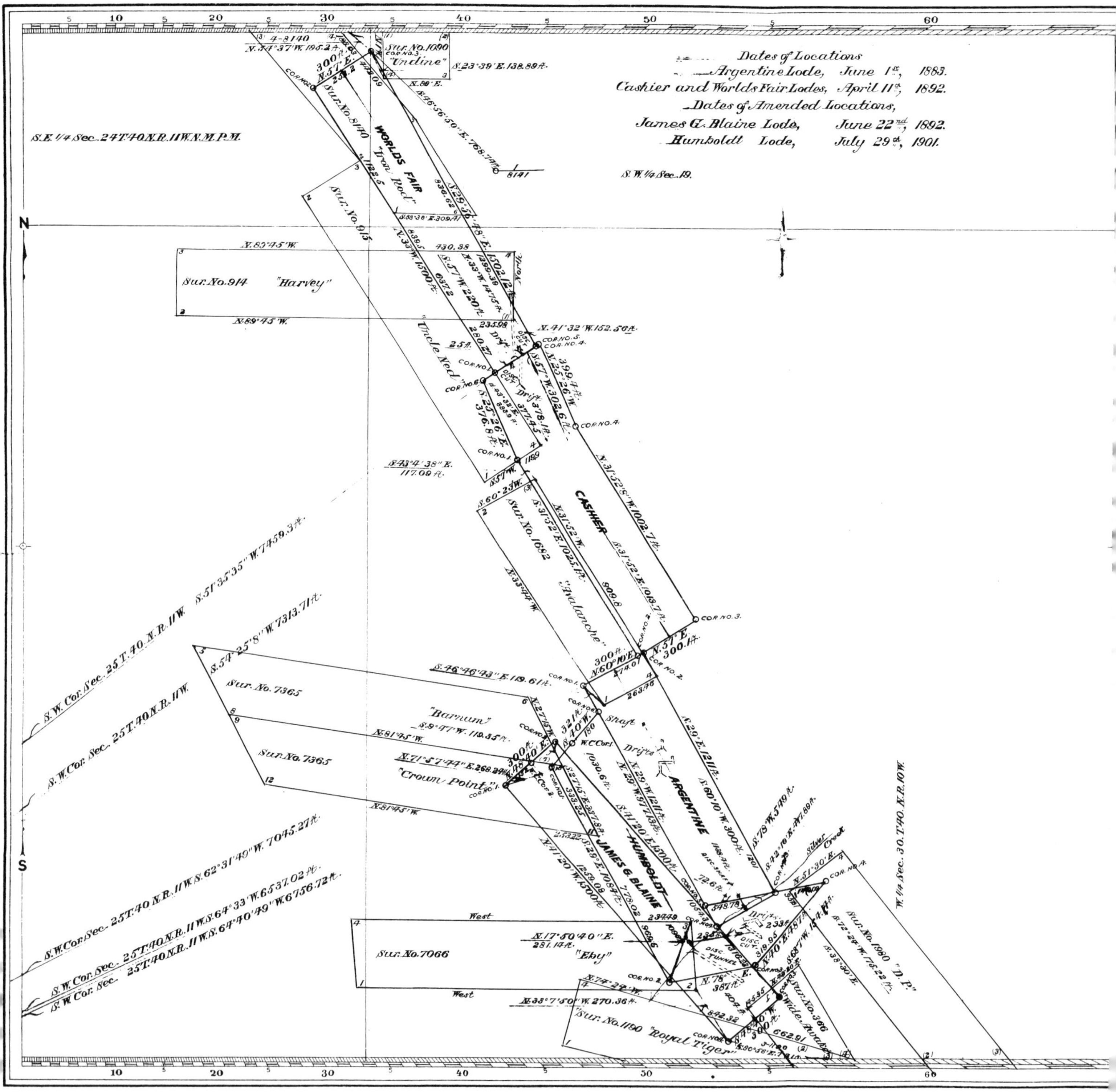

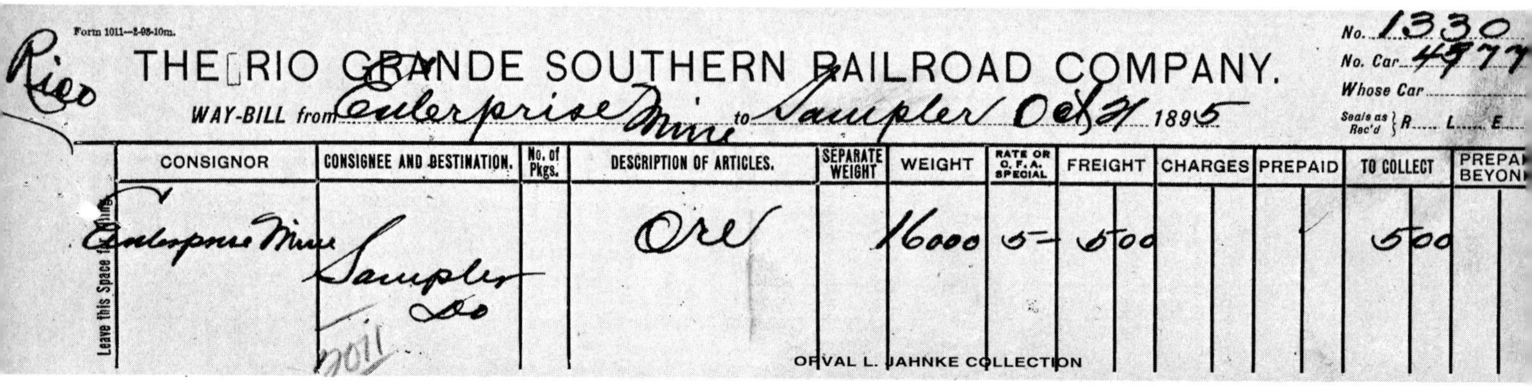
Form 1011—2-95-10m.

THE RIO GRANDE SOUTHERN RAILROAD COMPANY.

WAY-BILL from Enterprise Mine to Sampler Oct 31 1895

Rico

No. 1330
No. Car 4977
Whose Car
Seals as Rec'd R L E

CONSIGNOR	CONSIGNEE AND DESTINATION.	No. of Pkgs.	DESCRIPTION OF ARTICLES.	SEPARATE WEIGHT	WEIGHT	RATE OR C.F.A. SPECIAL	FREIGHT	CHARGES	PREPAID	TO COLLECT	PREPAID BEYOND
Enterprise Mine	Sampler Do		Ore		16000	5—	500			500	

Leave this Space for Stamp

ORVAL L. JAHNKE COLLECTION

A SHIPMENT FROM THE ENTERPRISE MINE

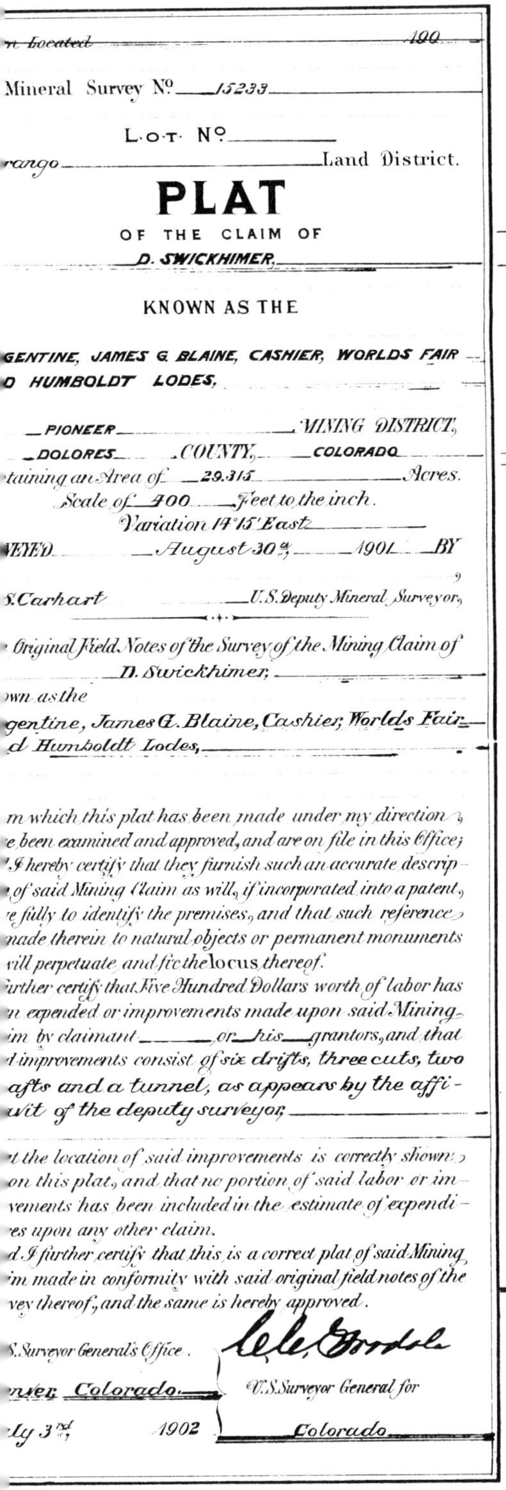

...n located ... 190...

Mineral Survey No. 15233

L·O·T· No. ______

...rango ______ Land District.

PLAT

OF THE CLAIM OF

D. SWICKHIMER,

KNOWN AS THE

...GENTINE, JAMES G. BLAINE, CASHIER, WORLDS FAIR ...
...D HUMBOLDT LODES,

PIONEER ______ MINING DISTRICT,
DOLORES ______ COUNTY, ______ COLORADO
...taining an Area of 29.315 ______ Acres.
Scale of 400 ______ Feet to the inch.
Variation 14°15' East ______
...VEYED ______ August 30th ______ 1901 ______ BY
...S. Carhart ______ U.S. Deputy Mineral Surveyor,

... Original Field Notes of the Survey of the Mining Claim of
D. Swickhimer,
...own as the
...gentine, James G. Blaine, Cashier, Worlds Fair
...d Humboldt Lodes,

...m which this plat has been made under my direction, ...e been examined and approved, and are on file in this Office; ...I hereby certify that they furnish such an accurate descrip... ...of said Mining Claim as will, if incorporated into a patent, ...e fully to identify the premises, and that such reference ...nade therein to natural objects or permanent monuments ...vill perpetuate and fix the locus thereof.
...urther certify that Five Hundred Dollars worth of labor has ...n expended or improvements made upon said Mining ...im by claimant ______ or his ______ grantors, and that ...d improvements consist of six drifts, three cuts, two ...afts and a tunnel, as appears by the affi... ...vit of the deputy surveyor;
...t the location of said improvements is correctly shown ...on this plat, and that no portion of said labor or im... ...vements has been included in the estimate of expendi... ...es upon any other claim.
...d I further certify that this is a correct plat of said Mining ...im made in conformity with said original field notes of the ...vey thereof, and the same is hereby approved.

...S. Surveyor General's Office. ...nver, Colorado. ...ly 3rd 1902

C. C. Goodale
U.S. Surveyor General for
Colorado

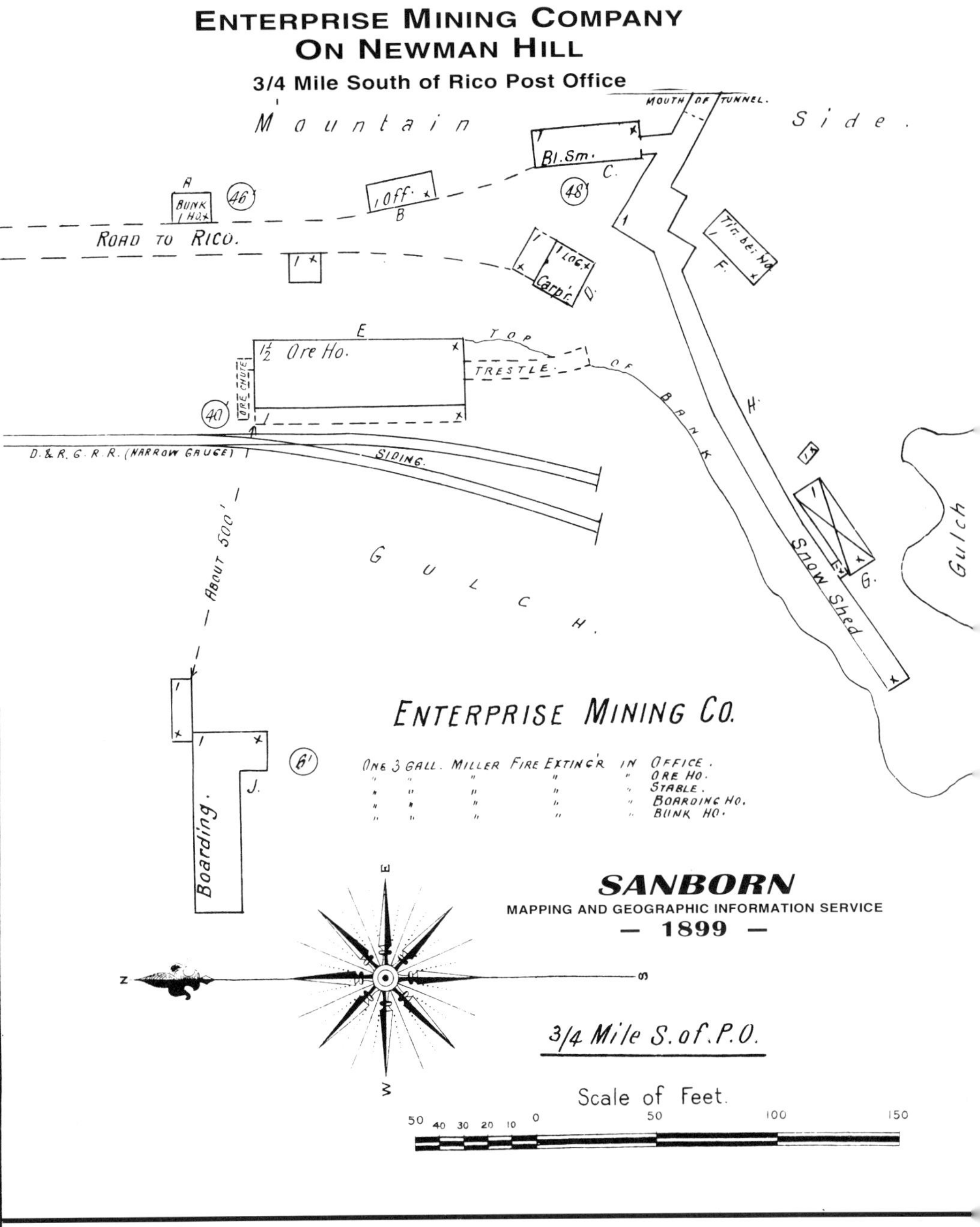

COURTESY KEN HAZEN

THE OFFICIAL PLAT of the mining claim of David Swickhimer, dated August 30, 1901, is reproduced on the page at the left. These valuable claims were strung out across Newman Hill, east of Rico, the lower portion of Dolores Mountain. These claims are up Silver Creek, in the area of the Rico-Argentine mill.

OVERLEAF: The Enterprise ore house was in the left-hand portion of this view taken during the period between 1892 and 1899. A string of empty boxcars was on the tail end of the track at the upper end of the Enterprise Branch. Several loading chutes were located on the western wall of this structure, and the boxcars were rolled down by gravity to be loaded there. Another chute is visible at the south end of the building, leading down from the trestle that supported trackage for the mine's railway. This chute would have had a hinged extension that would have been able to reach into a boxcar's doorway. Workers would then have to shovel the heavy ore into each end of the car, in order to center the load over each truck. The covered snowshed, at the right, protected the mine's outdoor trackage during the winter months.

LARRY PLEASANT COLLECTION

ported that Rico's population doubled or tripled, with 5,000 people living in the district and 2,000 men working in the mines. It was reported that 599 mines were operating in 1892. Many new substantial buildings were constructed, some of which are still standing. These new buildings included the Burley Block, the Dolores County Courthouse (a brick structure built in 1893), the Dey building, the Masonic Lodge building, the public school, the RGS depot (built in 1891) and the first enginehouse (a two-stall wooden structure), the Rico Mercantile Store, the Rico State Bank and the Silver League Theatre.

Glasgow Avenue became Rico's main business street early in the town's history. The present-day state highway (Colorado Highway 145) passes through Rico on Glasgow Avenue. A relatively small red-light district was centered around River Street. Although bordellos were illegal, the so-called "Female Boarding House" was not harassed by local authorities.

The Enterprise Branch of the RGS was built in 1892 to reach Rico's two most productive mines, the Enterprise, on Newman Hill, and the Black Hawk, located on the Silver Creek watershed. The Enterprise Mine originally was owned by David Swickhimer (and other partners), and it was the discovery of rich ore at the Enterprise that had set off the mining boom in Rico in 1887. The Black Hawk Mine had been claimed in 1878 by R. C. Darling, John Glasgow and E. A. Robinson. Otto Mears was also an investor in

DELL A. McCOY PHOTOS

IN THE EARLY PART of June, as aspen buds began to leaf out, a hike was taken along the old Enterprise Branch grade to the end-of-track at the mine opening. The shed covering the mine track was still standing, at the right. The ore bins, at the left, were in ruins.

THE ENTERPRISE MINE entrance shed was photographed in 1966, showing the north side, as the photographer stood next to the abandoned narrow-gauge grade of the Enterprise Branch. The old structure that was the ore house was almost unrecognizable.

the Black Hawk mining operation.

The Enterprise portal site, also known as the site of the Group Tunnel, looks down on the town, about 400 feet below. The nearby Rico-Aspen Mine was owned by the Rico-Aspen Consolidated Mining Company, which had been formed in 1891, and whose president was the prominent banker and railroad builder, David Moffat.

Rivalry between the Enterprise and the Rico-Aspen became violent in a disputed underground area, near the property boundaries. Enterprise miners set off sulfur fires in the disputed tunnels, but not before the Rico-Aspen miners had already extracted most of the ore. The Enterprise eventually won the dispute, after lengthy litigation. Both mines were good producers, and after completion of the spur, the Enterprise Mine regularly shipped about 150 cars a month, and the Rico-Aspen Mine was reported to have shipped about 60 cars a month.

(Continued on Page 223)

DAVID SWICKHIMER AND THE ENTERPRISE MINE

By William A. Graves

THE ENTERPRISE MINE became the dominant mine in Rico, Colorado, in the late 1880's, and this mine sparked the biggest mining boom that the town has ever had. In three years, annual silver production went from less than 100,000 ounces to over 600,000 ounces. As a result, Rico's population increased rapidly, and all kinds of things were scarce.

CHARLES M. ENGEL COLLECTION – COURTESY RACHEL A. HARDWICK

THE CORNET BAND of Rico was posed in the aspen grove near the Enterprise Mine before presenting a concert. The accompanying newspaper account from THE RICO DEMOCRAT tells about the occasion for the concert.

FOND FAREWELL BID TO THE SWICKHIMERS

August 14, 1891—The Rico Concert Band on Monday evening tendered a serenade to Mr. and Mrs. Swickhimer, that being the day upon which the control of the Enterprise Mine passed from their hands. The incoming management [people] were not left out, as the band rendered a musical discourse at the Enterprise, where Mr. J. B. Farrish, the representative of the new owners, has rooms.

A Rico handyman, saloonkeeper, and a one-time town lawman and county sheriff, David Swickhimer could not resist prospecting whenever he had time for it. In 1881, he became interested in a mining claim on Newman Hill called the Enterprise, and with financial help from two partners, David purchased the claim and sank a shaft. The shaft did not reveal any ore, and at a depth of only 35 feet, work on the shaft was stopped for lack of money.

The claim was abandoned in 1883, but Swickhimer could not forget the Enterprise. The location and findings at several of the other nearby mines made David believe that there was rich silver ore at the Enterprise. Gradually, he was able to accumulate some working capital, mostly with the help of his wife, Laura, who ran a boarding house and who gave him strong encouragement to try again. Repurchasing a controlling interest in the Enterprise Mine in 1886 took most of his funds, but he started work on the shaft again in March of 1887. However, before long, he was bankrupt, and he sold his saloon to get money to keep working the shaft. No ore was found, and Swickhimer found himself going deeper and deeper into debt, with local businessmen. Finally, his creditors notified him that they would not extend his credit. Thereupon, his crew of miners quit. At this time, when it seemed that all hope was lost, Laura received a big check in the mail. An enclosure explained that she was a lottery winner! She had bought a \$1.00 Louisiana lottery ticket from a transient who stayed at her boarding house. The Rico newspaper claimed the amount Laura won was \$5,000; however, old-time Rico residents placed the amount between \$2,500 and \$4,000.

In any case, work on the Enterprise shaft resumed—but David still had no success. And the lottery money Laura had given to him for the mining venture ran out, and he decided that he would have to give up on the project. However, Laura talked David into running one more work shift at the mine. On October 6, 1887, at a level of 262 feet, the final blast uncovered a 15-inch thick blanket of \$400-per-ton gold-and-silver ore!

MOST HONEST MAN IN COLORADO, DAVID SWICKHIMER, DIES IN CITY

Discoverer of the Enterprise Mine, Who Put Rico on Map, Finally Succumbs After Long Struggle To Recuperate His Fortune

By JOSEPH EMERSON SMITH
(From a Denver newspaper)

DAVID SWICKHIMER, pioneer, and who had the reputation of being the most honest man in Colorado, since he gave every cent of his fortune of nearly a million dollars to reimburse dollar for dollar, with 6-percent interest added, the depositors in the Rico [State Bank, the] bank he had founded, and which went down in the Panic of 1907, died early Thursday morning in comparative poverty after a determined struggle to recoup his fortunes. Death came from a cancerous growth in the lungs that could not be operated on. Mrs. Swickhimer was at the bedside at the [family] residence, 75 Downing Street.

DAVID SWICKHIMER
Old-Time Mining Man, Who is dead

It is a remarkable story, one stranger than fiction, the career of this widely known Coloradan, who would have been 67 years old next April. He was born in Ohio and worked on a farm in Missouri. Coming to Colorado in his early manhood, he went to the Western Slope and was one of the men who made the rich mining camp of Rico. He was known for years as "the backbone of Rico."

When Posey and Dayley bought his famous Enterprise Mine from him 25 years ago, Swickhimer passed through the receiving-teller's window at the First National Bank, Denver, the largest check that had been drawn here up to that time. It was for $1 million.

Lottery Ticket Prize Put Them on Their Feet

The discovery of the Enterprise at Rico made him. His first wife—later, Mrs. G. Smith—who at one time owned the Tabor Block, but who lost it on a second mortgage, had invested a few dollars in a lottery ticket. Times were hard with the Swickhimers, but the ticket put them on their feet, for it was among the lucky numbers and brought $5,000 [a large sum of money in those days]. Swickhimer, with his wife's consent, took this money and went back to Rico, where he started prospecting. Every cent was put into the Enterprise, and the last shot from the last stick of dynamite uncovered the ore body. Some years later, Mr. and Mrs. Swickhimer were divorced.

Before this, he had trudged the San Juan country looking for mineral. He reached Colorado in 1876, and for several years was on his "uppers." The discovery of the Enterprise put Rico on the map.

In the fall of 1896, he went back to Missouri and returned with a bride, who was Miss Annie Shear. Mr. and Mrs. Swickhimer for years lived at the Metropole Hotel, but when the Panic of 1907 came, and his bank at Rico closed its doors, he lost everything he possessed, and dollar for dollar, all depositors were paid and were given 6-percent [interest] on their money.

His Name at Stake He Told His Friends

This ruined him. He was told by his friends that it was quixotically foolish, but his reply was that his name was at stake.

With his wife, he went to live at the Hamilton Apartments, and then, despite all his efforts to recoup in business, the downward curve grew sharper, so the couple went to one room at the Plymouth Hotel. This was followed by the removal to 75 Downing Street [in Denver]. During the oil excitement, Swickhimer had hopes of gaining back some of the fortune he had relinquished, but the subsidence of the excitement and the comparative flattening of the market again left him stranded.

In the early years at Rico, Swickhimer served as sheriff [of Dolores County, having been elected to that office on November 6, 1883, and he served as a respected officer of the law in Rico prior to that time.... [Swickhimer] brought much money into the state, and was always among the first to give assistance to any move that would build up his community or state. He was retiring by disposition, but had a strong sense of duty, from which he could not be swerved. His reputation for integrity and honesty was statewide. Practically all his time in his earlier days was spent in mining and prospecting. He was vigorous and active up to a few weeks ago.

The funeral will be held from the Nash Mortuary [in Denver] at 11:00 o'clock Saturday morning.

— LARRY PLEASANT COLLECTION

WENDE PERINA STUART PHOTO

FROM AN OPEN AREA near Beamis Flats, a local photographer snapped this picture of Rico. She was pointing her camera toward the east for this 1994 view. This scene includes Nigger Baby Hill, on the left, the Silver Creek drainage, flanked by Newman Hill and Blackhawk Mountain, rising on the skyline.

The Swickhimers soon became millionaires. They operated the Enterprise Mine very successfully for several years, and then, they sold it for $1,250,000 in cash. Other details of the sale are subject to debate. The year of the sale is variously stated in five different accounts as being 1890, 1891 or 1892. An article published in the Colorado Historical Society's ***Colorado Magazine,*** written by Carl F. Mathews (in the issue of January, 1951), states that Swickhimer sold the Enterprise to two men from New York and Wisconsin, who had previously arranged for the sale of three Red Mountain mines to a British syndicate. The date Carl Mathews gives for the sale of the Enterprise is May of 1891. A rumor circulated at the time of the sale was that the same British syndicate was interested in buying the Enterprise.

This bonanza at the Enterprise Mine rejuvenated the Rico prospectors, and soon, new large ore bodies were discovered. Rico soon experienced boomtown growth and prosperity.

Unfortunately, prosperity did not stay with the Swickhimers very long. Laura lost her money in Denver real estate dealings. And mining at Rico had an unbelievably bad time of it during the Panic of 1907. As a result, the Rico State Bank, of which David was president, went into receivership. His personal fortune was used to pay off the bank's depositors, and the Swickhimers eventually divorced. Later, David remarried, but he never regained his wealth. (See the story about his death in the article from a Denver newspaper, which appears elsewhere in this volume.)

The Enterprise Mine paid to have a five-mile railroad branch constructed by the Rio Grande Southern, so that ore could be shipped directly from the mine. The Enterprise Branch had switchbacks and 5.0-percent grades, and it terminated near the mine's tunnel portal. The Enterprise was a major shipper for the RGS, and it often provided five carloads of ore per day. The last year of operation at the mine is not known; however, freight cars are shown on the branch in an 1899 photograph. In any case, the reported date for the abandonment of the RGS trackage to the mine is 1900.

THE STORY OF RICO AND THE PIONEER MINING DISTRICT

By William A. Graves

(Continued From Page 219)

1893 Prosperity in Rico reached a peak early in 1893. Driven by the Federal government's purchases of silver, over $2-million worth of silver was produced that year. In addition, the production of gold and lead also peaked in 1893. However, this over-production of silver at Rico was

(Continued on Page 226)

C. W. PURINGTON PHOTO – U.S. GEOLOGICAL SURVEY COLLECTION

THIS PICTURE WAS TAKEN during the summer of 1896. It documents the flourishing mining activity that took place near the base of Dolores Mountain—on the slopes of Newman Hill, just prior to the turn of the century. The same photograph has been enlarged and printed on the following two pages, in order to better show the detail in this scene. The Enterprise Mining Company's Group Tunnel was shipping five boxcar loads of ore per day at the time this photograph was taken. Looking at the ore bins, two loaded boxcars had been rolled down from the ore chutes, and one remained at the tipple for loading. At the same time, two empty boxcars were at the end of "empty track," waiting to be eased down for loading. The Vestal shaft was the highest location in this illustration, and the Jumbo was a short distance below it. Midway in this picture are huge mine dumps from the Newman Group of mines. At the right-hand edge of the view is the Syndicate mining property, with its support structures.

also taking place throughout Colorado and other Western states, and over-production caused the price of silver to drop. The Federal government's large horde of silver also was seen as a threat to the gold standard of the Federal Reserve System. Bankers began to lose faith in paper currency, and hundreds of banks in the U.S. began to close their doors. This crisis was blamed on the inflation of the silver price, and the so-called "Silver Panic of 1893" caused the Sherman Silver Purchase Act to be repealed on October 31, 1893. The end of the government purchases was a great disaster for Rico. Hundreds of mines in the Rico area closed their operations, and the miners departed. Many merchants also closed their businesses and left town. Although Rico survived, it had a greatly reduced population, and it never could recover its earlier booming vitality.

1894 The price of silver fell to 63 cents per ounce in 1894, and ore production in the Rico mining district dropped to less than one-half of what it was during the previous year. Traffic on the RGS also decreased to less than one half of what it was in 1893, and many railroad employees were laid off or had their wages cut.

1895 Production of ore continued to decrease, and the output in 1895 was only about one-third of 1894's production. Mining had also slowed because the best ore deposits at the Enterprise, Rico-Aspen and other Newman Hill mines

(Continued on Page 244)

COLORADO MILLING & CONCENTRATING COMPANY WET PROCESS MILL

See Pages 232, 234 and 236

SANBORN

MAPPING AND GEOGRAPHIC INFORMATION SERVICE

— 1899 —

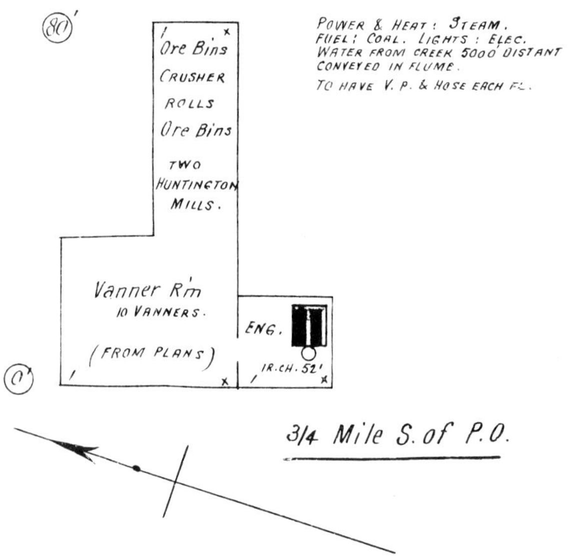

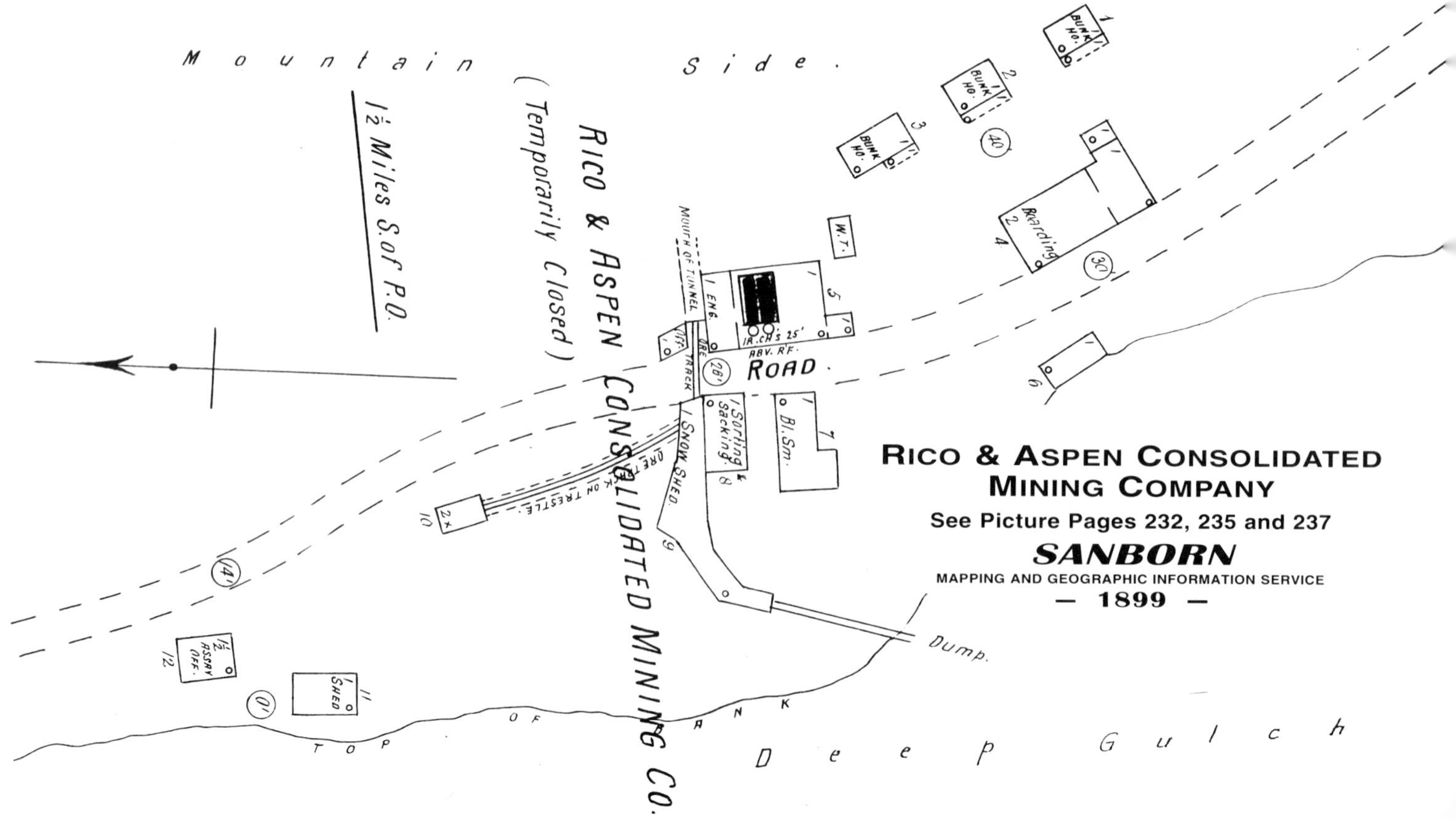

RICO & ASPEN CONSOLIDATED MINING COMPANY

See Picture Pages 232, 235 and 237

SANBORN

MAPPING AND GEOGRAPHIC INFORMATION SERVICE

— 1899 —

HOWE PHOTO – W. CROSS COLLECTION, U.S. GEOLOGICAL SURVEY

THE U.S. GEOLOGICAL SURVEY describes this picture as follows: "Looking up Silver Creek, from bench west of Rico. Dolores Mountain on the right: Telescope Mountain and Nigger Baby Hill on the left. Rico quadrangle, Dolores County, Colorado. July 1899." The original glass-plate print was reduced to 85 percent in the illustration printed above.

SANBORN

MAPPING AND GEOGRAPHIC INFORMATION SERVICE

— 1899 —

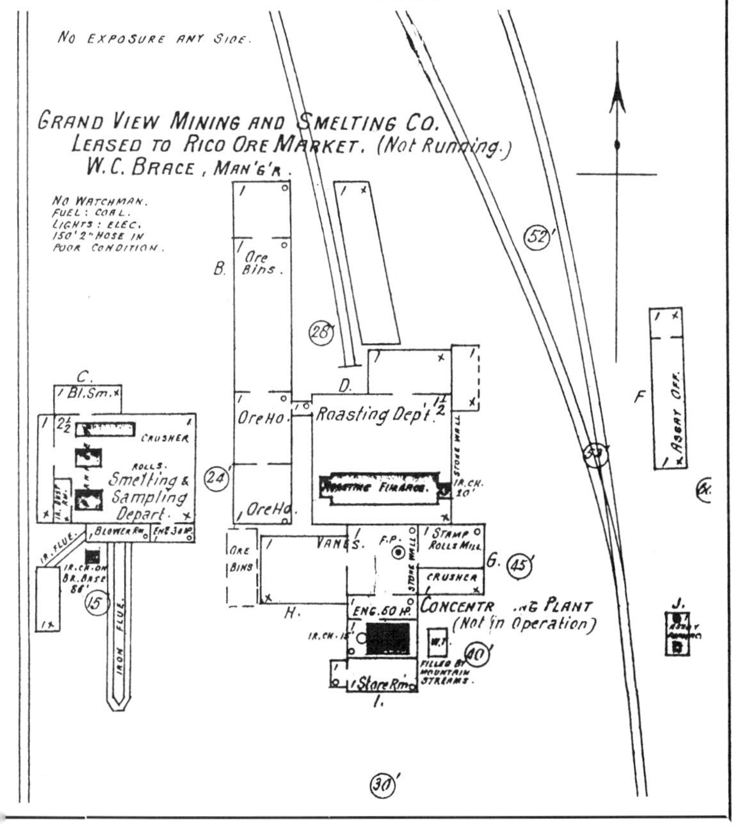

Final Stage of Development—See Page 238

THE GLASS-PLATE PRINT has been reproduced as a four-page illustration, enlarged 400 percent on the pages that follow. The Grand View Mining & Smelting Company works are in view at the northern outskirts of Rico. By carefully studying the grade of the Enterprise Branch, it is possible to follow along as the RGS track climbed the steep 5.0-percent grade, zigzagging up the lower slope of Nigger Baby Hill. Four switchbacks were needed to gain the required elevation, and they are shown on the first two pages that follow. Notice the abandoned funicular railway that once had been used to haul ore down to the Grand View smelter from the Ben Harrison claim. The cable railway was abandoned when Enterprise Branch trackage cut through the inclined railway in three places.

THE THIRD AND FOURTH PAGES of this enlargement portrays the end-of-track on the Enterprise Branch. A number of empty boxcars had been spotted at the end-of-track, on the grade slightly above the Enterprise ore bins. The snowshed sheltering the mine adit and tramway track to the dump appears on the right, and this mine track had been lengthened since the previous pictures had been taken. The Enterprise shaft, in which the original body of ore was discovered, was above and to the right of the tunnel and loading facilities. The shaft house was the small structure at the left end of the enlongated dump. The tunnel of the Golden Fleece Mine was located below the grade of the Enterprise Branch, above the southern town limit of Rico. Notice that these four pages portray the entire Enterprise Branch.

F. L. RANSOME PHOTO – U.S. GEOLOGICAL SURVEY COLLECTION

THIS PHOTOGRAPH was dated 1900, one year later than the preceeding one, and it shows the Enterprise Mining Company's Group Tunnel, at the left, and the Syndicate Tunnel, with its mine buildings, at the right. The mine in the extreme right-hand edge of the view was named the Stephanite. The new Colorado Milling & Concentrating Company's wet-process mill building can be seen at the foot of the Enterprise dump.

F. L. RANSOME PHOTO – U.S. GEOLOGICAL SURVEY COLLECTION

THE SAME PICTURE, enlarged even greater, shows additional details from this glass-negative print. Four empty boxcars were at the end of the Enterprise Branch, spotted above and alongside the ore house of the Enterprise Mine. The mine dump had grown considerably since the earlier view, and a new ore-reduction mill of the Colorado Milling & Concentrating Company had been built at the base to extract and recover values from copper, lead, iron and zinc. Silver and gold were of the primary importance during the earlier years, and the other metals were not sought. The Vestal shaft was the highest structure in this view, at an elevation of 9,861 feet. The Jumbo shaft house was below it, located at 9,706 feet. Farther to the left was the old Enterprise shaft dump and hoist house remains. The mine dump in the center of the view was at the mine named the Newman Group. At the far right was the Rico-Aspen Consolidated Mining Company. Late in 1891, a road was constructed from the Rico Smelting Company to the Newman mines. This road is visible at the lower right. Several empty boxcars were on the spur at the Rico Sampling Works, along with a string of flatcars. Wagons dumped ore into the chute at the back of the main structure.

MONTE BALLOUGH PHOTO – COURTESY MARGARET BALLOUGH PALMER

ON THE WEST SLOPE of Newman Hill, ca. 1902, the large Enterprise ore house and tailings dump were plainly visible, but it probably was not shipping ore over the Enterprise Branch by the time this scene was photographed. The Colorado Milling & Concentrating Company's mill, located below the Enterprise dump, had begun operating, as evidenced by the dump on the west side of the building. The large group of boarding houses, bunkhouses and assorted mine buildings, at the lower right of center in this scene, belonged to the Rico & Aspen Consolidated Mining Company. In the lower corner of the far right, you can see the Stephanite Mine, next to Ballough's credit. Higher on Newman Hill, the highest mines (left to right) were: the Union Carbonate and Buckhorn. The Laura shafthouse and dump were near the center top, with the Enterprise shafthouse directly below. The Jumbo was directly above the Montezuma, and the Vestal was at the extreme right.

NEWMAN HILL.
BALLOUGH PHOTO.

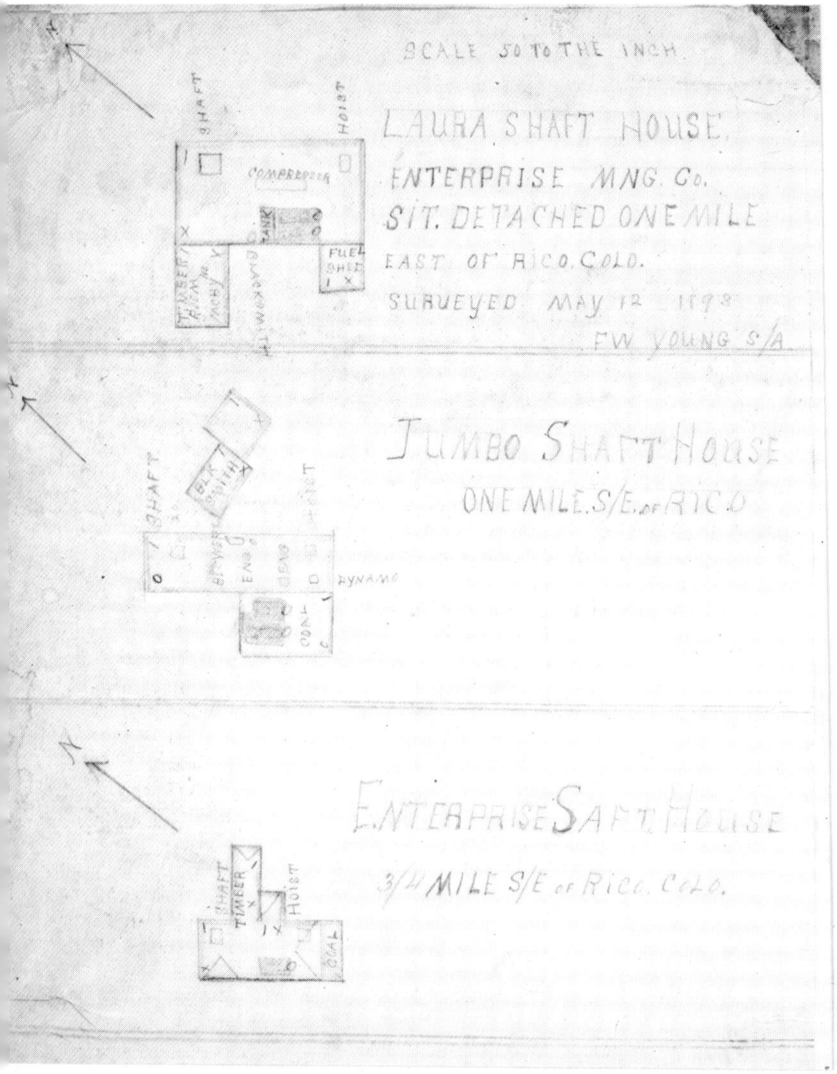

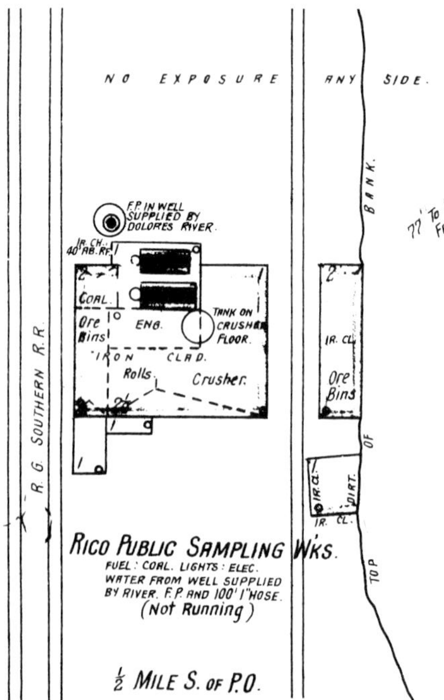

SANBORN
MAPPING AND GEOGRAPHIC INFORMATION SERVICE
— 1899 —

CHARLES MILLER ENGEL COLLECTION –
COURTESY RACHEL A. HARDWICK

LOCATED AT 10,200 FEET in elevation, this is believed to be the Union Carbonate Mine, located 1,912 feet below the summit of Dolores Mountain. The Union Carbonate Mine was the highest operation on Dolores Mountain. Sandstone Mountain may be the peak in view in the mist beyond the men. The mine's headframe had been in place for some time, and a shaft house was being constructed around it for winter opera-

tions. Hard-rock mining was backbreaking work, especially at this high altitude, and the mining men often would stop working to catch their breath in the thin mountain air. The miners in this scene were standing on the mine's railway track, which connected the mine tipple with the dump area. A mining engineer, with his transit or level mounted on a tripod, was near the center of this photograph.

FROM THE RICO DEMOCRAT:

April 24, 1891—The shaft on the Union Carbonate is down about 320 feet, the work progressing finely. Development work thus far has shown ore in large bodies.

FROM THE RICO SUN:

December, 1891—The boarding house at the Union Carbonate is nearing completion. The force on this mine will be increased at once.

C. W. PURINGTON PHOTOS – U.S. GEOLOGICAL SURVEY COLLECTION

C. W. PURINGTON'S SKILL as a photographer was paid for by the U.S. Geological Survey, and this picture was taken above Rico in 1896, as part of his job. The dark ridge of forested hillside, near the center of the view, was named Nigger Baby Hill. This photograph could have been taken from the shaft structure of the Logan Mine, showing the Mountain Spring Mine, at the lower right. The Logan Mine was near an altitude of 9,800 feet above sea level.

ENLARGED EVEN GREATER to the left, this view of C. W. Purington's picture makes it possible to identify the more prominent structures in the town of Rico. Most notable on Glasgow Avenue was the large Hotel Enterprise. Smoke was drifting from the smokejacks of the RGS enginehouse, next to the Dolores River. The darkened object upriver (to the right of the enginehouse) was a coal-storage shed. Notice the large number of houses along Riverside Street, at the lower right. THE RICO SUN printed an article, dated December 12, 1891, which stated, "Although her population will not justify it, the number of Rico's suburbs is such as to lead the uninformed to believe we are multitude of thousands." The newspaper listed such names as "...Piedmont, New Boston, Bellevue, Horse Gulch, Westport, Breezy Point and Mearsville." At the south end of the RGS trackage in this scene stood the Pasadena Smelter (run by the Adams Concentrating Works) that had four or five boxcars waiting on the empty track. This was near the RGS wye, which crossed the river twice, on Bridges 67-AS and 67-ASS. On October 30, 1891, the Rico newspaper stated that a switch was being spiked down by the RGS, in order to lay track to the Adams Concentrating Works. This spur can easily be seen taking off from the RGS mainline, between the two switches to the wye—just this side of the boxcars waiting on the empty track.

Form 3600

LOCAL WAY-BILL. **THE DENVER AND RIO GRANDE RAILROAD CO.**

From DENVER to Rico Date Sept 28 189_

Way-Bill No. ___ Car Initials ___ Car No. ___ | TRANSFERRED INTO: Initials ___ No. 1947 At ___ | WEIGH THIS CAR AT ___ | SHOW GROSS, TARE AND NET WEIGHT IN THIS SPACE. | ROUTE. Via ___ Station, and ___ R. R. | AGENT AT DESTINATION STAMP HEREIN THE DATE RECEIVED.

CONSIGNOR.	CONSIGNEE AND DESTINATION.	No. of Pkgs.	DESCRIPTION OF ARTICLES.	SEPARATE WEIGHT.	WEIGHT.	RATE.	FREIGHT.	CHARGES.	PREPAID.	TO COLLECT.	PREPAID BEYOND.
[illegible]	Enterprise Hotel Co										
		1	Bx Sar Chips		50	275	138				
		1	Lp Cheese	70							
		1	Pail Cocoanut	75	95	255	242				
		1	Cs Shredded Wheat	45							
		2	Can Gds	100							
		1	Bx Sausa Smo	50							
		6	Kits Jelly	100							
		1/2	Bbl S Fish	115	310 (410)	190	589 (779)			969 (1159)	
Do 2	Otto Koster										
		1	Bbl Fish Pails								
		1	" " " Corn		40	412	165				
		1	Cs Mustd		30	255	77				
		1	Bx Am Sausa	60							
		1	" Hams	65							
		1	Cs Can Gds	65	190	190	361			603	
Do 3	[illegible]										
		1	Dr D Beef	100							
		5	Cs Eggs	245	345	255	880				
		1/2	Bbl S Fish	120							
		2	Bxs S Sausa	65	185	190	352			1232	
					1245 (1345)		[illegible] (2994)			2804 (2994)	

Corr. #241 Oct 3

D&RG LOCAL WAYBILL

— DENVER to RICO —

For the Enterprise Hotel and Other Consignees

ORVAL L. JAHNKE COLLECTION

NEWS-SUN.

SATURDAY, JANUARY 1, 1898. VOL. 3. NO. 27.

PULLIAM,

ATTORNEY.

n 5, First National Bank Block.

RICO, COLORADO.

Hogg. F. W. Ingersoll.

HOGG & INGERSOLL,

TORNEYS AT LAW.

Practice in all the courts.

ver Postoffice, - Telluride, Colo.

HOMFELD & CO.

Rico Assay Office.

ork warranted to be strictly correct. s received by mail or otherwise will prompt and careful attention.

Opposite Hotel Enterprise.

- - - - - Colorado

J. P. NORTON,

NOTARY PUBLIC,

REAL ESTATE AGENT,

OFFICE: Two Doors North Hotel Enterprise.

. DAWSON,

YSICIAN AND SURGEON.

ffice in Bank of Rico Building.

COLO

IBBS, M. D.

ysician and Surgeon.

Rooms in Burley Block.

ICO - - COLO.

. HESS,

l Estate and Insurance.

Notary Public.

Office in Burley Block,

- - - - - - COLO.

E. W. HUNT,

Deputy Mineral Surveyor

Office on Glasgow Avenue.

ico, Colorado.

. J. TARSNEY,

orney at Law.

ANGO COLORADO.

co Beer Hall

hindelmeyer, Proprietor.

Domestic and Imported Wines, Liquors and Cigars.

r in

s Beer Always on Tap.

The Rico State Bank,

RICO, COLORADO.

Authorized Capital, - - - $200,000.
Paid in Capital, - - - - 50,000.

OFFICERS.	DIRECTORS.
L. D. SMITH, President.	L. D. SMITH,
DAVID SWICKHIMER, Vice-Pres.	D. SWICKHIMER,
JOHN EDER, Secretary,	JOHN LITTLE,
JOHN LITTLE, Cashier.	JOHN EDER.

Interest allowed on time deposits. Sell Drafts payable in any part of the world. Fire and burglar-proof safety deposit boxes for rent We solicit your patronage.

THE DEIS

Meat and Produce Co.,

DEALER IN

FRESH AND SALT

MEATS

Vegetables, Fish and Game

Butter, Eggs, Etc.

Koster's Old Stand. RICO, COLORADO.

The

Hotel Enterprise

H. DUNTON, Prop.

The Only First-Class Hotel in Rico.

RATES--$3 per day. Table board $7 per week. Sample rooms free to commercial men.

Bar and Billiard Room in Connection.

THE RICO NEWS-SUN
March 26, 1898

City Livery

FRANK HALL, Prop.

Fine Turnouts and Saddle Horses.

DEALER IN

Hay, Grain and Coal!

Ore Hauling, Transfer and Express.

BUSSES - MEET - ALL - TRAINS

Colorado Electric Light and Power Co.

Plant is now in condition to furnish lights to all.

For rates or repairs of any kind call at Rico State Bank.

THE RICO NEWS
November 5, 1898

COURTESY MARGARET MATZICK

MONTE BALLOUGH PHOTO – COURTESY MARGARET BALLOUGH PALMER

THIS PHOTOGRAPH, taken by one of Rico's resident photographers, was shot in 1900, prior to the construction of the Pro Patria mill. This large mill was later constructed at the end of Mantz Avenue.

TO ENHANCE the details in this picture, it has been enlarged by 300 percent and reproduced on the next three pages. The northeast section of Rico appears in the first frame and shows the homes of that time. The schoolhouse built in 1892, along with two outhouses, is at the lower right. The home of Charles Engel, Sr., stands alone near the center of the view.

THE STORY OF RICO AND THE PIONEER MINING DISTRICT

By William A. Graves

(Continued From Page 226)

were just about mined-out. The RGS transferred the majority of its stocks and bonds to the D&RG, and in return, the D&RG acquired much of the Rio Grande Southern's indebtedness. These moves made the D&RG the principal owner of the RGS, and the receivership ended on December 1, 1895.

1896 Recovery of zinc from Rico ore was reported for the first time in 1896, and 30,000 pounds were produced. In later years, zinc became the mainstay of production for the district.

1897 The price of silver was down to 63 cents per ounce. Some small-scale mining and two stamp mills kept Rico alive.

1898 Mining had a brief revival, and several mines were reopened. About a dozen businesses and several professional offices kept their doors open. Unfortunately, the Silver League Theatre, located on Glasgow Avenue, was destroyed by fire in 1898. The theater had been used by traveling road shows, as well as a local performing company, to put on plays, musicals and variety shows (the very popular "Gay Nineties" vaudeville performances, which featured lively songs, dances, skits, acrobatic acts, pretty girls in colorful costumes, etc.).

1899 This was a slow year for Rico, as the 1898 mining revival slowed down.

(Continued on Page 254)

Rico Colo.

MONTE BALLOUGH PHOTO – COURTESY MARGARET BALLOUGH PALMER

ON THESE PAGES you can see a rare view of houses built along River Street (shown below, at the center, and extending upward to the right). As of this date, uptown Rico did not have many vacant lots available for expansion along Glasgow Avenue. The Enterprise Mine is at the upper left, while the newer recovery mill is in view down the embankment, on Newman Hill. The RGS water tank stands at the far right, in a patch of sunlight, with shadow covering the freight yard, which was filled with cars. A two-story industrial building (emitting smoke) is believed to be the Rico Bottling Works, standing along the RGS stub spur at the lower right.

MONTE BALLOUGH PHOTO – COURTESY MARGARET BALLOUGH PALMER

THESE LAST TWO PAGES of this picture, enlarged by 600 percent, show more details. Three empty boxcars are in view at the Enterprise Mine's ore house, waiting to be loaded. Various mine-support structures, including bunkhouses, are visible around the mine and below the dump. This photograph affords one of the few views that show the entire block of store fronts along the east side of Glasgow Avenue in Block 14. Store buildings located in this block appear on the Sanborn Map, dated 1899, on Pages 364-365. The final remodeled Hotel Enterprise is included. Skylight windows placed on the rooftop (shown at the lower right) provided sunlight at strategic locations within the building.

DELL A. McCOY PHOTO

SNOW ACCUMULATION had not yet built up to any depth when this scene was photographed in the winter of 1990. This is Rico's Glasgow Avenue—Colorado Highway 145—and you are looking toward the north. On the left you can see the old Masonic Hall, Baptist Church, Rico Fire Department, Burley Block, movie theater, an old 1880's building and the first brick structure. Beyond Mantz Avenue, the 1892 Dey Building is visible. The familiar bay window over the sidewalk and the one-time Obendorfer Grocery are at the right, on the east side of the main street.

WILLIAM A. GRAVES PHOTO

THE ENTERPRISE CAFE and Lounge, in the old Dey Building, normally is open. However, the owner—Ken Jahnke—was taking refuge on a warm beach during February, as the sign states.

FROM "THE RICO ITEM," March 29, 1912, at right:

COURTESY MARGARET MATZICK

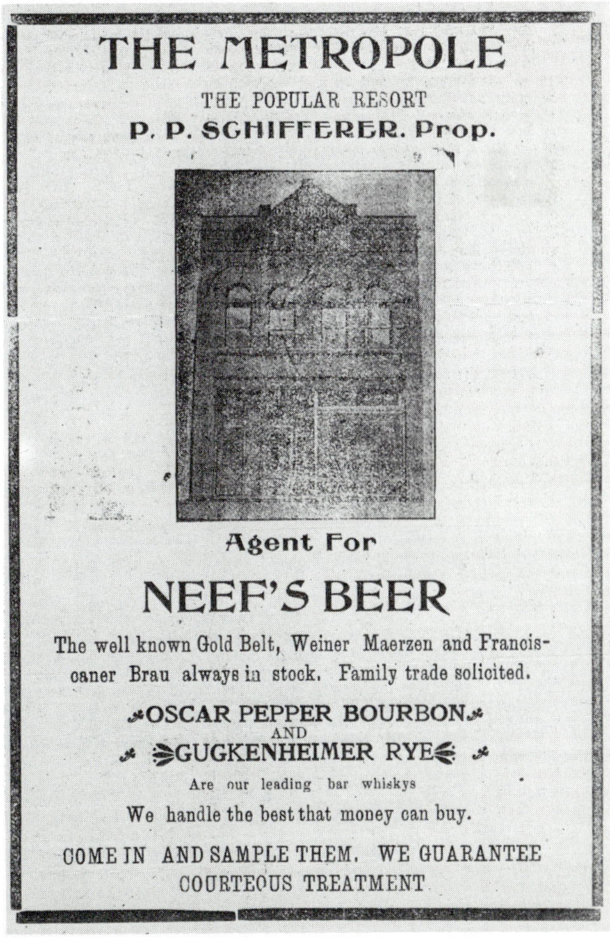

THE METROPOLE

THE POPULAR RESORT

P. P. SCHIFFERER. Prop.

Agent For

NEEF'S BEER

The well known Gold Belt, Weiner Maerzen and Franciscaner Brau always in stock. Family trade solicited.

OSCAR PEPPER BOURBON
AND
GUGKENHEIMER RYE

Are our leading bar whiskys

We handle the best that money can buy.

COME IN AND SAMPLE THEM. WE GUARANTEE COURTEOUS TREATMENT

KEN JAHNKE COLLECTION

THE METROPOLE SALOON occupied the Dey Building, constructed in 1892, on Glasgow Avenue. Ken Jahnke presently owns the building, which houses a restaurant and lounge (as of 1996). This photograph illustrates how the establishment looked after electricity had arrived in Rico. Light bulbs are in view at the entry way, as well as in the window.

KEN JAHNKE COLLECTION

GOLD BELT BEER, brewed by Keef Brothers, was being served at the Metropole Saloon in Rico. The board sidewalk was showing signs of wear, as patrons lounged for a portrait in the morning light.

Coors' Golden Beer!

Always on Draught and in Bottle. The Beer that Won Medals now Famous, 5 Cents a Glass! : :

The Rico Beer Hall.

M. SCHINDELMEYER, Prop.

COURTESY MARGARET MATZICK

FROM "THE RICO NEWS-SUN," March 26, 1898

BILL PLUNKETT COLLECTION

BOTH RICO AND TELLURIDE set up a flagpole in the middle of the busiest intersection in their towns and mounted a bandstand around the pole, for the Fourth of July and other celebrations. Rico's flagpole stood in the center of the intersection of Mantz and Glasgow avenues. According to the Sanborn Map of 1899, the furniture sign advertised what is believed to be E. Carey & Company Furniture Dealers. This was in the Masonic Hall, a brick building located at 218 Glasgow Avenue, and a saloon was south of it. All lots along the east side of Glasgow Avenue were occupied by one- and two-story buildings. Since power poles were scattered throughout the business district, shopkeepers apparently were making use of this new and convenient form of illumination.

RICO BOTTLING WORKS,

TRADE MARK.
ANHEUSER, BUSCH BREWING ASS'N
ST. LOUIS, MO.

I. N. SIMPSON,

Proprietor.

Manufacturers of

-o- Soda, Birch,
o Ginger Ale,
-o- Sasaparilla and
—o— Strawberry soda
-o- Orange Cider
o and Mineral
-o- Waters.

Ph. Zang and Anhauser, Draught and Bottled Be

RICO LIVERY, FEED AND SALE STABLE

Transfer and Delivery

Hay, Grain, Coal

FINE SADDLE HORSES

GARDNER & McGEOCH - - - - - PROPS.

COURTESY MARGARET MATZICK

FROM "THE RICO ITEM"
November 26, 1909

DELL A. McCOY PHOTO

THE RICO TOWN HALL currently occupies the former Dolores County Courthouse (as of 1996). In this winter view, taken in 1990, the old Congregational Church still stands on Mantz Avenue, and a glimpse of the former Charles Miller Engel home is beyond Commercial Street, near the fire plug.

1900 Rico's population and ore production continued to decline. According to the official 1900 census, the town's population was down to 811 people. The yearly total value of ore production was only $173,817, a drop of 90 percent in seven years. The only ore produced in the district was mined by small-scale operators, who leased small areas of the big mines.

1901 A major consolidation of Rico's mine holdings came about in 1901. A newly formed corporation, known as the United Rico Mines Company, had its main offices in New York City. The new company acquired most of the major properties in the district, including the Atlantic Cable, Cross, Enterprise, Grand View, Hope and Rico-Aspen claims. New ideas for the milling of lead-and-zinc ores helped to keep the district busy. The Rico Mining & Milling Company constructed a zinc-concentration mill, which used 20 stamps, 6 Wilfley tables and some newly developed devices called magnetic separators. These separators were intended to divide the ore into its lead, copper, iron and zinc components.

(Continued on Page 264)

WILLIAM A. GRAVES PHOTO

THE FORMER People's Congregational Church, now owned by the Presbyterian congregation, was photographed during the winter of 1995 by Bill Graves, a narrow-gauge model railroader and railfan photographer from Fort Collins, Colorado. Dell A. McCoy, publisher, was chatting with the former mayor of Rico, Ken Hazen. The old 1970 Ford Bronco continues to roll up the miles on picture-gathering missions. (See this building when it was new on Page 119.)

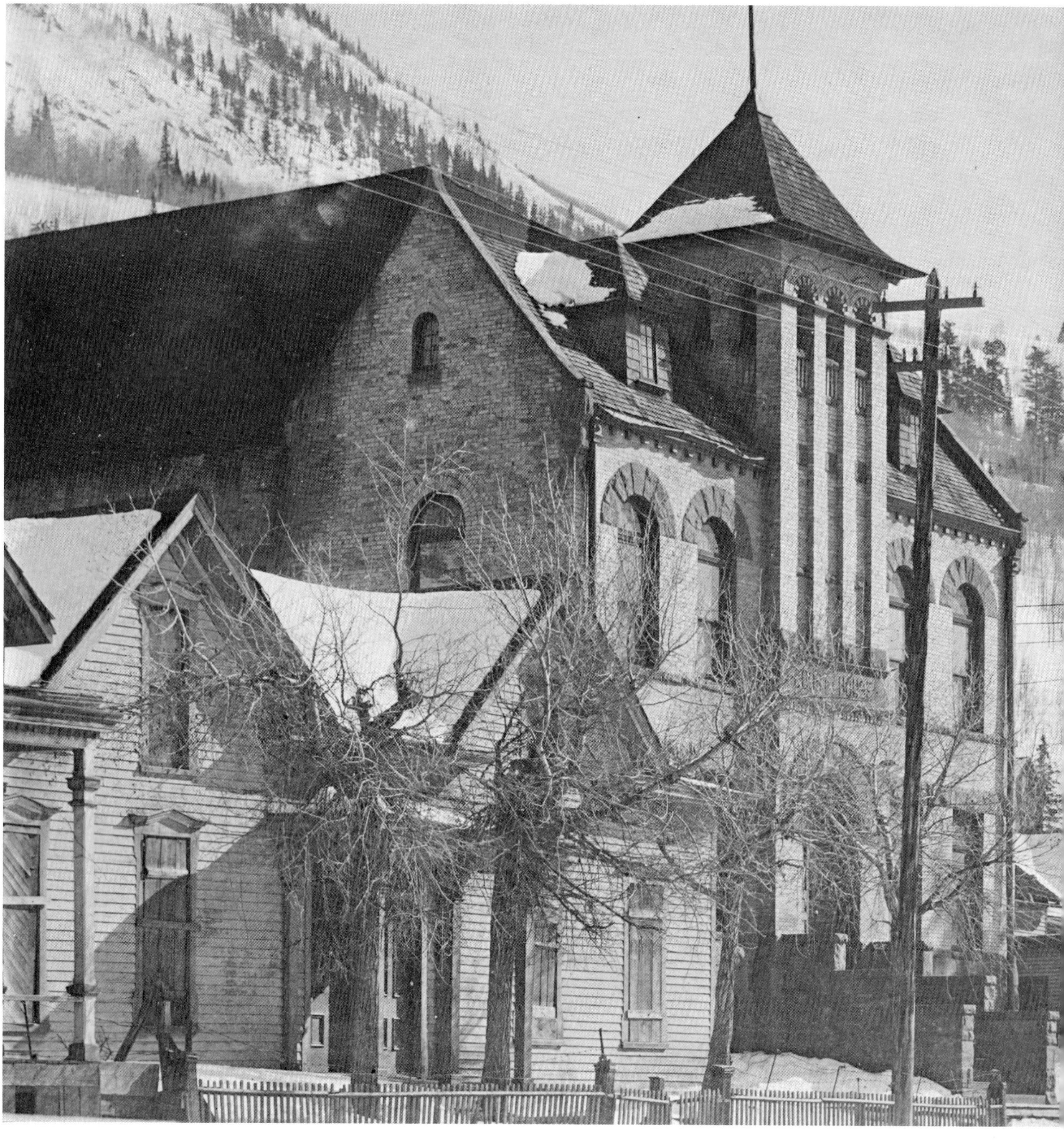

CHARLES MILLER ENGEL COLLECTION –
COURTESY RACHEL A. HARDWICK

THE FIRST PHOTOGRAPH of the new Dolores County Courthouse, which was built in 1893, appears on Page 180. In this mid-winter view, taken during 1915, an ore bucket was hanging from the "wire rope" (as mining men referred to tramway cable), next to one of the Pro Patria aerial-tramway towers. This tram line had been placed directly in the middle of Mantz Avenue. The west and north walls of the Dolores County Courthouse are shown here, along with the duplex home on the adjoining lot. Notice that the windows of the duplex were boarded up, and the

trees in the front yard had been topped so their branches would not interfere with the overhead electric lines near the sidewalk. An ice-skating pond had been created in front of the courthouse entrance, making a fun place for Rico children to enjoy sledding or skating.

COUNTY COURT PROCEEDINGS IN THE RICO DEMOCRAT...

August 25, 1891—A change of venue has been granted in the case of the People vs. John McCullough, Adam Embling and Walter Brown, charged with conspiracy and malicious mischief in blowing up the building of Grigsby & Hall last spring. The case will be tried at the next term of the district court for San Juan County [in Silverton].

—The case of Charles McCullough, et al, which was taken to Silverton on change of venue, came up for trial on Tuesday. After waiting for more evidence most of the day, the prosecution entered a *nolle pros*, and the case was thrown out of court.

September 18, 1891—The A. P. Adams Hose Company did good work last Tuesday at the jail fire. They had two streams of water running at full pressure three minutes after the alarm.

—Acting Coroner Justice McLaughlin held an inquest on the body of Sullivan, the man burned in the jail fire on Tuesday evening. The jury returned a verdict of "Death caused by suffocation in one of the cells in the county jail; cause of fire unknown."

December 19, 1891—A steel cage, manufactured by the Detroit (Mich.) Safe Co., arrived Thursday, and was neatly put in place by Geo. Tabor and one or two assistants. The cage is 5 x 7 feet in size, and it is thought of ample strength for the safe-keeping of the most desperate criminal. The cost of this much-needed contrivance, laid down in Rico, was $375.00, which does not include the cost of erection.

—A moment's gaze at Dolores County's new steel cage will probably induce the reformation of law breakers.

December 26, 1891—I. N. Simpson, junior member of the firm of Morgan & Simpson, proprietors of the Rico Bottling Works, was arrested the first of the week at the instance of the town board, charged with selling beer without a wholesale license. The trial occurred on Tuesday before Justice McLaughlin. W. A. Wallace conducted the defense, and H. W. Strickler represented the prosecution. The [Colorado] state law makes no provision for wholesale licenses, and as the city ordinance failed to cover the ground sought to be protected by the town, defendant was discharged, and the costs of the suit fell upon the town.

NOTE: Above articles refer to happenings in the original courthouse.

MONTE BALLOUGH PHOTO – LARRY PLEASANT COLLECTION

WHAT COULD BE FINER than a ride through Rico on a pleasant summer day? This group of Rico's finest church ladies had mounted their burros and horses for a ride about the town on a Sunday afternoon. The Dolores County Courthouse stands behind them. Sun bonnets kept their complexions white, while garter hose (candy-striped on the lady in the foreground) provided protection against mosquito bites.

THE BRICK COURTHOUSE for Dolores County was constructed in Rico during 1892, facing on Commercial Street. The Dolores County seat was moved to the town of Dove Creek in 1944, due to the lack of population in Rico. Mantz Avenue is at the right, with a view of the Congregational Church up the block. The duplex home (at the left) occupied Lots 23 and 24 in Block 15 of the Sanborn map.

CHARLES MILLER ENGEL COLLECTION
– COURTESY RACHEL A. HARDWICK

THE DOLORES COUNTY COURTHOUSE, in Block 15, is the subject of this commercial photograph, taken at the intersection of Commercial Street (at the left) and Mantz Avenue (in the foreground). This was another view taken during 1915, and it is valuable due to the fact that the houses at the left are now gone.

CHARLES MILLER ENGEL COLLECTION
– COURTESY RACHEL A. HARDWICK

Rico, Colo.

CHARLES MILLER ENGEL COLLECTION –
COURTESY RACHEL A. HARDWICK

THIS PUBLIC SCHOOLHOUSE in Rico was built during 1892, and it replaced the small, cozy frame structure built in the 1880's. The new building fronted on Commercial Street and contained four classrooms. This was the winter of 1918, when log cribbing protected the schoolyard by preventing soil from sliding into Silver Creek during its occasional rampages. No inside plumbing was provided at the school; however, a teacher's outhouse was over the embankment (at the left), and the boys and girls had separate entrances on their outhouse (at the right). The school building was razed during the late 1940's or early 1950's, apparently due to pressure applied by the mine owners who ran a tunnel beneath the schoolhouse. This bore was from the Atlantic Cable Tunnel to the Van Winkle shaft.

ONE OF THE FIRST close-up photographs of the school kids in Rico was taken on March 27, 1903. The setting was at the front entrance of the school, which was located on Commercial Street, in Block 21. It was to the north of Silver Creek and Soda Street.

CHARLES MILLER ENGEL COLLECTION
– COURTESY RACHEL A. HARDWICK

FROM THE RICO DEMOCRAT:

August 7, 1891—The school board have purchased ground upon which to build, and before the season closes we will have a school building that will be a credit to the town.

—Slow progress is made in getting the new school house underway. We understand the delay is with the architects who have not yet furnished the plans.

August 14, 1891—George L. Sheldon agent for the Union School Furniture Company, of Battle Creek, Michigan, has closed a contract with the Rico school board to furnish the new school building with their goods. – The plans for the Rico school house are found to be too elaborate for the amount of money available for building, and at present it seems not unlikely that the construction will be deferred for some time.

August 28, 1891—Our public schools commenced Monday. Mrs. Quigley in charge of the high school, has 30 scholars, and Miss Kilpatrick in charge of the primary department with 50.

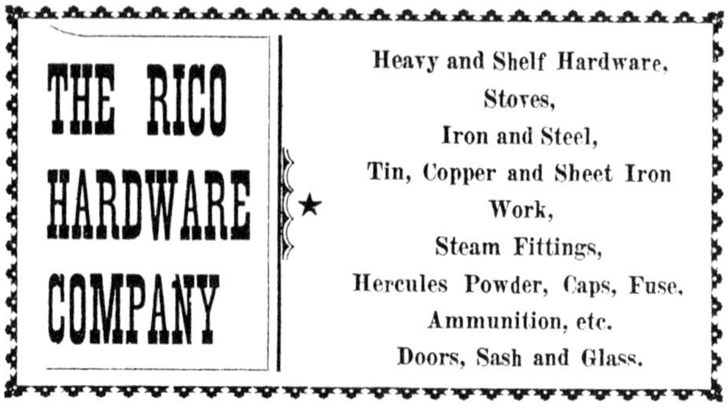

COURTESY MARGARET MATZICK

CHARLES J. LEE PHOTO –
NORMA L. BUTLER COLLECTION

RICO SCHOOL CHILDREN were on an outing near Rico, and they paused in their journey for a rest stop. The boy in front carried a back pack for their lunch, which they enjoyed beside the creek. The young lady at the left appears mature enough to have been one of the teachers.

Rico, Colo, 1914-15

CHARLES MILLER ENGEL COLLECTION – COURTESY RACHEL A. HARDWICK

THIS RICO CLASS was photographed during 1912, and the student's names are listed below (from left to right). Back Row: George Stanley, Frank Pinkerton, Jim Pinkerton, Charles Lee and George Hicks. Third Row: Grace Lincoln, Mary Columbo, Angela Lafquist, Edna Clark, Julia Knoll and Vera Krantz. Second Row: Frank Knoll, Charles Brittian (sticking his tongue out), Charles Engel, Gladys Epperson, Edna Corn and Leland Anderson. First Row: Vinton Hay, Vergie Crandall, Laurella Moore, Blanche Lincoln, Mabel Clark and Ramona Anderson.

DAVID C. BUTLER COLLECTION

RICO'S CLASS OF 1914–1915 was photographed at the west, or back wall, of the public school building. The teachers probably were the young ladies at the left (in the top and middle rows). A coal-chute door had been built along the basement wall, at the right. Steam provided heat, as long as the janitor shoveled enough coal into the furnace.

WITH CHEERLEADERS like these girls at the Rico Public School, the boys must have won every game. This snapshot was taken during the years 1935–1937. From the left, the girls are: Ivabelle Speer, Helen Drummond, Emma Sturdevant, Marge O'Driscoll and Ann Heidt (Ferando).

ANN AND JAMES FERANDO PHOTO

CHARLES MILLER ENGEL COLLECTION – COURTESY RACHEL A. HARDWICK

THE LAST SCHOOLHOUSE to be constructed in Rico is used today as the Rico Visitor Center. It was built in 1953 as the public grade school. Older students were bused to Dolores.

THE STORY OF RICO AND THE PIONEER MINING DISTRICT

By William A. Graves

(Continued From Page 254)

1902 David Swickhimer's Pro Patria Company also was interested in the new type of separators, and a new concentration mill was built to use them. This mill was located at the foot of Mantz Avenue, and it was connected to the mine portal by a 3,600-foot-long aerial tramway. The RGS built a spur to serve the new Pro Patria (Spanish for "For Fatherland") mill. Although production at the new mill was started late in 1902, the milling results were not satisfactory, and the mill was little used until 1908.

1903 Rico's only major fire burned the Hotel Enterprise to the ground in 1903. This hotel was a three-story building, and it measured about 60 feet in width and 75 feet in length. The grand Hotel Enterprise was the largest hostelry in town, and it had 62 rooms, not counting rooms in the Hotel Annex or other adjacent hotel buildings. The Enterprise had been the host hotel for the big banquet that celebrated the arrival of the RGS in Rico 12 years earlier. Unfortunately, this historic hotel was never rebuilt after the fire. Several smaller buildings nearby also caught fire and burned down at the same time. (See the complete story of the disastrous Hotel Enterprise fire in THE RICO NEWS article printed elsewhere in this volume.)

The Hotel Enterprise had been a favorite meeting place for Rico residents, and its loss lowered the town's morale. Ore production was down, the town's population was down, and over 60 percent of Rico's commercial businesses had closed their doors—and then, the Hotel Enterprise burned down.

1906 Rico's economy began to depend more on the production of base metals. For the first time, the annual value of zinc production was greater than the value of silver output.

1907 The yearly value of metal production in the Rico area was only $56,283, and it was the lowest annual amount since 1880.

David Swickhimer's Rico State Bank failed as a result of the severe "Panic of 1907." Ignoring the advise of business associates, Swickhimer used his personal fortune to pay off the bank's depositors, a loss that financially ruined him. He was called Colorado's "most honest man" as a result of using all his wealth to pay back the Rico State Bank's depositors. Following David Swickhimer's bankruptcy, the United Rico Mines obtained a controlling interest in the Pro Patria property; however, this company went into bankruptcy itself four years later.

1908 The Pro Patria mill started production again and helped to boost the output of lead and zinc concentrates in the district. However, the mill was destroyed by fire during October of 1908, and it remained inactive for the next five years.

1909 Heavy rains during September of 1909 caused the Trout Lake dam to fail, and the San Miguel River washed out a lot of the Rio Grande Southern's track. This kept Telluride isolated for months, and it meant that through train service was interrupted.

1910 This was another slow year for Rico. The town lost about half of its population during the 10-year period between 1900 and 1910. In the same period, ore production also dropped about one half.

(Continued on Page 333)

GOLD FILIGREE PASS FROM OTTO MEARS
MORRIS ABBOTT COLLECTION – COURTESY E. J. HALEY

THE PRO PATRIA MILL

By Gabe Bradford

RICO'S LANDMARK MILL...

"How do I get to the Pro Patria mill?"

No. 7

BACK IN THE GOOD OLD DAYS, any resident of Rico, Colorado, would have replied… "Go down Glasgow Avenue, then west on Mantz Avenue… You can't miss it." Completed by the end of 1902, this mill was the collection point for ore mined on the western slope of Dolores Mountain, east of the little mining town of Rico. [Many of the] mines on the mountain were owned by David Swickhimer, a banker, and one of Rico's leading citizens. Dave Swickhimer continued his activity in the mining industry with the digging of the Swickhimer Tunnel—later known as the Pro Patria Tunnel. [Dave obviously named his mining venture in a patriotic gesture, since "Pro Patria" means "For Fatherland" in Spanish.]

The Pro Patria Tunnel tied together most of the upper-slope workings on Dolores Mountain. From this central point, the ore was carried by an aerial tramway down the mountain and through Rico to the Pro Patria mill. Unfortunately for Mr. Swickhimer, his Rico State Bank failed during the Panic of 1907, and he was forced to liquidate all his mining interests to repay his depositors.

The United Rico Mines Company, established in 1901 and headquartered in New York City, purchased the entire Pro Patria operation from Mr. Swickhimer. United Rico Mines Company already operated another mill in Rico, so the Pro Patria became its second concentrating plant in Rico. Unfortunately, this acquisition added further to the United Rico's problems, and the company was forced into bankruptcy in 1911. During 1912, the company was reorganized as the Rico Mining Company.

In 1908, one year after the financial panic, the Pro Patria mill was damaged in a fire. By the time United Rico's successor, the Rico Mining Company, finished remodeling the mill, it operated at capacity for only one year—1913. However, as it turned out, 1913 was a banner year for metals' production, exceeding the best years of the 1890's.

MONTE BALLOUGH PHOTO – COURTESY MARGARET BALLOUGH PALMER

MONTE BALLOUGH, a resident photographer in Rico, probably stood on the Enterprise Mine's tailings dump to focus his camera lens across Rico during the winter of 1902. Elliot Mountain was covered with snow (on the center skyline), while Sandstone Mountain slopes off to the right.

THE SAME PHOTOGRAPH has been enlarged and reproduced on the following three pages to show changes that were taking place in Rico. At the right, the RGS station grounds had a number of gondolas and boxcars spotted on yard tracks. The roadmaster's office was inside the old RPO car body, sitting at a right angle to the depot—where a locomotive was emitting steam. You can see that a room extended off the back of the wooden enginehouse. Houses in south Rico are in view (below).

YOUR WINTER TRIP

To

CALIFORNIA AND
THE PACIFIC COAST

Will Be Most Enjoyable If You Use

The

Denver & Rio Grande

"Scenic Line of the World"

THROUGH STANDARD AND TOURIST
SLEEPING CARS WITHOUT CHANGE

SUPERB DINING CARS--SERVICE, a la
CARTE--ON ALL THROUGH TRAINS

For further information address

S. K. HOOPER, General Passenger and Ticket Agent, - - - Denver, Colorado

COURTESY MARGARET MATZICK
FROM "THE RICO ITEM," NOVEMBER 26, 1909

The recovery did not last—and after just two years of operation, the Rico Mining Company filed for bankruptcy. Nevertheless, during its two years of operating the Pro Patria, the Rico Mining Company assembled the largest group of Rico mining claims ever held by one company.

(Continued on Page 270)

THIS TWO-PAGE SPREAD shows uptown Rico, with store fronts on the west side of Glasgow Avenue in view. The south wall and front of the Hotel Enterprise appears along this main street. The newly constructed Pro Patria mill, with white-framed windows, stands at the end of Mantz Avenue. The industrial section of Rico had reached its full development. Distant houses line Riverside Street in neighboring Piedmont, visible at the top of this view, beyond the Dolores River. The Rico Power & Light Company had constructed a wooden flume that ran across the eastern face of Expectation Mountain. The inlet for this flume was at Burns, upstream on the Dolores River, about 2.5 miles north of Rico. The trestle crossing the river carried water from this flume into the hydroelectric plant, located adjacent to the Pro Patria mill (to the north). At the far right is a glimpse of the Grand View Smelter, beyond the Dolores County Courthouse.

MONTE BALLOUGH PHOTO – COURTESY MARGARET BALLOUGH PALMER

RICO WAS PHOTOGRAPHED again during the winter of 1902, shortly after the Pro Patria mill had been constructed. Monte shot two frames to cover the town from north to south.

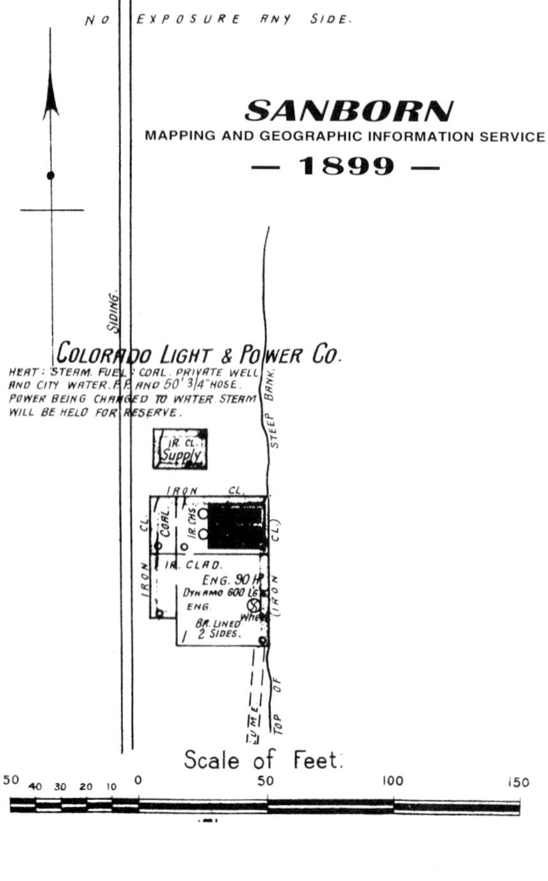

Early in 1916, the Syndicate Mining & Milling Company bought the interests of the Rico Mining Company. During this period, they remodeled the power plant that served the mill. This coal-fired plant and its successors were called "lighting plants," even though they generated electricity primarily for the mills. As life continually teaches, timing is everything. In 1915, silver was quoted at 50 cents per ounce. Then, during World War I, the price of not only silver, but zinc, lead and copper, all climbed—with silver hitting $1.00 per ounce by 1918. Unfortunately for the local workers and the com-

(Continued on Page 289)

AN ENLARGEMENT of this same two-part picture has been produced to enhance detail, and it is printed on the next six pages. The first page reveals the Sampling Works, with an RGS stub spur near the front door (shown at the lower left). A string of coal chutes had been constructed as a long building next to the siding at the lower left. The following page offers a splendid study of houses fronting on River Street. Notice the extensive addition that was built across the alley from the Hotel Enterprise, with bridge connections.

THE NEXT TWO PAGES of these illustrations contain the newly constructed Pro Patria mill, where a flatcar load of lumber and an RGS boxcar had been spotted on the spur at the west wall of the outer building. Notice that the mill had been neatly painted, probably with oxide red and white trim. Mantz Avenue was designed to be 80 feet wide, perhaps with the idea that wooden towers would someday be built to carry an aerial tramway up to the Pro Patria tram house. The Dolores County Courthouse stands out prominently, at the intersection of Mantz Avenue and Commercial Street. The RGS water tank is in view (at the lower right).

THE LAST TWO PAGES of this display center on the RGS depot, painted in its original darker paint scheme. The roadmaster's office was in the old ex-D&RG RPO car, set on the ground at right angles to the depot. A string of gondolas, loaded with large chunks of coal, stood near the old car body. These coal cars were the old low-side gondolas. South Rico contained a number of larger houses, located below the Enterprise Mine. A flume had been constructed that ran almost on the level below the Enterprise Mine. This was the flume from Silver Creek that appears in the picture taken of the curved trestle (Bridge 66-AS) from the Iron Mine dump. This flume also appears on Pages 196–197, in Chapter 6. Bridge 66-AS carried the Enterprise Branch across Silver Creek from Nigger Baby Hill to Dolores Mountain.

MONTE BALLOUGH PHOTO – COURTESY MARGARET BALLOUGH PALMER

RICO

COLO,
TUDE 8720
"BALLOUGH" PHOTO
02

STATION MAP OF RICO
1902
DRAWN BY JIM KEY

3. SAMPLING WORKS
4. RICO P&L WAREHOUSE
5. S. J. HOOD LUMBER YARD
7. COAL SHED
16. RGS WATER TANK
17. RGS DEPOT
21. COAL GONDOLA
22. SAND HOUSE
23. COAL DOCK (FOR ENGINES)
24. FIRST ENGINE HOUSE (WOODEN)
25. SECTION HOUSE
26. BUNKHOUSE
27. COAL SHED
28. WOODEN BOARDWALK
34. RPO BODY FOR ROADMASTER
37. DEPOT OUTHOUSE
39. TRACK SCALE
40. SECOND WYE
41. RGS BRIDGE 67-AS
42. RGS BRIDGE 67-ASS
45. LOCATION OF FIRST WYE
46. RICO SMELTING CO. (ADAMS)
47. PASADENA REDUCTION CO. (TRACT)

JACKSON & SHARP PHOTO – E. J. HALEY COLLECTION

THE ROADMASTER'S OFFICE was set up in a railroad car similar to the one in this picture of a Denver & Rio Grande Railway RPO-baggage car, first used in the early 1870's. This car was built at the Jackson & Sharp Car Builders in Wilmington, Delaware, in 1871. Perhaps this car was obtained after the old car had been abandoned for some 25 years. No doubt it was hauled to Rico on a flatcar. One end of the car featured a platform door. A similar car body rests in Silverton, Colorado, the Silverton Railroad's baggage-express car purchased by Otto Mears, ca. 1891. This car is 8 feet-1 inch wide, the roof is 30 feet long, and the body is 26 feet long.

WALL ST SUMMARY. JULY 2, 1907.

NOTICE

THE NEW YORK OFFICES AND TRANSFER DEPARTMENT OF

THE UNITED RICO MINES CO

ARE LOCATED AT

9 AND 11 WALL STREET (ROOMS 19 AND 20)

REGISTRAR

NEW YORK TRUST COMPANY, 24 BROAD ST., NEW YORK

FOR INFORMATION REGARDING

THE UNITED RICO MINES CO.

OR SUBSIDIARY COMPANIES, FORMERLY KNOWN AS

The Rico Smelting and Refining Co.
The Colorado Milling & Concentrating Co.
The Rico Mining & Milling Co.
The Atlantic Cable Consol'd Mining Co.
The Rico-Aspen Consol'd Mining Co.
The Enterprise Mining Co.
The Futurity Mining Co.
The Grand View Mining Co.
The The Black Hawk Mining Co.
The Hope & Cross, and others.

Apply at Offices of the Company at Above Address.

LISTING OF UNITED-RICO MINES COMPANY'S SUBSIDIARY MINING FIRMS

MARGARET MATZICK COLLECTION

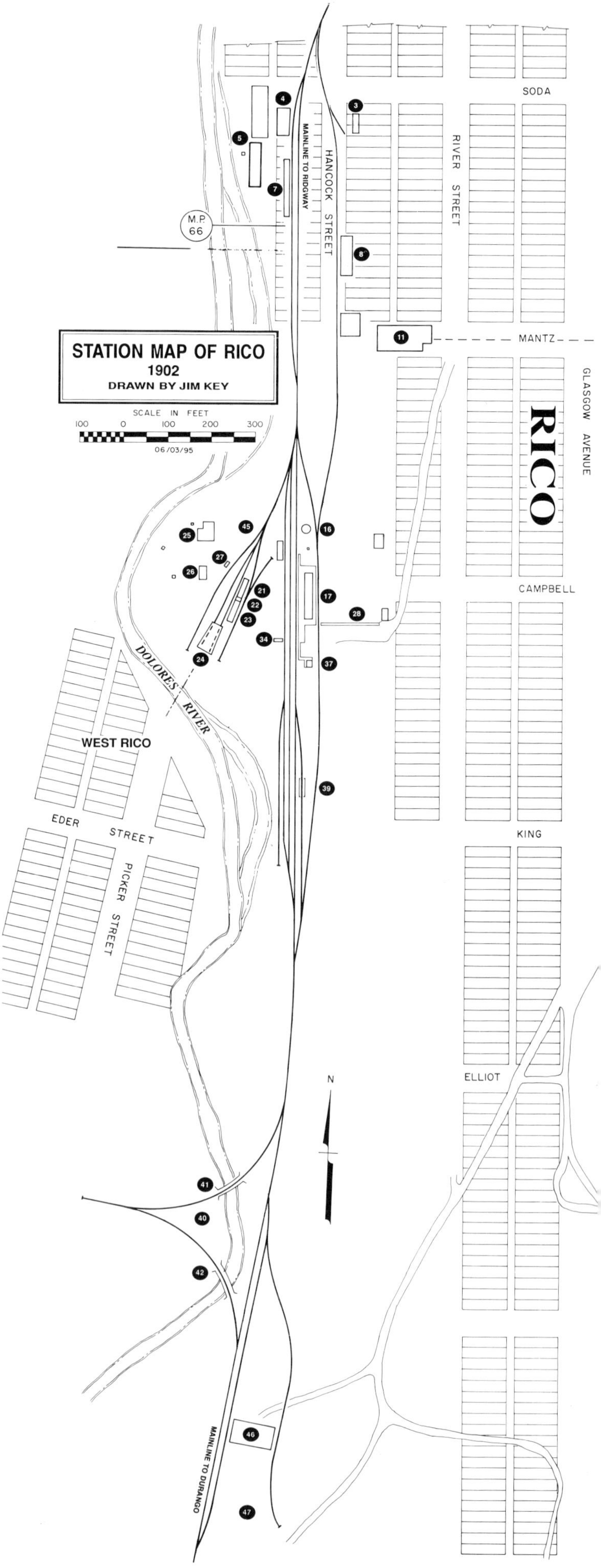

MONTE BALLOUGH PHOTO – MARGARET MATZICK COLLECTION

THE UNITED-RICO MINES Company called upon Monte Ballough to use his panoramic camera to photograph the Rico landscape from north to south. Supposedly, this view was produced in the spring of 1902.

OVERLEAF: This panoramic view has been reduced slightly from the original, which measured 36 inches in width. It has been reproduced on the next four pages. Homes lined Riverside Street in the "suburb" of Piedmont, located toward the northwest of Rico (at the far left). The Grand View Mining & Smelting Company's buildings had reached their maximum extent by this time. The old funicular railroad's scar is visible, where it climbed Nigger Baby Hill to the Grand View Mine. Industrial buildings that warehoused building materials lined the RGS spur at the northern end of Rico. Also included is a large unnamed mining company building, located beside the spur that led to the Sampling Works. A string of work cars were on the coal-chutes siding, just south of the shed. The Colorado Light & Power Company's powerhouse (with a tall smokestack) had a penstock that carried water across the Dolores River on a trestle. Since the Rico Power & Light Company's powerhouse has a tall stack, it had been equipped with a steam boiler that could be put into service to operate a generator and produce power when water was not available from the Dolores River. The Pro Patria building, next to the RGS mill spur, had received a roof change, with ventilator cupolas placed on top. The aerial tramway was 3,600 feet long from the mill to the Pro Patria Mine, visible at the dump on Dolores Mountain. The United-Rico Mines firm acquired other major ore producers in the Rico district, including the Atlantic Cable, Cross, Enterprise, Grand View and Hope, as well as the Rico-Aspen.

THE LAST TWO PAGES of this broad panoramic view show most of the company holdings across Newman Hill, above the south end of Rico. The RGS station grounds laid within the mineral rights of the town, which is shown by the border lines on the picture. Notice how the Dolores River, crossed by a bridge that led to West Rico, bends around the station grounds. A hospital, complete with nurses' quarters, had been built in this area (at the far right). The second RGS wye crossed the Dolores River, visible at the far right. The Rico Sampling Works occupied space beside a spur located next to the RGS mainline leading southward toward Dolores and Durango.

A SMALL SECTION of Monte Ballough's panoramic photograph was enlarged to 300 percent of its original size in order to provide details in the RGS station grounds, as shown on Pages 284–285. This scene was recorded on film during the spring of 1902. No noticeable activity was present, but a few boxcars lined the house track beyond the depot. An unidentified shed occupied space in front of rotary snowplow No. 2. Alongside No. 2's tender what appears to be a gondola containing lump coal had been placed on ties. The sand house sat south of the gondola, and farther to the right is what is believed to be a coal dock used for fueling locomotives. The original wooden enginehouse is farther to the right. Between the coal dock and depot outhouse stands the old RPO car body, in use by the roadmaster. In the foreground (at the left) the larger building (with a kitchen lean-to) housed the section foreman and his family. A large bunkhouse was located to the right. Both of these buildings were quite close to the Dolores River. Residences ranging from shacks to three-story houses had been located on top of the embankment east of (and above) the RGS depot and yard.

MONTE BALLOUGH PHOTO – COURTESY MARGARET BALLOUGH PALMER

U.R.M.Co
Properties
of
The United Rico
Rico, Colo

Rico

FROM THE RICO NEWS

Rico, Colorado, October 24, 1903

GREATEST FIRE IN RICO'S HISTORY

HOTEL ENTERPRISE A TOTAL LOSS

Loss of Over $53,000 Covered by Insurance of Only $7,500

ON TUESDAY evening last, at about 7:00 o'clock, a vast cloud of black smoke was discovered arising from the rear portion of the third story of the Hotel Enterprise, and an alarm of fire was sounded by the bell at the hose house and the whistle at the Atlantic Cable shaft almost simultaneously.

The volunteer firemen responded promptly, but before the first line of hose could be laid, the blaze was breaking through the roof, and within a few minutes thereafter, every window on the second and third stories was belching forth a mass of flame, and it became evident at once that there was no possibility of saving the building, and the firemen were confined to keeping the fire as low as possible and preventing it from spreading to surrounding buildings, and this they succeeded in doing for about 30 minutes, by which time the heat of the fire had become very intense and spread through the connections to the kitchen, which was a frame building across the alley. As soon as this building, which was a two-story structure and very inflammable, caught fire, it was evident that others would go, and in fact, it was thought that it would be impossible to save a single building on either side of the alley south of the hotel, and had there been any breeze other than that caused by the flames, this would have undoubtedly proven to have been the case.

The flames spread from the hotel kitchen to the rear of Ed McGill's Saloon, then to the hotel cottage, Hall's Ice House, the Cranston cottage and stables, the McGill cottage and those in the rear of the Gault Meat Market... [caught fire] in quick succession, notwithstanding that two streams of water under a pressure of 120 pounds were kept playing on them constantly. In the meantime, the flames had spread to a one-story building north of the hotel owned by G. T. Myers, and it was soon made a wreck by the flames, notwithstanding it was protected by the brick wall of the hotel and Mr. Myers kept a constant stream of water from a small hose playing on it. The United Rico Mines Company's building, the old **NEWS** office and C. F. Middaugh's building across Glasgow Avenue commenced to smoke, and the paint was blistering from the intense heat, and had the water not been turned upon them at once, there is no doubt but that they would have caught afire, and as it was, nearly all the glass was broken or cracked in the United Rico Mines Company building. The north wall of the Hotel Enterprise was the first to fall, it being only a brick veneering, but it helped to smother the flames, and was quickly followed by a portion of the south wall, which was of stone and 18 inches thick, which fell on McGill's building, crushing in the burning rafters.

During the time the firemen were fighting the fire in the main street, under supervision of Chief John Shaver of the M&S Hose Company, and in the alley, under the command of Chief George E. Moore of the A. P. Adams Fire Company, Bert Blodgett was doing some good work with the hose from the Colorado Electric Light & Power Company's plant, and to him must be given the credit of saving numerous cottages on River Street, as well as the livery stable and Pro Patria mill. The company's water plant is entirely separate from that of the city, and has a pressure nearly as great. Theo. Barlow, G. E. Smith and Arthur Story kept the fire brands from igniting buildings south of Mantz Avenue by the use of small garden hose, and the bucket brigade did good work in many instances.

As soon as the alarm was given, a score or more of men rushed to the hotel and carried the chairs out from the office and a piano out from the hall, which was just being opened for the evening by Hall's Comedy Company, and assisted others in removing the stock and fixtures from the bar room, which was located in the south lower room of the brick portion of the building, and from the cellar beneath it.

Owing to the fact that he thought the heavy stone wall of the Enterprise would protect his building, Ed McGill refused to allow any of his goods or fixtures to be removed until fully 30 minutes after the fire started, and of consequence, [he] lost a great many of his fixtures and all the liquors stored in the cellar.

A portion of the stock of T. R. Williams, all of J. D. Diehl's stock not in the vault, and the furniture from Mrs. Rohde's rooming house in the Bank of Rico building, was moved into the street on Mantz Avenue, and the household effects were removed from Mrs. Garren's rooming house on River Street and from several cottages in close proximity.

The fire was under complete control at about 10:30 o'clock, after having damaged property to the extent of $53,125... divided as follows:

Hotel Enterprise Company	$45,000
Ed McGill Building	5,000
J. M. Hall's Ice House	1,000
Judson Cranson residence	1,000
G. T. Myers	500
John Gault's Stables	100
Gasser & Schendelmeyer building	25
Damage to other buildings by fire and breakage of glass	500
Total	$53,125

MARGARET MATZICK COLLECTION

THE GAUNT SKELETON of what once had been Rico's finest hostelry, the Hotel Enterprise, jutted up from among its smaller neighbors. All that remained after the devastating fire in 1903 were one tall chimney and most of one masonry wall. The photograph probably dates from just after the fire. A shanty town had been built up at the rear of the Rico Bottling Works, which shared the same spur with the Sampling Works (the small building to the right). The modest houses that lined River Street can be seen in the foreground, this side of the hotel ruins.

All the Mining and Local News of the District appears in these columns, up-to-date and authentic. Keep posted.

THE RICO NEWS

DEVOTED TO THE MINING AND MATERIAL INTERESTS OF DOLORES COUNTY

Mines can be Made in this District at a minimum of cost, because the rock is soft, timber handy and power cheap.

VOLUME 25. RICO, DOLORES COUNTY, COLORADO, SATURDAY, OCTOBER 24, 1903. NUMBER 7

D. SWICKHIMER, President. J. H. GOODKNIGHT, Asst. Cashier.

THE RICO STATE BANK

Interest allowed on time deposits. Sell drafts payable in any part of the world. Fire and burglar proof safety deposit boxes for rent. We solicit your patronage.

PAID IN CAPITAL, $50,000.00

PEACHES

Pears, Apricots, Apples, Grapes, Plums, Etc., for canning and preserving. Fruit is ripe and prices are right.

Koster's Market

FREE DELIVERY

City Livery

Fine Turnouts AND Saddle Horses.

V. H. LEE & Co., Proprietors.

DEALERS IN

HAY, GRAIN, COAL, LIME AND LUMBER

Ore Hauling, Transfer and Express.

BUSSES - MEET - ALL - TRAINS.

Leonard Andeway. S. O. Burgwin.

Andeway & Burgwin

— DEALERS IN —

Dry Goods, Groceries

Gents' Furnishings

Boots, Shoes, Etc.

GENERAL OUTFITTING SUPPLIES

FOR

Miners and Campers

DUNTON - - - COLORADO.

PERSONAL AND LOCAL

Mayor Rider returned from Durango Sunday afternoon.

Dr. McNeill returned to Dunton, Wednesday morning.

Rev. Carney returned from a visit to Dunton this morning.

Additional local news will be found on the fourth page.

J. C. Williamson was an arrival from Ophir, yesterday morning.

John Dennison returned from a week's visit in the valley, Thursday.

Richard Keller was a south-bound passenger yesterday morning.

Mrs. Rohde lost a silver syrup pitcher during the fire. Finder will please return.

James Lenton purchased all of the piping and metal in the ruins of the Hotel Enterprise.

Roy Lewis came up from Red Rock the first of the week and is the guest of his father, George Lewis.

Mike Schendelmeyer came over from Telluride yesterday morning to size up what the fire had done for him.

The delinquent tax list for the years 1901 and 1902 will be found on the second and third pages of this issue.

J. D. Burgwin, Leon Eggers and Jos. Samide were those from the West Fork who were doing jury duty in this city the first of the week.

Leonard Andeway, of the Dunton firm of Andeway & Burgwin, came over Thursday afternoon and returned to his town yesterday morning.

Uncle John Eder came over from his Fish creek ranch Wednesday to view the remains of Rico's hotel. He returned home yesterday morning.

The city council should see that the fire department is furnished with axes and lanterns at once. And another thing, stop the loaning of fire aparatus.

Mrs. Reik and daughter, Mrs. Joseph A. Irving, mother and sister respectively of Mrs. W. C. Carley, arrived Tuesday last and are guests at the Carley home on Soda street.

Jim Hall's bill file was carried out with other stuff from the Hotel Bar last Tuesday night. It is supposed to be in a galvanized bucket. He would like it returned, if found.

Allen Moore departed for Denver, Sunday afternoon, to meet a brother and sister, and arrived back in Rico Thursday morning. This is what is generally termed a flying trip.

J. B. Blair and wife, of Kansas City, Mo., were visitors in this camp last Sunday. Mrs. Blair is a daughter of the late John J. Graham, and was here to look up matters regarding any estate her father might have left. Mr. Blair is superintendent of terminals for the C., R. I. & P. railway at Kansas City.

The president has issued a proclamation calling the fifty-eight congress in extraordinary session on November 9 at 12 o'clock. The proclamation states that the purpose of the session is to consider a commercial convention between the United States and Cuba which requires the approval of Congress.

District court convened Monday at 10 o'clock and adjourned at 9:15 Tuesday morning. Those present from outside points were: Judge J. L. Russell, Stenographer Mary I. Mesecre, Attorneys H. M. Hogg, S. R. Fitzgarrald, B. F. Ritter and Reese McCloskey. The cases disposed of consisted of a judgment in favor of the Rico Hardware company against the Pro Patria Mining and Milling company; a judgment in favor of the Industrial Building and Loan association against Dr. F. A. McNeill; the refusing of a new trial in the case of Pringle vs. Ditchfield, and the dismissal of the cases against Frank Worthing.

RED SEAL LYE

Is not like old-style concentrated lye. It is granulated; comes in handy *sifting-top* cans, and comes in handy for every house-hold use.

DISINFECTANT

CLEANSER,

WATER SOFTENER

It not only removes the dirt you can see, but dirt you can't see—the dangerous dirt. Kills all germs. Keeps off contagious disease.

SOLD EVERYWHERE.

Insist on getting RED SEAL Lye

P. C. TOMSON & CO.

Philadelphia, Pa.

GREATEST FIRE IN RICO'S HISTORY

Loss of Over $53,000 Covered By Insurance of Only $7,500.

On Tuesday evening last, at about 7 o'clock, a vast cloud of black smoke was discovered arising from the rear portion of the third story of the Hotel Enterprise, and an alarm of fire was sounded by the bell at the hose house and the whistle at the Atlantic Cable shaft almost simultaneously.

The volunteer firemen responded promptly, but before the first line of hose could be laid the blaze was breaking through the roof and within a few minutes thereafter every window on the north and west sides of the building on the second and third stories was belching forth a mass of flame, and it became evident at once that there was no possibility of saving the building, and the firemen were confined to keeping the fire as low as possible and preventing it from spreading to surrounding buildings, and this they succeeded in doing for about 30 minutes, by which time the heat of the fire had become very intense and spread through the connections to the kitchen, which was a frame building across the alley. As soon as this building, which was a two story structure and very inflamable, caught fire, it was evident that others would go, and in fact, it was thought that it would be impossible to save a single building on either side of the alley south of the hotel, and had there been any breeze other than that caused by the flames this would have undoubtedly proven to have been the case.

The flames spread from the hotel kitchen to the rear of Ed McGill's saloon, then to the Hotel cottage, Hall's ice house, the Cranston cottage and stables, the McGill cottage and to the stables in the rear of the Gault cottage and those in the rear of the Gault meat market stand in quick succession, notwithstanding that two streams of water under a pressure of 120 pounds were kept playing on them constantly. In the meantime the flames had spread to a one-story building north of the hotel owned by G. T. Myers and it was soon made a wreck by the flames, notwithstanding it was protected by the brick wall of the hotel and Mr. Myers kept a constant stream of water from a small hose playing on it. The United Rico Mines company's building, the old NEWS office and C. F. Middaugh's building across Glasgow avenue commenced to smoke and the paint was blistering from the intense heat and had the water not been turned upon them at once, there is no doubt but that they would have caught afire, and as it was, nearly all the glass was broken in or cracked in the United Rico Mines company building. The north wall of the Hotel Enterprise was the first to fall, it being only a brick veneering, but it helped to smother the flames, and was quickly followed by a portion of the south wall, which was of stone and 18 inches thick, which fell on McGill's building, crushing in the burning rafters.

During the time the firemen were fighting the fire in the main street under supervision of Chief John Shaver, of the M. & S. Hose company, and in the alley under command of Chief George E. Moore, of the A. P. Adams Fire company, Bert Blodgett was doing some good work with the hose from the Colorado Electric Light and Power company's plant, and to him must be given the credit of saving numerous cottages on river street as well as the livery stable and Pro Patria mill. The company's water plant is entirely separate from that of the city and has a pressure nearly as great. Theo. Barlow, G. E. Smith and Arthur Story kept the fire brands from igniting buildings south of Mantz avenue by the use of small garden hose and the bucket brigade did good work in many instances.

As soon as the alarm was given a score or more of men rushed to the hotel and carried the chairs out from the office and a piano out from the hall, which was just being opened for the evening by Hall's Comedy company, and assisted others in removing the stock and fixtures from the bar room, which was located in the south lower room of the brick portion of the building, and from the cellar beneath it.

Owing to the fact that he thought the heavy stone wall of the Enterprise would protect his building, Ed. McGill refused to allow any of his goods or fixtures to be removed until fully thirty minutes after the fire started and of consequence lost a great many of his fixtures and all the liquors stored in the cellar.

A portion of the stock of T. R. Williams, all of J. D. Diehl's stock not in the vault, and the furniture from Mrs. Rohde's rooming house in the Bank of Rico building, was moved into the street on Mantz avenue, and the household effects were removed from Mrs. Garren's rooming house on River street and from several cottages in close proximity.

The fire was under complete control at about 10:30 o'clock, after having damaged property to the extent of $53,000, which is divided as follows:

Hotel Enterprise company (insurance $7,150)	$45,000
Ed. McGill, building, fixtures, stock and cottage	5,000
J. M. Hall, ice house contents etc.	1,000
Judson Cranson, residence and contents (insurance $350)	1,000
G. T. Myers	500
John Gault, stables, etc.	100
Gasser & Schendelmeyer, building	25
Damage to other buildings by fire and breakage of glass	500
	$53,125

Ed. McGill's loss, which is given at $5,000, is complete, as he carried no insurance.

* * *

Mrs. Story kept open house for any one who desired to store goods or get warm.

* * *

Mike Schendelmeyer donated $5 to the fire boys for refreshments and says he is going to do more.

* * *

Robt. Gibbs worked hard then made it "free for all" who had an idea that they would catch a cold.

* * *

Andy Schrader had a little water works and hose of his own and put out several dangerous fire brands.

* * *

All of the cottages burned were in good condition and their loss will mean the repairing of other buildings.

* * *

Lee & Royce had their transfer teams out ready to move goods to a safe distance should the flames get beyond control.

* * *

T. R. Williams showed his appreciation of the good work done by the firemen by making each company a present of $25.

* * *

Mrs. J. D. Diehl and her female rescue corps came in for a good share of praise. Even our women, young ladies and children are "fire scrappers."

* * *

Old fixtures were carried out with the utmost care at several places and glassware and other breakables were thrown into the street for safe keeping.

* * *

John H. Goodknight, who was the only one occupying a room at the hotel, saved his trunk and a portion of his wearing apparel, but lost nearly all his pictures of sweethearts—past, present and future.

* * *

Bill Kneidler helped move the sluice table, etc., out of Ed. McGill's, but forgot to move his own bedding and "keyster" which were stored there. This wasn't a case of unselfishness—but just plain forget.

* * *

Manager Rider, of the United Rico Mines company, called all the men from the Atlantic Cable workings to fight fire, then gave them Wednesday's shift off. Another case of harmony between men and manager.

* * *

As soon as possible after the fire started and it was seen that the hotel could not be saved, J. M. Hall wired J. A. Pulliam, who was at Durango, and who is administrator for the Dey estate, for possession of the Metropole block and by Wednesday evening was doing a refreshment business at that stand.

* * *

Mrs. Pulliam organized a rescue corps of men and women and succeeded in landing the Judge's law library in the Rico Hardware company's brick store across the street, in short order. Bet she can't lift one end of a desk to-day that we saw her lift out of the way alone on the night of the fire.

* * *

The Hotel Enterprise as it stood before the fire, cost the Rico State Bank and David Swickhimer for its construction and maintenance, close to $100,000. It was a big money maker in 1892 and up to the panic in 1893, since which time it has paid taxes and maintenance but a portion of the time. The insurance of $7,150 would tot pay for the furniture in the building.

* * *

Richard Keller, manager of the Emma mine, at Dunton, had all his winter clothing stored in room 110 of the Hotel Enterprise, in fact had all his clothing there except a small amount which he carries with him when going out of town. There will be business for some gents' furnishing goods man as a result, as nothing was removed. Mr. Keller grieves more over the loss of a number of trinkets than his clothing.

H. OBENDORFER

Dealer in Bakery Goods, Stationery, and Fancy

GROCERIES

Finest Line in the City

H. OBENDORFER

LIST OF NOMINATIONS

To Office, as certified to the County Clerk of Dolores County, Colorado, by the Secretary of State, for the office of JUDGE OF THE SUPREME COURT, together with the name and address of each candidate to be voted for in the State of Colorado and County of Dolores, at the General Election to be held on Tuesday, November 3d, 1903.

TO VOTE A STRAIGHT PARTY TICKET, WRITE WITHIN THE BLANK SPACE IMMEDIATELY HEREUNDER THE NAME OF THE PARTY YOU WISH TO VOTE FOR.

I hereby vote a Straight ________ Ticket,

Except where I have marked opposite the name of some other candidate

*If you have voted a Straight Ticket above, and place a cross mark (**X**) opposite any name below, such cross mark (**X**) will be counted for that candidate, and the vote cast for the candidate for the Straight Ticket for the same office will not be counted.*

FOR JUDGE OF THE SUPREME COURT. (Vote for One.)		MARK IN THIS COLUMN.
JOHN CAMPBELL, Postoffice address, Capitol Building, Denver, Colorado. Residence, 1520 North Tejon Street, Colorado Springs, Colorado. Place of Business, Capitol Building, Denver, Colorado.	Republican.	
I. A. KNIGHT, Postoffice address, Pueblo, Colorado. Residence, 1707 West 17th street, Pueblo, Colorado. Place of Business, Pueblo, Colorado.	Socialist Labor.	
FRANK W. OWERS, Postoffice address, Leadville, Colorado. Residence, Court House, Leadville, Colorado. Place of Business, Court House, Leadville, Colorado.	Peoples Party.	
CHANNING SWEET, Postoffice address, Denver, Colorado. Residence, 1445 Gaylord street, Denver, Colo.	Socialist.	
ADAIR WILSON, Postoffice address, Williams Block, Durango, Colorado. Residence, Durango, Colorado. Place of Business, Durango, Colorado.	Democrat.	

STATE OF COLORADO, } ss.
COUNTY OF DOLORES.

I, Edward B. Clark, County Clerk of Dolores County, Colorado, do hereby certify the above and foregoing to be a true and correct list of nominations as certified to me by the Secretary of State of the State of Colorado, to be voted upon at the General Election to be held November 3rd, A. D. 1903.

Witness my hand and official seal this 23rd day of October, A. D. 1903.

[SEAL.]

EDWARD B. CLARK,
County Clerk.

NOTICE.

By virtue of authority from the Board of Directors of the Enterprise Mining Co., a corporation, I will offer for sale to the highest cash bidder, at my office in Rico, Colorado, on Tuesday, November 10th, 1903, at 4 p. m., all of the Ute Lode Mining Claim, Survey No. 6796, situate in Pioneer Mining District, Dolores County, Colorado.

PERCY S. RIDER, Agent.

Don't leave town without subscribing for the NEWS.

HOGG & HENRY

The undersigned, Herschel M. Hogg formerly of Telluride, and Lyman I. Henry, of Ouray, announce that they have formed a partnership for the general practice of law, under the name and style of Hogg & Henry, with offices at 7 and 8 Pope block, Pueblo, Colo.

The business of the firms of Hogg & Watson at Telluride, and of Henry & Sigfrid at Ouray, will be conducted at each place as it has been.

HERSCHEL M. HOGG,
LYMAN I. HENRY.

Pueblo, Colo., Oct. 1st, 1903.

JOSEPH E. BYERS PHOTO – RICHARD A RONZIO COLLECTION

JOSEPH E. BYERS was a photographer based in Telluride, and he had ventured over Lizard Head Pass one day about 1905, to see and shoot pictures of Rico. The ruins of the Hotel Enterprise stood, but the tall chimney had toppled. After a long winter, summer weather had reached Colorado's high country. On the mountain slopes above the mining town, aspen leaves were quivering on the trees with white bark. Joseph had pointed his camera lens toward Silver Creek for this historic view.

pany stockholders, in 1920 the mill was again forced to close. The Syndicate Mining & Milling Company and the Rico-Argentine Company were battling each other in the courts, and this had completely crippled all mining activities in Rico.

During 1923, while the legal battle continued, a new company was formed in Rico. Robert and Elizabeth Pellet created the Pell-Eyre Mining & Milling Company, with the support of the International Smelting & Refining Company, a subsidiary of Anaconda Copper. Their formation of this company would later prove to be the salvation of the Pro Patria mill.

However, in 1926, new ownership for the Pro Patria was close at hand—the Rico-Argentine Company finally prevailed in the courts. And again, the Pro Patria mill was about to bounce back. With the help of the United States Smelting & Refining Company, Rico-Argentine renovated the old Pro Patria mill.

(Continued on Page 302)

LARRY PLEASANT COLLECTION

THE RICO NEWS, dated October 24, 1903, told the shocking story of the terrible fire that consumed the Hotel Enterprise. The actual sheet size of the newspaper's front page is 18 x 24 inches.

JOSEPH E. BYERS PHOTO – RICHARD A RONZIO COLLECTION

ON THESE TWO PAGES you will see an enlarged print of Mr. Byers' photograph, which illustrates most of the major structures of Rico. These buildings had survived the Silver Panic of 1893, a severe economic depression. The stacked lumber down near the RGS depot had arrived by train for a contractor to erect a 125-foot-long brick enginehouse. The old wooden enginehouse had proved to be inadequate for servicing engines and bad-order cars, especially during winter, when cars had to be moved inside for repair work. From this photograph, you see the south end of Rico from a different angle. It looks as though a string of homes was at the far right, without doors or windows—also in view on Page 282.

PROCEED TO PAGES 292-293 for a closer look at the RGS station grounds and the business district of Rico, along Glasgow Avenue—reproduced at 470 percent of the original view. At the left, above the RGS mainline, are two boxcars spotted at the United Rico Mines Company's Pro Patria mill. Farther along to the right is the RGS water tank and depot, complete with a fenced lawn and adjoining outhouse. The ex-D&RG RPO being used as the roadmaster's office was still standing across the mainline from the depot, resting on the ground, next to large timber stacks. A couple of gondolas await unloading. A coal shed for servicing locomotives had been set up next to the siding beyond RGS Rotary Snowplow 2. Two more gondolas and three boxcars were positioned on the two spur tracks next to the original enginehouse. In the foreground, houses can be seen on this side of the Dolores River, adjacent to the yard, on Picker Street in West Rico.

A LOG RETAINING WALL had been constructed to hold back the embankment along the access road leading down to the depot and yard from Mantz Avenue. Along Glasgow Avenue, at the far left (at the edge of the view), stood the Atlantic Cable Mine's head frame and hoisthouse. Farther down the street, to the right, stood the forlorn ruins of the Hotel Enterprise, destroyed by fire in 1903. This hotel had been Rico's pride and joy, where the formal celebration was held after the RGS had reached Rico. Crossing Mantz Avenue and continuing to the south, the next significant structure—identified by the skylight on the large brick building on this side of Glasgow Avenue—housed the Burley Block, still standing, as of 1996. This building contained a drugstore, jewelry shop and gent's clothing store. The Dolores County Courthouse is in plain view, at the intersection of Commercial Street and Mantz Avenue, while the prominent Congregational Church stands beyond the brick courthouse.

IN ALL LIKELIHOOD, the Rico depot's original paint scheme matched that of the Telluride depot. It was discovered that this color scheme was oxide red (cherry red) on the lap siding, pine-tree green on the corner posts and other trim work (around windows, doors, etc.), while the lower wainscot siding was Tuscan red (a dark reddish brown). The shingle trim at the upper peaks of the depot had been painted a buff color (straw yellow). This finding came from paint chips removed from the Telluride depot before this historic building was renovated in 1991.

JOSEPH E. BYERS PHOTO – RICHARD A RONZIO COLLECTION

All the Mining and Local News of the District appears in these columns, up-to-date and authentic. Keep posted.

THE RICO NEWS

Mines can be Made in this District at a minimum of cost, because the rock is soft, timber handy and power cheap.

DEVOTED TO THE MINING AND MATERIAL INTERESTS OF DOLORES COUNTY

VOLUME 24. RICO, DOLORES COUNTY, COLORADO, SATURDAY, JANUARY 31, 1903. NUMBER 20

D. SWICKHIMER, President. J. H. GOODKNIGHT, Asst. Cashier.

The Rico State Bank,

RICO, COLORADO.

Authorized Capital, - - - $200,000.
Paid in Capital, - - - 50,000.

Interest allowed on time deposits. Sell Drafts payable in any part of the world. Fire and burglar-proof safety deposit boxes for rent We solicit your patronage.

The Finest

{And most complete line of

PIPES

And smokers' articles in the city.

A nice Pipe makes a very suitable present for your smoker friends Call and examine my stock and I will make prices suitable.

HENRY OBENDORFER.

City Livery

Fine Turnouts AND Saddle Horses.

V. H. LEE & Co., Proprietors.

DEALERS IN

HAY, GRAIN, COAL, LIME AND LUMBER

Ore Hauling, Transfer and Express.

BUSSES - MEET - ALL - TRAINS.

Leonard Andeway. S. O. Burgwin.

Andeway & Burgwin

— DEALERS IN —

Dry Goods, Groceries
Gents' Furnishings
Boots, Shoes, Etc.

GENERAL OUTFITTING SUPPLIES

FOR

Miners and Campers

DUNTON - - - COLORADO.

Koster's Market

OTTO KOSTER, Prop.

Fresh, Salt and Smoked Meats.
• • •
Fish and Game in Season.

Canned Goods of all Kinds.
• • •
Butter, Eggs and Vegetables

Free Delivery to all Parts of Town.

MODERN LIFE INSURANCE.

In connection with the commemoration of the founding of American life insurance on February 1, 1843, the Canadian Society of New York is putting up at 56 Wall street a tablet in honor of Morris Robinson, the first president of the Mutual Life Insurance Company. Robinson was a native of Nova Scotia, though connected by birth and marriage with powerful New York families; but he came back to this country when a young man and became one of its leading financiers in the first quarter of the last century.

The branch of finance which he started with practically no capital except good judgment, honesty and conservatism has steadily prospered until it has reached a position of the first importance. The volume of business of this single company in the first year was scarcely a million and a half; to-day 103 American companies have nine billions of insurance in force and the money now in their hands from premises amounts to two billions. Ten thousand dollar policies, which used to be sought only by the very wealthy, have become a necessity in comfortable middle-class homes, and million-dollar policies are not unheard of. Two such have already been written by the Mutual; one for George W. Vanderbilt of New York, who pays an annual premium of $35,000; the other for the late Frank H. Peavey of Minneapolis, well-known as the "Elevator King." The latter was paid about two years after it was taken out, Mr. Peavey dying of pneumonia in December, 1901; but in accordance with the conservatism that has done so much to make modern life insurance possible, as an extra precaution, his policy had been reinsured by the Mutual to the extent of $750,000 in twenty other companies. That the Mutual was not seriously disturbed by payment of the $250,000 for which it stood individually responsible, so to speak, can be understood from the fact that at the date of settlement this company had on hand $20,000,000 in cash, while its total assets were $350,000,000.

In the exercises commemorative of its 60th anniversary the broadened scope of life insurance is likely to receive not less attention than its bigness. As a science it has progressed wonderfully since 1843. When the New York Mutual started it wrote only "straight" insurance for life or for a period of years; now policies have been devised which enable a man to husband resources against the future in the easiest and most efficacious way, and from a mere form of protection it has been developed into a great plan of investment, directed by the most expert financiers, while the protective advantage has been in no way sacrificed. Beginning in young manhood, the policy holder protects his family and makes his maximum savings at a time when his earnings are at their highest without pledging himself to carry any heavy burden after middle age has been passed.

And not only for families but for estates and business interests life insurance has come to be used as a means of protection. It has become very common for two or more partners to insure their lives for the benefit of the survivor or survivors. Some firms likewise are taking the precaution of insuring the lives of leading officials; and the custom of insuring individual lives for the benefit of religious, charitable or educational institutions grows steadily from year to year, affording a unique means by which people of moderate means can help along causes in which they are especially interested by making several yearly appropriations, within their means, in the form of life insurance premiums, with the knowledge that eventually they will become benefactors to a considerable amount.

Of the original policy-holders in the first American mutual company three are still alive. All of them took out their policies in the first year and two of them in its first week, and an interesting circumstance is that all three are living in the state of New Jersey and that the oldest living policy-holder—Charles H. Booth, who is approaching his 100th birthday—and the holder of the oldest policy—Mr. William E. Shepard—both are residents of the town of Englewood.

The present head of the Mutual is Richard A. McCurdy, who, as its president, has come into the front rank of American financiers. With him are associated as trustees—the Mutual has no salaried board of directors—such representative men as Cornelius Vanderbilt, Stuyvesant Fish and William Rockefeller of New York; Effingham B. Morris of Philadelphia; William Babcock of San Francisco, and Justice Rufus W. Peckham of the United States Supreme court.

The site on Wall [illegible] in which the Mutual company began business in 1843 is famous as having been occupied in the latter part of the seventeenth century by Captain William Kidd. The present building on this plot, erected in 1836, and recently remodeled into one of the fine buildings of this famous thoroughfare, is now the property of Senator George Peabody Wetmore of Rhode Island, and on its front is affixed the Canadian Society's commemorative tablet of bronze.

X-MAS IS OVER!

We thank you for your past patronage and wish to have a continuance of the same, We have some exceptionally good bargains to offer you in the way of Fancy Crockery, Silverware, Watches and Jewelry.

Don't forget that we do first-class Repair Work in all its branches, and all work is positively guaranteed.

Our name stands for reliable quality and fair price.

J. D. DIEHL,

Licensed Watch Inspector for the Rio Grande Southern Railroad Company.

A SILVER-COPPER TRUST.

The Standard Oil Company has plans well under way to make the United Metals Selling company the sole dealer of the world in copper and silver.

This plan was disclosed through a circular issued by James Lewis & Sons of Liverpool.

The movement accounts in a way for the bearish attitude of the Standard Oil people on the stock market, this being to break in the price of stocks on metal producing companies so that control may be bought.

It was discovered that the United Metals Selling company, which is the selling agent for the Amalgamated Copper company, and controlled by the Rockefellers, the agents for which are the Lewissohns, is seeking to make long time contracts at present prices with independent companies for their copper. This is being done in the face of the alleged fact that there is an overproduction of copper and prices are to break.

The company has contracts with the Greene Consolidated company, and has an insidious hold on the great Calumet & Hecla company that will force it to terms at any moment desired.

"The object of this peculiar method of trade," say the Lewissohns, "is evidently to get control of the United Copper company mines of the Heinzes, either by purchase or successful litigation."

It is also said that the American Smelting company is marked as prey for the United Metals Selling company. This is the silver trust.—Mining Review.

Letter List.

Following is the list of letters remaining uncalled for in the post office at Rico, Colo., for the week ending Jan. 31, 1903.

DOMESTIC

S. A. McMurray.
Jas. J. Merrison.

Parties calling for the above letters will please say advertised and give date of this list. C. R. ELLIS, P. M.

E. W. Grove

This signature is on every box of the genuine **Laxative Bromo-Quinine** Tablets the remedy that cures a cold in one day

Fifty Years the Standard

DR. PRICE'S CREAM BAKING POWDER

Awarded Highest Honors World's Fair Highest tests U.S. Gov't Chemists

PRICE BAKING POWDER CO. CHICAGO

PERSONAL ** AND LOCAL

Below zero several nights this week.

F. A. Smith left for Denver Wednesday.

M. W. Tozer was over from Dunton Monday.

P. S. Rider left for Durango Wednesday morning.

The dancing club gave the usual hop Saturday night.

Will J. Scoutt returned to Rico Thursday from a trip to Telluride.

Dr. McNeill came over from Dunton the first of the week for a short visit with his family.

Dr. Dawson was sick a few days this week—probably took some of his own medicine by mistake.

Richard Keller, manager of the Emma mine at Dunton, was here Wednesday on his way to Durango.

The stockholders of the Pro Patria M. and M. company held their regular monthly meeting last night

A special with the new superintendent and other Rio Grande officials, on a tour of inspection, passed through northward bound yesterday.

George Ditchfield will return to Rico this morning to look after his divers and various interests, leases, co-partnerships and merchandise.—Wednesday's Durango Democrat.

At the business meeting of the members of the Congregational church Friday evening it was decided to engage the Revs. Carmichael for the period of of one month, and longer if mutually satisfactory.

Among those registered at the Hotel Enterprise during the week we find the following: Frank Thaler, E. A. Smith, Ferd Weinland, Chas. Greenclay, Denver; V. C. Rocho, Omaha; B. Burton, Sioux City, Iowa.

Geo. Hutt secured a few days' furlough from the senate "after the ball was over" and visited Rico from Monday morning until Wednesday afternoon. He is looking as though life at the capital agrees with him.

Mrs. E. L. Short and children departed on Monday last for Denver and points in Kansas and Nebraska to visit relatives and friends, and will be absent three or four weeks. Meanwhile E. L. is keeping late hours and learning to frog.

The snowstorm of Wednesday, with the high wind that raged, caused lots of trouble for the railroad company, the snow drifting badly in many places. The flangers were needed to buck the drifts near Lizard Head. About eight or ten inches of snow fell.

Anticipation

A life insurance policy is usually for a long period. The record of the company in which you insure, therefore, becomes of first importance. The Mutual Life Insurance Company of New York gives you the best security for the future.

Its Assets, larger than those of any other life insurance company in the world, exceed

$352,000,000

It has paid Policy-holders over

$569,000,000

which is more than any other life insurance company in existence has disbursed.

Write for "Where Shall I Insure?"

THE MUTUAL LIFE INSURANCE COMPANY OF NEW YORK

RICHARD A. MCCURDY, President.

JOHN L. STEARNS, Mgr., Denver, Colo.
F. LEE FORMAN, District Manager.

HENRY M. TELLER ELECTED U. S. SENATOR

Hill and Madden, Ill, Put In Appearance at Joint Session Saturday, Completing Quorum Sufficient to Elect.

After several days of unavailing attempts on the part of the Democratic members of the joint session of the senate and house to secure the attendance of all their members in order to have enough present to make the election of a United States senator legal, success crowned their efforts last Saturday afternoon, January 24. The absence of several members by reason of sickness during the last [illegible] nounced: "Senator Henry M. Teller having received a majority of the votes of the members elected to the Fourteenth General assembly, I hereby declare him elected senator for the term in the United States senate beginnin March 4, 1903."

Then the enthusiasm [illegible] loose, and a wild yell was fo[illegible] by shouts and cheers, and of handkerchiefs an[illegible]

DAVID C. BUTLER COLLECTION

THIS SAMPLE of THE RICO NEWS is dated January 31, 1903, and both the front and back pages are reproduced here. See the back page for a Colorado & Southern story about the Denver-to-Georgetown line. There is a multitude of railroad articles, including RGS stories. The Pro Patria stockholders' meeting is in Column 4. The back page has a story of the Iron Clad Gold Mine, telling that an RGS spur had been laid to the mine. There is a display advertisement for the Dunton general store.

Rev. Neil Carmichael and wife, who have been staying at the Hotel Enterprise during the past week, expect to secure a furnished house the first of next week and settle down to housekeeping. It is hoped they will like Rico and decide to remain permanently.

Major A. B. Litchfield and wife left Telluride Wednesday morning for their future home in the South African gold regions, where the Major has accepted the management of a large mining property at a handsome salary. He is an old timer in the San Juan and a mining man of wide experience, and his many friends are well pleased at his success.

William Callan, a San Juan pioneer who went to California forty years ago and has been in the San Juan since the earliest days, who once, in the early eighties, had the Rico and La Sal Mountain mail contracts out from Ouray, is interesting some parties to go into Southern Nevada on the line of Clark's railroad, into the gold fields in that section, and Mr. Callan is a safe man to tie to.—Wednesday's Durango Democrat.

A New Summer Resort.

Ground is being prepared and a tunnel has been run into the mountain following the flow of water from a hot spring that will rank with some of the best in the state in medicinal properties. The spring is located at Placerville, about twelve miles northwest of Telluride, and is owned by C. P. Lemon He expects to have the ground cleared by spring and a fine hotel and bath house erected. The springs are not over 100 yards from the Rio Grande Southern depot and are reached by a foot bridge crossing the San Miguel river.

Rich Ore In the Portland Group.

C. C. Koenneker, who is the controlling factor in the Portland group of mines at La Plata, is highly elated with the prospects of developing the Portland into another Neglected, says the Silverton Standard. The six claims that comprise the Portland group are within 800 feet of the side lines of the famous Neglected, and the report that a richer and greater quantity of ore has just been encountered in the shaft at the 175-foot level, the deepest yet attained, is pleasant news for Koennecker and partners. An option to sell the property at the sum in the neighborhood of $50,000 was recently given to W. W. Parshall of Durango, who has money men on the string.

To Cure a Cold in One Day

Take Laxative Bromo Quinine Tablets. All druggists refund the money if it fails to cure. E. W. Grove's signature is on each box. 25c.

MINERAL POINT MINE SOLD.

F. H. Butler, a man well known, is the latest to make a stake in the mines, says the Ouray Herald, and he departed last week for New York, the old home, to spend the balance of his life a little easier. Mr. Butler had prospected in the San Juan mountains for 22 years and has been up against is as hard as few men get but never lost faith.

He located and owned the Nettie group at Mineral Point, just over the line in San Juan county, and concluded a sale last week to a New York company for $125,000, of which $25,000 was cash down. He says he has earned every cent of it and would not go through the same experiences again for that amount.

We understand the company will push work on the property the coming summer, which will largely increase the present force. New mining machinery and a mill are also contemplated. With a good road from Ouray to Mineral Point much of the trade from that and adjacent properties might be brought to Ouray.

Strike in the Silver Queen.

Another important strike has been made during the week, this time in the Silver Queen, a property in the upper Animas district, operated under the direction of B. D. Smith, for the Silver Queen Mining company.

The new body or vein of ore is found at a depth perpendicular at 175 feet by a crosscut tunnel 200 feet long, and while the foreman, Johnnie James, believes the vein to be the Silver Queen proper, others claim that it is the extension of the Sound Democrat vein. Be it one or the other, the find is considered by all who have inspected the same to be par excellence. There is at the present time five feet of the material—a quartz heavily impregnated with lead and copper—and a product which, if not rich enough to ship in its crude state, will certainly mill. The Silver Queen is situated in that section into which Otto Mears contemplates extending his railroad.—Silverton Standard.

CHURCH SERVICES.

Morning service	10:30	a. m.
Sunday school	11:30	"
Junior C. E. society	3:00	p. m
C. E. society	6:30	"
Evening service	7:30	"

Communion services will be held to-morrow after the regular morning service.

Stops the Cough and Works off the Cold.

Laxative Bromo Quinine Tablets cure a cold in one day. No cure, no pay. Price, 25 cents.

THE RICO NEWS

GEO. H. HUTT, ED. AND PROP.
B. W. FITCH, MANAGER.

PUBLISHED EVERY SATURDAY.

SUBSCRIPTION TERMS.

One Year.......................$2 50
Six Months.....................1 25

INVARIABLY IN ADVANCE.

Rico Sun........Vol. 5, No. 31.
News-Sun.......Vol. 6, No. 5.

SATURDAY, JANUARY 31, 1903.

RIO GRANDE SOUTHERN TIME CARD.

DEPART.

No. 8—Passenger for Ridgway and intermediate points.......... 7:00 a. m.
No. 5—Passenger for Durango and intermediate points ....10:30 a. m.
No. 6—Mixed train for Telluride and intermediate points.......... 2:45 p. m.

ARRIVE.

No. 7—Passenger from Ridgway and intermediate points.......... 7:20 p. m.
No. 5—Mixed train from Telluride and intermediate points ....10:22 a. m.
No. 6—Passenger from Durango and all points south .... 2:35 p. m.

OFFICIAL DIRECTORY

COUNTY OFFICERS.

Commissioners, J. D. Clarke, Chairman; Wm. Pearce, Charles A. Goble.
County Judge, George W. Hunter.
County Clerk, E. B. Clark.
County Treasurer, V. J. Kraft.
County Sheriff, Joseph Theno.
County Assessor, A. V. Gorla.
County Supt. of Schools, Lizzie Kelly.
County Surveyor, Pendleton Hunter.
County Coroner, Wm. Sherry.

TOWN OFFICERS.

Mayor, Percy S. Rider.
Treasurer, Jesse E. Crecelius.
Clerk, Elizabeth Muncaster.
Trustees, N. J. Bradley, W. B. Hess, Otto Koster, W. A. Miller, John Muncaster, S. M. Ransom.

CHURCHES.

People's Congregational Church. Regular services at 10:30 a. m. and 7:45 p. m. every Sunday. Sabbath school at 11:30 a. m. Regular prayer meeting on Wednesday evening at 7:45.

Chapel of Immaculate Conception. Services second Sunday of each month

LADIES' AID SOCIETY. Meets on the second Thursday afternoon of each month. Mrs. E. Muncaster, Pres.; Belle Dawson, Sec'y.

SOCIETIES.

Cards under this head will be charged for at the rate of 25c. per month. When lodge cards are inserted resolutions of condolence will be published free of charge.

SILVER CRESCENT LODGE No. 40, I. O. O. F. Meets every Wednesday evening in Odd Fellows' hall. Visiting brothers always welcome. J. G. Begue, N. G.; L. E. Brown, Sec.

RICO LODGE No. 79, A. F. and A. M. Meets at Masonic hall, the second and fourth Tuesday of each month. John Anderson, W. M.; J. T. Carnahan, Secretary.

RICO CHAPTER No. 31, Order of Eastern Star. Meets at Masonic hall first and third Thursday of each month. Mrs. Oakes, W. M.; J. Anderson, W. P.; Eva Ramsey, Secretary.

THREE STAR REBEKAH LODGE No. 56, I. O. O. F. Meets every Thursday evening in lower hall of Odd Fellows' building. Visiting members always welcome. Aldrett Misch. N. G.; Agnes Hall, Secretary.

RICO LODGE No 48, A. O. U. W. Meets every Friday evening in Odd Fellows' hall. Visiting brothers always welcome. V. H. Lee, M. W.; E. B. Clark, Financier; Geo. S. Hicks, Recorder.

RICO LODGE No. 9, D. of H., A. O. U. W. Meets every Tuesday evening in lower hall of Odd Fellows' building. Visitors always welcome. Margaret V. Clark, C. of H.; Abi Quine, Recorder; Charles Engel, Financier.

HAZEN POST No. 63, G. A. R. Meets in Odd Fellows' hall second Sunday of each month, 7:30 p. m. S. H. Burghardt, Commander; Jacob Winkfield Adjutant.

LAFAYETTE COUNCIL No. 15, Jr. O. U. A. M. Meets every Saturday evening at 7:30 o'clock, in Masonic hall. Visiting members always welcome. Geo. L. Garren, C.; M. G. Ballough, R. S.; J. T. Carnahan, F. S.

RICO CAMP No. 284, W. of W., meets on the first Monday evening of each month at the Odd Fellows' hall. L. E. Brown, C. C.; C. F. Middaugh, Clerk.

RICO CIRCLE No. 12, W. of W. Meets 1st and 3d Monday evening of each month in lower hall of Odd Fellows' building. Mrs. E. Muncaster, G. N.; Isabelle Dawson Clerk.

BUSY BEE HIVE NO. 19, LADIES OF THE MACCBEES, meets every second and fourth Friday in each month at I. O. O. F. Hall. Visitors welcome. Alice Herr, L. C.; Kate Obendorfer, R. K.

SILVER GLANCE LODGE No. 82, K. of P., meets every Thursday evening at Masonic hall. Visiting brothers always welcome. John F. Stevenson, C. C.; John Garland, K. of R. & S.

A. E. ARMS,
U. S. Deputy Mineral Surveyor
Special attention given to underground work.
RICO, - COLORADO.

J. A. PULLIAM,
ATTORNEY-AT-LAW,
Rooms 5 and 6, Post Office Building,
RICO, COLORADO.

P. A. LARSH,
Assayer and Chemist
Assays Every Day.
At Merdith Stand.
Bring Samples Before Noon, If Possible.

W. B. HESS,
Real Estate and Insurance.
Office in Rico Furniture Store
RICO, COLO.

F. A. McNEILL, M. D.
Physician and Surgeon.
DUNTON, COLO.

H. H. DAWSON,
PHYSICIAN AND SURGEON.
Office Rooms, front Burley Block,
RICO, COLO.

Confectionary
We have a large stock of Fresh Candies and nuts of all kinds always on hand.
Books.
Periodicals.
Fine Stationery.
STORY'S

Vade, Baron Wolcott.

The Baron of Wolhurst failed to make good.

There seems to be an epidemic of terrible railroad wrecks at the present time.

Our legislators seem to have a pretty good idea of what the people want, after all.

Now comes a Chicago physician with the assertion that pneumonia is highly contagious.

A bill for female suffrage in Kansas was defeated in the senate of that state Wednesday.

Senator Henry M. Teller will look after our interests at the National capital for awhile longer.

Colorado has had a narrow escape from Wolcottism, and she should offer up a prayer of thanks.

Most of the people of Colorado drew a deep breath of relief when the news came that Teller was elected senator.

Dave Day is a fighter from 'way back. The Typographical union of Durango and its friends are still boycotting the Democrat, and Dave seems to enjoy the scrap.

A combination of great corporations of the state is now lobbying in the senate to defeat the eight-hour law. The capitalists are against everything that benefits labor.

The constantly recurring squabbles and disgraceful scenes in various legislatures throughout the country over the election of United States senators, with the plain evidence in many instances of boodle in connection therewith, must inevitably lead, sooner or later, to a change in the constitution providing for their election by popular vote. The people are becoming thoroughly disgusted with the present method of electing them.

THE IRON CLAD GOLD MINES COMPANY.

On Saturday, January 17, 1903, the articles of incorporation of the Iron Clad Gold Mines company were filed with the secretary of state.

The company is capitalized for $200,000 under the laws of Colorado, and the stock is fully paid and non-assessable. One-half of this stock has been placed in the treasury to be sold for the purpose of developing the Iron Clad mine, which is patented, and covers 1500 feet on what is known as the Iron Clad fissure.

The company owes no debts and the property is free from all incumbrances.

The vein on the Iron Clad ground is a true fissure showing a pay-streak of ore fully two feet wide, which has been exploited by two tunnels, 100 and 200 feet in depth, respectively, and by a winze from the second level for a perpendicular depth of 25 feet below the tunnel level. During this development of the property something over $25,000 worth of ore has been extracted and shipped to the smelters at Pueblo and Durango.

The vein averages about two feet in width throughout the entire workings and the ore is a copper pyrites, carrying values in copper, gold, silver, lead and zinc.

The property is most admirably situated for its economical working, as ores from the second level can be dumped into railroad cars at the mouth of the level, and a spur from the Rio Grande Southern railway has been constructed from the main track for a distance of about 300 yards for that purpose.

The officers of the company are T. R. Williams, ex-mayor of Rico, president and manager; Florence E. Pulliam, vice-president; V. J. Kraft, county treasurer of Dolores county, treasurer; W. A. Smith, secretary; Jas. A. Pulliam, district attorney Sixth Judicial district, attorney.

The property has a most excellent record for the amount of development that nas been done, and as its principal officers are men of vast experience in mine development and operation, and are reliable business and professional men, there is no doubt that the Iron Clad will soon be listed as one of the heaviest dividend paying propositions in Colorado.

BILL TO REGULATE SMELTER CHARGES.

A bill of vast and far-reaching importance to all mining men of Colorado was introduced in the house on Wednesday by Representative Samuel Belford. It is House Bill No. 134, and in substance provides for an amendment to the constitution classifying all ore handling concerns under the head of "Business of public interest," and empowering the state to fix and regulate all smelting charges for the reduction of ores. It is the hardest blow ever aimed at the smelter trust in Colorado, and if passed it will be of great benefit to the producers of ore and will cut down the enormous profits of the smelters to a reasonable figure. The smelter trust will put up a hard fight against its passage.

Another Short Line From Denver West.

It is now rumored that Harriman intends to change to broad gauge the Colorado and Southern branch running from Denver to Georgetown, complete and make use of the old Atlantic and Pacific tunnel under Gray's Peak, and build on through northwestern Colorado and connect with the Union Pacific at either Rawlins or Granger. This will open up different territory from the new Moffat road.

We are agents for the White sewing machine. Catalogue and full particulars free. J. B. BEGUE & CO.

Old Papers For Sale.

Suitable for laying under carpets, wrapping bundles, etc. Twenty-five cents for 100, 15 cents for 50. This office.

On and after July 1st coal will be sold at following rates:

One-fourth ton—500 lbs...........$2 00
One-half ton—1000 lbs.............3 50
One ton—2000 lbs..................6 50

Special rates on carload lots.

V. H. LEE.

WANTED—FAITHFUL PERSON TO TRAVEL for well established house in a few counties, calling on retail merchants. Local territory. Salary $1024 a year and expenses, payable $19.70 a week in cash and expenses advanced. Position permanent. Business successful and rushing. Standard House, 334 Dearborn St., Chicago.

Proceedings of Board of County Commissioners

Of Dolores County, Colo., at their Meetings held January 5, 6 and 13, 1903.

A resolution was adopted authorizing the County Treasurer to assign to J. D. Adams the tax sale certificates for ¼ Chicago Lode and ¼ South Park Lode for the sum of $20.00 and the payment of the accrued taxes thereon.

Also authorizing the County Treasurer to assign tax sale certificates for Lot No. 5, in Block 17, to Joseph Holzner for the sum of $7.00 and the payment of the accrued taxes thereon.

James A. Pulliam, Esq., was appointed County Attorney.

Howard H. Dawson, M. D., was appointed County Physician.

George A. Rule was appointed Janitor of the County Court House.

THE FOLLOWING BILLS AGAINST THE COUNTY WERE ALLOWED:

NAME.	AMOUNT CLAIMED	AMOUNT Allowed	FUND	ACCOUNT.
M. W. Tozer	$156 00	$156 00	Road	Road work Dist. No. 2
Pen Hunter	30 00	30 00	do	do do
V. H. Lee	4 00	4 00	do	do do
S. R. Robinson	99 00	99 00	do	do 3
C. A. Goble, Co. Com.	26 20	26 20	O. C. R.	Salary and mileage
Otto Koster, do	10 00	10 00	do	do
W. J. Muncaster, do	15 00	15 00	do	do
V. H. Lee, do	5 00	5 00	do	do
E. B. Clark, Co. Clerk	76 85	76 85	do	Clerk of Board, etc.
V. J. Kraft, Co. Treas.	278 25	278 25	do	Making assignments, etc.
Rico Furniture Co.	1 75	1 75	do	Glass and putty
Jos. Theno, Co. Sheriff	7 20	7 20	do	Fees
J. W. Ridlen & Co.	36 43	36 43	do	Coal
Out West Co.	26 41	26 41	do	Blanks for Co. offices
Geo. A. Rule	77 00	77 00	do	Janitor work, Nov. & Dec.
Wm. Sherry	5 00	5 00	do	Putting up election booths
A. V. Gorla, Co. Assessor	350 00	350 00	do	Salary
B. N. Griffing	36 10	36 10	do	Lights, Oct., Nov. & Dec.
Grant Morgan	75 50	75 50	do	Painting Court House
Robert Howell	24 18	24 18	do	Coal
J. A. Pulliam, Co. Attorney	75 00	75 00	do	Salary
J. D. Diehl	1 25	1 25	do	Ink for Co. offices
Out West Co.	15 00	15 00	do	Record book, Co. Treas.
V. H. Lee	62 50	62 50	do	Coal
E. E. Tremble	27 00	27 00	Poor	Boarding pauper
Sarah Kinnish	27 00	27 00	do	do
H. H. Dawson, Co. Physician	21 00	21 00	do	Med. attendance pauper
Rico Drug Co.	11 25	11 25	do	Medicines for do
V. J. Kraft, Co. Treas.	121 20	121 20	Co. Treas. Fee	Fees, Oct., Nov. & Dec.
S. F. Thomas, Clerk Dist. Court Montezuma Co.	50 60	50 60	D. C. Ex.	Fees, Chesebro et al. case

STATE OF COLORADO, } ss.
COUNTY OF DOLORES. }

I, Edward B. Clark, County Clerk in and for said County, in the State aforesaid do hereby certify the above and foregoing to be a true copy of the Proceedings of the Board of County Commissioners of said County, as therein set forth.

Given under my hand and official seal this 27th day of January, A. D. 1903.

[SEAL.]

EDWARD B. CLARK,
COUNTY CLERK.

Wood box stoves, all sizes, at Begue & Co.'s.

Clocks, watches and jewelry at cost.
BEGUE & CO.

New invoice—iron bedsteads cheap at J. B. BEGUE & CO.'S.

We have a few more boxes of W. D. Cammon, Jr., & Co.'s infant and complexion powder, @ 15c per box.
BEGUE & CO.

Through Tourist Sleeping Cars to Kansas City and St. Louis via Rock Island System, the Shortest and Quickest Route

Leaves Denver every Saturday 3 p. m. Arrives Kansas City every Sunday 9:30 a. m. Arrives St. Louis every Sunday, 6 p. m.

MAKE YOUR MONEY EARN MONEY.—A safe investment, weekly dividends, sums of $20 up received; your money always at your command, and all money to your credit can be withdrawn on demand. We have earned and paid between 4 and 8 per cent. each and every week; this is no gambling but a legitimate business proposition; best of references throughout the United States; write for our booklet, free. Address H. Brolaski & Co., 356 Dearborn St., Chicago, Ill.; Denver office, 406 Cooper bldg., W. C. Lape, Mgr. 2-7

...THE DENVER REPUBLICAN

Is Clean, Truthful, Reliable and Progressive.

It prints more news than any other paper in Colorado. It stands for the best interests of the state and enjoys the confidence and esteem of all intelligent readers.

It has the Biggest and Best Circulation At Home and Abroad.

As an advertising medium THE DENVER REPUBLICAN is superior to any other paper in Denver, because its readers are prosperous and progressive.

DAILY AND SUNDAY BY MAIL
Postpaid, Per Month,
75c
WEEKLY—Postpaid, Per Year,
$1.00

IT DOES NOT HAVE TO BE "YELLOW" IN ORDER TO BE READ

PIONEER BARBER SHOP
A. V. GORLA, Prop.
FIRST-CLASS WORK IN LATEST STYLES

FORFEITURE NOTICE.

To Christina Cohnan and Edward J. Cohnan, their heirs or assigns:

You are hereby notified that we have expended during the year 1902 $200, in labor and improvements, upon the Silver Swan and Condor lode mining claims, situated in Pioneer mining district, County of Dolores and State of Colorado, of which the location certificates are found of record in book 31, page 520, in the office of the recorder of Dolores County, Colorado, in order to hold said claims under the provisions of section 2324 of the revised statutes of the United States and amendment thereto approved January 22, 1880, concerning annual labor upon mining claims, being the amount required to hold said lode claims for the period ending on the 31st day of December, 1902.

If within ninety days after the last publication of this notice you fail or refuse to contribute your proportion of such expenditures as co-owners, your interest in the claims will become the property of the subscribers, your co-owners, who have made the required expenditures, by terms of said section and amendment thereto.

Dated at Rico, Dolores County, Colorado, this 16th day of January, 1903.

A. E. ARMS,
F. G. DAY,
F. A. W. DAY.

First pub. Jan. 17; last pub. April 18.

FORFEITURE NOTICE.

To Christina Cohnan and Edward J. Cohnan, their heirs or assigns:

You are hereby notified that we have expended during the year 1902 $500, in labor and improvements, upon the Mesaba, Tomale, Dawson, Sachem and Pine Knot lode mining claims, situated in Pioneer mining district, County of Dolores and State of Colorado, of which the location certificates are found of record in book 31, pages 521, 521, 527, 528, 528, respectively, in the office of the recorder of Dolores County, Colorado, in order to hold said claims under the provisions of section 2324 of the revised statutes of the United States and amendment thereto approved January 22, 1880, concerning annual labor upon mining claims, being the amount required to hold said lode claims for the period ending on the 31st day of December, 1902.

If within ninety days after the last publication of this notice you fail or refuse to contribute your proportion of such expenditures as co-owners, your interest in the claims will become the property of the subscribers, your co-owners, who have made the required expenditures, by terms of said section and amendment thereto.

Dated at Rico, Dolores County, Colorado, this 16th day of January, 1903.

A. E. ARMS,
F. A. W. DAY.

First pub. Jan. 17; last pub. April 18.

Through Tourist Sleeping Cars to Oklahoma City, Little Rock and Memphis via Choctaw Route. Best Line to Hot Springs, Ark.

Leaves Denver via C. & S. 9:30 p. m. every Sunday. Leaves Colorado Springs via C. & S. 12:05 a. m. every Monday Leaves Pueblo via C. & S. 1:20 a. m. every Monday.

Important! New Train Service.

The Chicago, Milwaukee & St. Paul railway has consummated an arrangement under which, commencing January 1, 1903, it will have the same train service over the entire Union Pacific system as that of the Chicago and Northwestern. This will include standard sleeping cars and chair cars to and from all points.

SHOES
For Men and Children!
Wear well.
Perfect fit.
Give satisfaction.
Best Shoes in market.
WM. GROEN,
The Exclusive Shoe Dealer

POHLE & PARMELEE
ASSAYERS AND CHEMISTS.
Special Attention to Control & Uupire Work. Ores tested to determine the best method of treatment. We have a new and thoroughly equipped Laboratory. Over 30 years of practical experience in Colorado. Prices and sample sacks free on application.
1627 Champa St., Denver, Colo.

NEW Through Train Service DAILY DENVER to CHICAGO
—VIA—
THE UNION PACIFIC
—AND—
CHICAGO, MILWAUKEE AND ST. PAUL LINE.
Through Sleeping Cars
Through Free Chair Cars
Through Day Coaches
ELECTRIC LIGHTED.
Elegant New Equipment, Superior Service, Quick Time.
ALL MEALS SERVED IN DINING CARS
THE BEST IN THE WORLD.
All trains arrive at UNION STATION, Canal and Adams streets, Chicago, in the heart of the city and convenient to all Eastern Line Depots. Ask your Agent about it, or apply to
J. E. PRESTON,
Commercial Agent,
1029 17th Street, DENVER, COLO.

FROM
RIDGWAY TO TELLURIDE, OPHIR RICO, DOLORES MANCOS, LA PLATA AND DURANGO.
Opening up the most magnificent scenery in the Rocky Mountains, heretofore inaccessable, and passing through the Famous
Montezuma and Shenandoah Valleys,
The great Agricultural Region of
THE DOLORES RIVER
AND THE
Famous GOLD FIELDS of San Miguel County and the La Plata Mountains.
This new line brings the Tourist within easy ride of the wonderful
HOMES OF THE CLIFF DWELLERS
This line in connection with the Denver & Rio Grande, forms a new and attractive
All Rail "Around the Circle" Trip.

...THE...
Rocky Mountain News
(Morning and Sunday)
The Denver Times
(Afternoon and Evening)
THE WEEKLY NEWS
AND
THE COLORADO WEEKLY TIMES
(COMBINED)
The Great Representative Newspaper of the Rocky Mountain States and Territories. All the News from All the World. Illustrations, Cartoons, Special Features, Etc.

SUBSCRIPTION RATES.

THE NEWS		THE TIMES	
Per Month,	$0 75	Per Month,	$0 45
Per Year,	9 00	Per Year,	5 20
Sunday Only, yr.,	2 50	With Sun. News, per month,	65
Weekly News and Colorado Weekly Times (combin'd) per year,	1 00	With Sun. News, per year,	7 80

ADDRESS
THE NEWS-TIMES PUBLISHING CO.
Denver, Colorado.

RICO BOTTLING WORKS
I. N. SIMPSON, Proprietor.
Wholesale Dealers in Zang's Pilsener, Anheuser's Faust, Pabst's Blue Ribbon.

THE ONLY ICE DEALERS IN THE CITY.
Manufacturers of All Kinds of Carbonated Goods.
BOTTLED BEER TO ORDER FOR FAMILY USE

THE Board OF Trade
We Study To Please
OLD STONEWALL WHISKIES.

THE DENVER & RIO GRANDE AND THE RIO GRANDE WESTERN
THE POPULAR LINE TO
COLORADO SPRINGS, LEADVILLE, CRIPPLE CREEK, PUEBLO, ASPEN, GLENWOOD SPRINGS, SALT LAKE CITY, GRAND JUNCTION, HELENA, OGDEN, BUTTE, SAN FRANCISCO, LOS ANGELES, TACOMA, SEATTLE, PORTLAND. ★ ★ ★ ★ ★
REACHES ALL THE PRINCIPAL TOWNS AND MINING CAMPS IN COLORADO, UTAH AND NEW MEXICO.

THE TOURIST'S FAVORLTE ROUTE
TO ALL MOUNTAIN RESORTS.
The Only Line Passing Through Salt Lake City Enroute to the Pacific Coast.

THROUGH SLEEPING CARS BETWEEN DENVER AND
CRIPPLE CREEK — SALT LAKE CITY
LEADVILLE — OGDEN
GLENWOOD SPRINGS — POTRLAND
GRAND JUNCTION — SAN FRANCISCO
LOS ANGELES
CHICAGO, ST. LOUIS AND SAN FRANCISCO
DINING CARS SERVICE A LA CARTE ON ALL THROUGH TRAINS

E. T. JEFFERY, President, DENVER, COLO.
RUSSELL HARDING, Vice-Pres. and Gen'l Mgr., ST. LOUIS, MO.
J. M. HERBERT, Manager, DENVER, COLO.
A. S. HUGHES, Gen'l Traffic Manager, DENVER, COLO.
S. H. BABCOCK, A. G. Traffic Mgr., SALT LAKE CITY, UTAH.
S. K. HOOPER, Gen'l Passenger and Ticket Agent DENVER, COLO.

Rock Island System
8000 MILES
THE MOST CONVENIENT LINE
To Chicago, Kansas City, Omaha, St. Paul, St. Louis, Hot Springs and all points in the East
Through Sleeping Cars to Chicago, Kansas City, Omaha, St. Louis, Memphis, Little Rock and Hot Springs, Ark.
Best Dining Car Service. Mid-day Luncheon, Fifty Cents.
C. M. VAN LAW, C. P. A., 800 17th St., Denver. W. W. WOOD, C. P. A., 2 Pike's Peak Ave., Colorado Springs. C. H. SCHELL, C. P. A., Third and Main Streets, Pueblo.
HAL S. RAY, General Agent, Denver, Colorado.

MINING AND SCIENTIFIC PRESS
FORTY-THIRD YEAR.
48 Pages : Weekly : Illustrated.
INDISPENSABLE TO MINING MEN.
$3 PER YEAR, POSTPAID.
SEND FOR SAMPLE COPY.
MINING and Scientific PRESS
330 MARKET ST., SAN FRANCISCO, CAL.

FOR GOOD JOB PRINTING
CALL AT THE NEWS OFFICE.

ANN AND JAMES FERANDO PHOTO

SHOT ON A WINTER DAY in the 1940's, this view—where the photographer was looking west from Glasgow Avenue—showed that the hoist house and headframe at the Atlantic Cable Mine had been kept in fairly good condition.

CHARLES MILLER ENGEL COLLECTION – COURTESY RACHEL A. HARDWICK

THE ATLANTIC CABLE MINE was located at the north end of Rico, on the west side of Glasgow Avenue. Silver Creek ran beneath the bridge that connected the hoist house, at the left, with the shaft house. The hoist house contained an air compressor that provided fresh air, while a steam boiler powered the hoist with a multi-strand wire rope (cable), which raised and lowered the ore

bucket in the shaft below the sheave wheels. The office and change room probably were in the building at the right. The miner at the far right had just pushed the small four-wheel ore car out of the headframe house and was about to dump its load into the ore bin. A wagon would then transport the ore to a mill to be processed into concentrate—before going to a smelter to be refined. A spare bucket for the mine shaft was standing beside the shaft house. Miners were lowered into the mine in these buckets to the level where the mine was being worked. Glasgow Avenue is in the foreground of this view. It should be noted that there were originally two sheave wheels atop the headframe, although the wheels cannot be seen in later pictures.

DELL A. McCOY PHOTOS

WITH HIS CAMERA pointed toward the west, Dell McCoy photographed the Atlantic Cable Mine's surface structures from Glasgow Avenue, with Expectation Mountain looming in the background. The headframe is in the foreground, with the hoist house on the opposite bank of Silver Creek. The four pictures on Pages 298–299 were taken during 1966.

CHANGING HIS LOCATION, Dell then shot the west and south sides of the hoist house, with Nigger Baby Hill in the background.

DELL A. McCOY PHOTOS

LOOKING SOUTHEAST from the bed of Silver Creek in Rico in 1966, the photographer shot this view of the west side of the Atlantic Cable Mine's headframe. The picture shows tho heavy timbering used in the construction of the headframe.

THE NORTH AND WEST sides of the Atlantic Cable's hoist house were photographed from the bed of Silver Creek, with the outhouse in the foreground.

ANN AND JAMES FERANDO PHOTO

DURING THE EARLY 1940's, the headframe (at the right) and a part of the hoist house of the Atlantic Cable Mine were photographed from Glasgow Avenue, as the camera faced toward the north. All of the buildings were still on the property, and two ore cars were positioned on the tramway track, beneath the headframe. The remains of the mine's footbridge is visible.

NEW ORE FOR RICO

JULY 8, 1907—Calamine, a mineral new to the Rico, Colo., district, has been found in the Atlantic Cable mine of the United-Rico Mines Company. It is found in the south upraise from the lowest level of the mine, which is silicate of zinc, such as is found in rich ores in Newman Hill and Dolores Mountain. The upper workings, containing probably the largest body of zinc blende in the United States, are all in "Black Jack" zinc, or zinc associated with lead and copper.

— *From The New York Sun*
COURTESY MARGARET MATZICK

DELL A. McCOY PHOTOS

FROM THIS ANGLE, the head frame and ore bin are shown, with the west and south sides of the structure in view, photographed during 1966.

THE HOIST HOUSE for the Atlantic Cable Mine was in a state of deterioration when this picture was taken during late September of 1966. This view shows the south and east walls (or elevations) of the hoist house and the south side of the headframe, with Silver Creek at the right. The headframe and ore bin appear at the right, in the lower view (at the left).

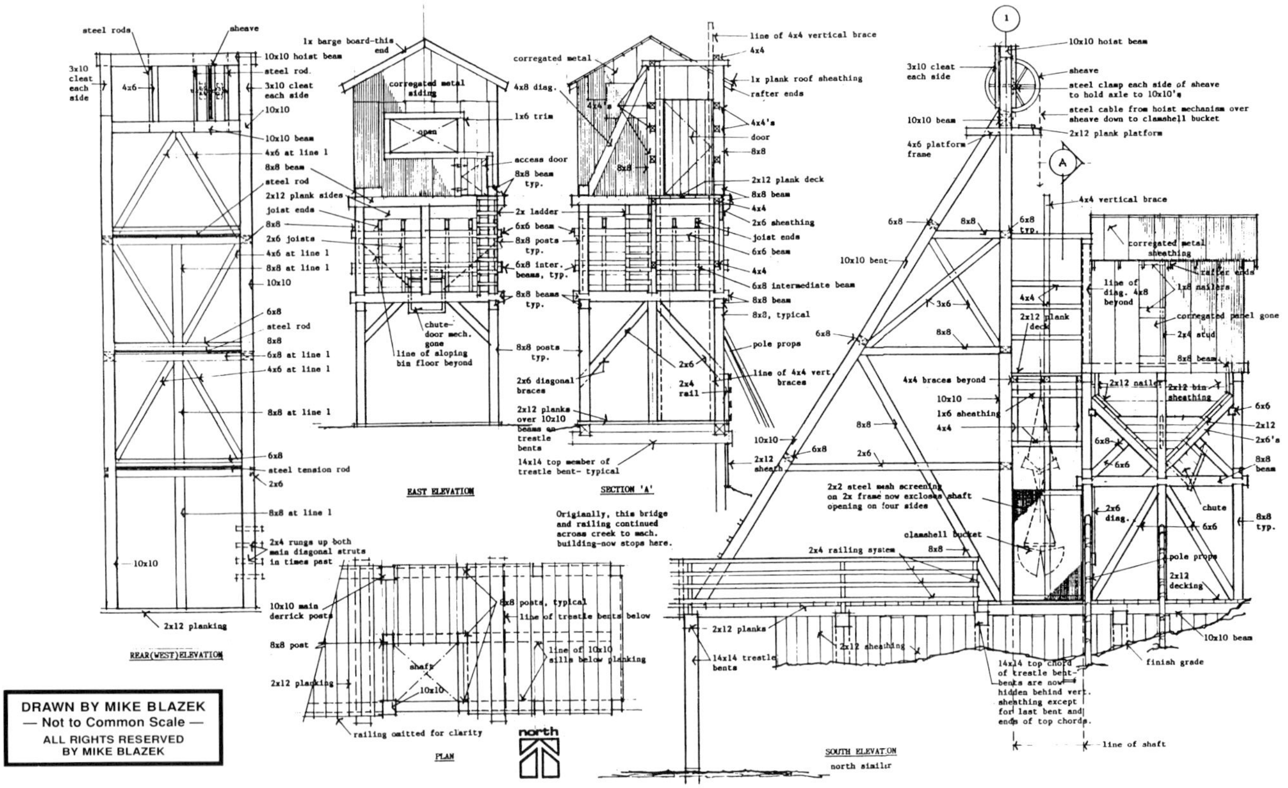

ATLANTIC CABLE HEADFRAME AND HOIST HOUSE

The 1926 renovation used many of the old buildings and foundations of the Pro Patria mill. All the machinery, however, was scrapped, and the latest state-of-the-art equipment was installed to increase the mill's efficiency to a higher level. Operating capacity was increased to 250 tons of ore a day, and the new mill was one of the most modern selective-flotation plants to be built in Colorado.

However, no sooner had the mill been renovated and put into operation than the Western Colorado Power Company completed an 11,000-volt transmission line to Rico. This made the old coal-fired power plant unnecessary (including the one that had been remodeled by the Syndicate Mining & Milling Company in 1916).

In 1929, the U.S. stock market plummeted. Shortly after that, metal prices collapsed again—and so did mining operations in Rico. The Great Depression was real, and only two mining-and-milling companies were left operating in Rico after the crash—the Pell-Eyre Mining & Milling Company and the St. Louis Smelting & Refining Company. Elizabeth Pellet had gone into politics (and she eventually was elected to the Colorado General Assembly, and later, she became a...[Colorado state] senator), but Robert continued in the mining business.

Exactly what happened has been lost in the mists of time, but somehow his operations came to include the newly renovated Pro Patria mill. It appears that he never owned it, but only had an option to operate it. He was backed by the International Smelting & Refining Company—and presumably, a fee or percentage of the profits went to Rico-Argentine, the owner of record. By exploiting a vein of high-grade silver ore, discovered just after the crash, Bob Pellet was able to keep the Pro Patria operating, and the workers on the job.

During 1937, with World War II looming on the horizon, metal prices began to rise. Copper, lead and zinc were in demand again, and Rico had all of them. Then, in 1941, the Rico-Argentine operation acquired most of the holdings of the International Smelting & Refining Company. In the process, Rico-Argentine got back full control of the Pro

(Continued on Page 307)

JOSEPH E. BYERS PHOTO – BILL PLUNKETT COLLECTION

A SEGMENT OF THE VIEW that follows on the next two pages has been enlarged to show more detail on RGS 2-8-0's, Nos. 12 and 17. Homes in Piedmont (just outside Rico) appear along the west bank of the Dolores River, while others were on the hillside.

OVERLEAF: The townsfolk of Rico no doubt were awakened on many occasions with all the locomotive whistles calling out signals in the railroad yard west of town. This interesting three-engine train was standing still while being photographed, as indicated by the smoke and the three men standing beside it. Photographed ca. 1908, this train was hauling 10 carloads of ore concentrate upgrade toward Lizard Head Pass, en route to Ridgway. Some of the boxcars may have been secondhand Rio Grande Western 4000-series cars; however, the first car behind Engine 17 was a newer 3000-series D&RG boxcar. RGS 2-8-0 No. 17 was the train's road engine, while 2-8-0 No. 12 was a helper (which would be cut off from No. 17 at the summit). Notice the flanged domes, diamond smokestacks and oil-burning headlights on the locomotives, all of which help to date this picture. Another helper engine was coupled to the rear of Caboose W-404. (The "W" stood for "way-car," a common railroad term in Colorado, referring to a crew car, or caboose, going "all the way" to the train's destination.) All of the engines were Class 60 Consolidations (later reclassified as C-16 2-8-0's). What later became Colorado Highway 145 was located on the hillside beyond the train on the west side of the Dolores River. It dropped down and crossed the Dolores River on a wooden bridge (behind the rear helper) and entered Rico via Glasgow Avenue. By 1880, several houses and out-buildings had been built at the northern end of Rico. This flourishing village was named "Piedmont." High on the slopes of Expectation Mountain was the Rico Power & Light Company's flume, which brought Dolores River water to the hydroelectric-generating equipment in the powerhouse at Rico.

JOSEPH E. BYERS PHOTO – BILL PLUNKETT COLLECTION

12

STEVE FRAZEE COLLECTION

ONE OF THE LAST VIEWS of the original two-stall wooden enginehouse in Rico, this shot was photographed prior to the construction of the larger brick enginehouse. It is believed this picture was taken during 1907, a short time before the Dolores River rampaged through the RGS yard—destroying ties and rail in the yard. Much to your delight (hopefully), an unknown photographer shot the northbound passenger train as it was standing at the depot in Rico. A breeze carried the locomotive's smoke toward the north, up along the mainline. Stacks of lumber were still in view, beyond the B&B work cars, which had been spotted on the spur beside the old wooden enginehouse. A drag flanger also was standing on the spur, and RGS rotary snowplow No. 2 was sitting on a spur track to the left of the enginehouse, just west of the passenger train. The back walls of the RGS bunkhouse and section house stand to the left (north), with their "water closets" (outhouses) next to the Dolores River.

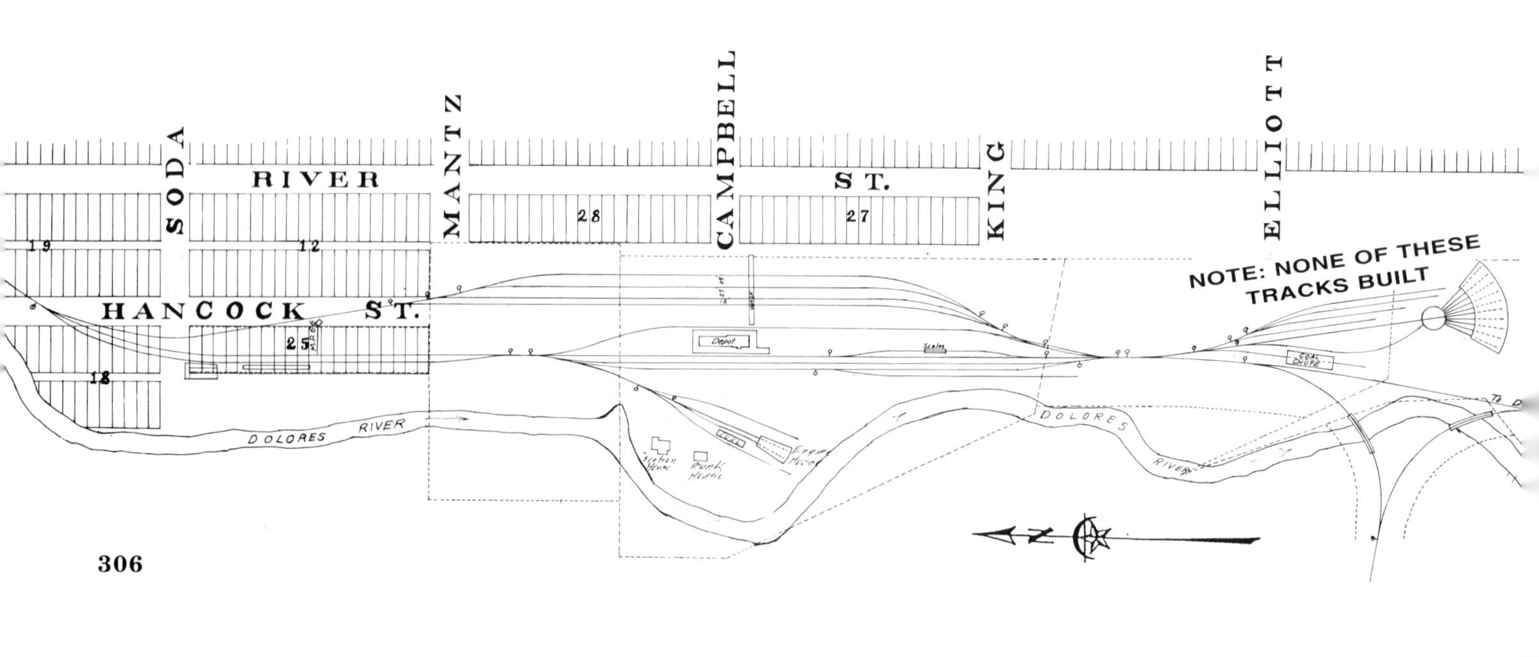

—— PROPOSED CHANGES ——

RICO STATION GROUNDS

RIO GRANDE SOUTHERN

CA. 1910

JOHN W. MAXWELL COLLECTION

THIS SKETCH shows proposed additions to the Rio Grande Southern's yard at Rico. Notice that a large 10-stall roundhouse had been planned for this division point.

Patria mill. Lead and zinc were the principal ores mined and processed in Rico during the war years.

The Pro Patria mill closed just after the end of World War II, but it may have reopened in 1950 for a brief period. In any case, by the mid-1950's, the mill was in an advanced state of decay, and only the memories remain today.

NOTE: This story was previously published in the ***Narrow Gauge and Short Line Gazette***.

CHARLES MILLER ENGEL COLLECTION – COURTESY OF RACHEL A. HARDWICK

OF GREAT INTEREST in this view, taken ca. 1910, is the old wooden two-stall enginehouse still in place beside the new brick three-stall enginehouse. Two enginehouses appear side-by-side in the upper center of this view. The RGS annual report for the year ending on June 30, 1910, lists an enginehouse and coal chute costing $12,614.51. (This information is available at the Colorado Railroad Museum.) The second RGS wye, located at the southern edge of Rico, was still in use, crossing the river twice in the process. In the foreground, track was still in place on the railroad's abandoned Enterprise Branch. This print was enlarged 500 percent from the small original print, which was not as sharp as it should have been. However, you can compare buildings in this picture with those drawn on the Sanborn Map of 1899, pinpointing older structures. Illustrated on the right-hand page, three lots remained vacant in Block 13, where the Hotel Enterprise stood before the fire of 1903. Farther north, on Glasgow Avenue, is the large wagon shop and horse barn, located in Block 14, Lots 38, 39 and 40. The second Rico school is visible in the right foreground.

ANDY PAYNE COLLECTION – COURTESY OF BILL PLUNKETT

THIS SNOWY SCENE at Rico station might well have been photographed during the winter of 1906. RGS Engine 19 had arrived with a plow-flanger and a caboose, running ahead of a passenger train to clear snow from the line. The train had plowed heavy snow over Lizard Head Pass, and the crew was glad to arrive at Rico. The little Rico depot must have been a busy place, with two trains spotted there simultaneously—in view in the distance, at the right. RGS No. 19 was a Baldwin 2-8-0 built in 1882 for the D&RG, and she had been acquired by the RGS in 1891. Notice that the caboose was carrying flags rather than marker lamps, a common practice at this time.

THE REVENUE AND PRO PATRIA mines were photographed from the south slope of Nigger Baby Hill, looking across the valley of Silver Creek toward Dolores Mountain. (The Revenue was above the Pro Patria.) Mine trackage, sheds, a bunkhouse, blacksmith shop and upper tramway house are all in sight at the Pro Patria.

ORVAL L. JAHNKE COLLECTION

COLORADO RAILROAD MUSEUM COLLECTION
COURTESY OF MICHAEL KOCH

THIS HANDSOME CLASS 60 (C-16) 2-8-0 had been serviced at the Rico enginehouse, located next to the RGS section house and other small buildings. Engine 15 may have been in service as a helper locomotive on the line up to Lizard Head Pass. In any case, the engine had encountered snow on a run during 1906. Automatic "Janney" couplers were now standard equipment on the RGS and D&RG narrow-gauge lines; however, this engine still had a D&RG-type oil headlight. No. 15 was another Baldwin 2-8-0 built in 1881 for the D&RG and acquired by the RGS in 1891.

BRIEF RGS NOTATIONS...

Class 60 2-8-0 Engines 13, 18 and 21 were dismantled in 1916, while Nos. 3, 6, 10, 15, 16, 17, 20 and 21 survived into the 1920's. Nos. 1, 2, 4, 5, 7, 8, 9 and 20 were out of service before the 1920's.

During September of 1900, the following RGS engines were in service on the northern First District: Nos. 3, 4, 12, 13, 15, 17, 18, 19, 20, 22, 25 and D&RG 176 (a 4-6-0). On the Second District, Engines 2, 4, 6, 15, 16, 18, 22 and 25 were in service. At the same time, Nos. 4, 12, 13, 17, 19 and D&RG 176 worked on the Telluride Branch. (Engines 17 and 19 were the ones most often used on this branch.)

October of 1900 saw the following engines in service on the First District: Nos. 3, 4, 13, 15, 17, 19, 20, 21, 22 and 25. Engines 2, 3, 4, 16, 18, 19, 20, 22 and 25 were commonly used on the Second District. On the Telluride Branch, Nos. 15, 17, 19, 21 and 22 were in use.

Other months in 1900 saw the same engines in general service on both the Ridgway–Rico district and the Rico–Durango district. Detailed rosters of RGS locomotives, as well as freight and passenger cars, will be printed in the last volume of this series.

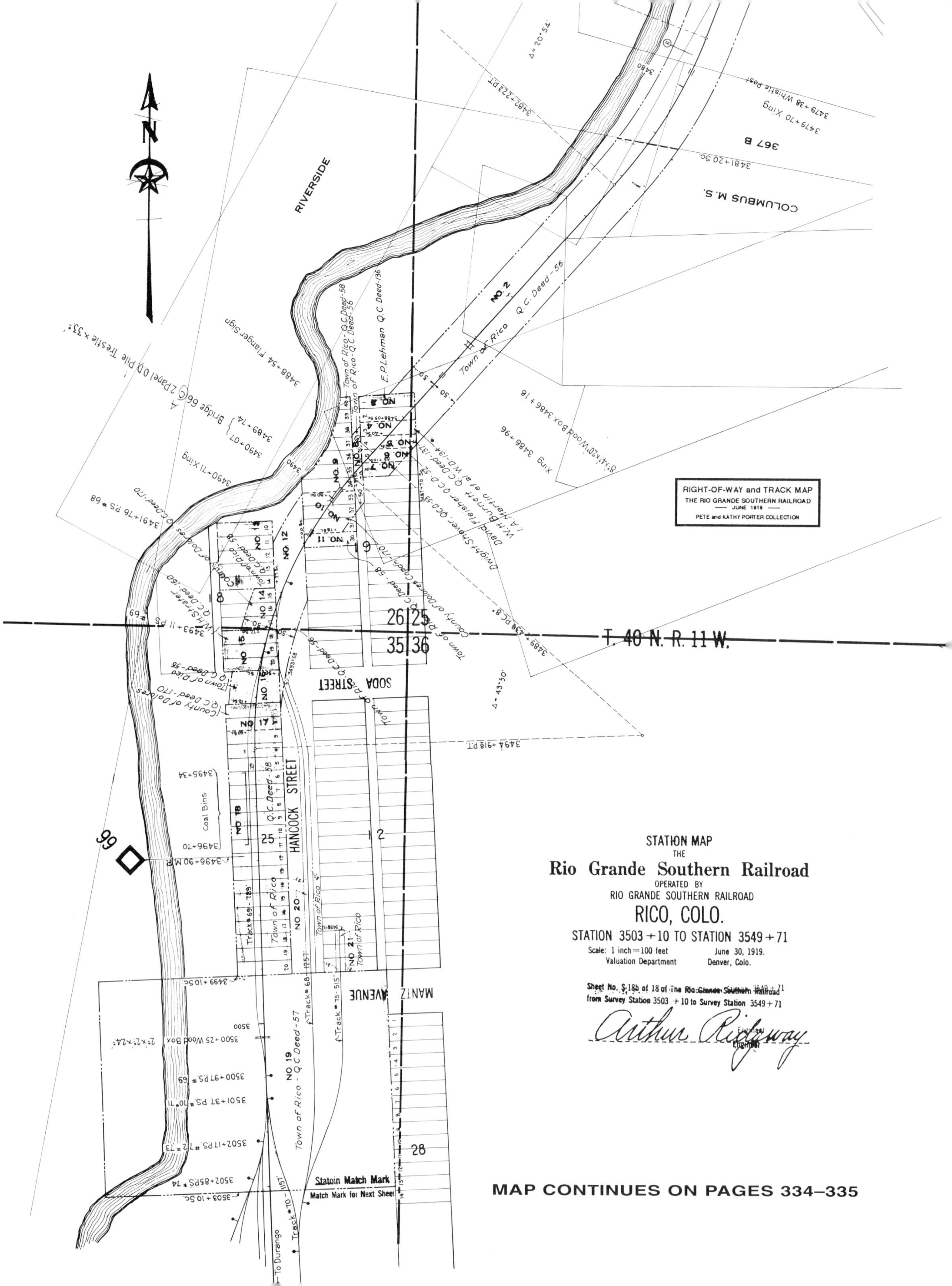

MAP CONTINUES ON PAGES 334–335

Mile by Mile to...

The Rico Division Point

The End of the First District and the Beginning of the Second District

By Russ Collman

The Big Flood in October of 1911

No. **8**

RICO—Milepost 66.2, Elevation 8,737 feet—originally **CARBONATE CITY**. The early days of this mining town and its little narrow-gauge railroad have been covered in earlier chapters of this volume. The Rio Grande Southern briefly thrived—as did Rico and the Pioneer Mining District after the arrival of the railroad. The freight and passenger business of the RGS reached a high point during the early part of 1893.

The fortunes of the RGS for the first half and the second half of 1893 are two totally different stories. Net earnings for the first six months of 1893 were reported to have been $340,486, almost triple what they had been for the same period a year earlier. However, when the mines in the Rico district closed down, the earnings disappeared. The "Silver Panic" also made it impossible for the RGS to sell many of its first-mortgage bonds, a holdover from the railroad's cost of construction. As a result, the railroad was forced to go into bankruptcy, and subsequently went into receivership on August 2, 1893. The Denver & Rio Grande Railroad had anticipated that the RGS would be a good moneymaker, and it had bought a large number of RGS bonds as an investment. Therefore, it was logical that E. T. Jeffery, president of the D&RG, was appointed as receiver. The RGS had operated as an independent railroad company for only 18 months. Although the receivership lasted for only 28 months, Otto Mears never again regained a controlling interest in the RGS—which proved to be a most unfortunate circumstance for the little railroad. The far-distant management of the D&RG cared little about their "step child" in the San Juans, and the condition of the RGS worsened over the years. However, local authorities of the RGS occasionally managed to show an interest in Rico. On December 10, 1898, **The Rico News-Sun** reported the management of the RGS contributed funds to build a fence around the public cemetery.

STATION MAP OF RICO
1910
DRAWN BY JIM KEY

3. SAMPLING WORKS
4. RICO P&L WAREHOUSE
5. S. J. HOOD LUMBER YARD
7. COAL CHUTES
8. RICO P&L POWER PLANT
11. PRO PATRIA MILL
16. RGS WATER TANK
17. **RGS DEPOT**
22. SAND HOUSE
24. FIRST ENGINE HOUSE (WOODEN)
25. SECTION HOUSE
26. BUNKHOUSE
27. COAL SHED
28. WOODEN BOARDWALK
33. SECOND ENGINE HOUSE (BRICK)
37. DEPOT OUTHOUSE
39. TRACK SCALE
40. SECOND WYE
41. RGS BRIDGE 67-AS
42. RGS BRIDGE 67-BS
43. SECTION-CAR SHED
45. LOCATION OF FIRST WYE
46. RICO SMELTING CO. (ADAMS)
47. PASADENA REDUCTION CO. (TRACT)
48. PENSTOCK FOR POWER COMPANY
49. PENSTOCK SUSPENSION BRIDGE
67. BURNETT & CLIFTON COALING FACILITY

A newspaper report in **THE SAN MIGUEL EXAMINER** early in 1909, stated that "...the Rio Grande Southern is to build a new enginehouse in Rico...." And the D&RG's annual report dated June 30, 1910, shows that the brick enginehouse (as well as the coaling facility south of town) had been constructed prior to that date, ***before*** the old wooden enginehouse was torn down—courtesy of information provided by Wally Maxwell, from the Colorado Railroad Museum's collection. A photograph printed in this volume shows this to be true. (Of interest is the fact that it was widely held that the second enginehouse replaced the first one after the original structure had been destroyed by a flood.) The second enginehouse had three run-through tracks, with the capacity to hold six engines. The three entrances on each end of the enginehouse were protected by large wooden doors—on both the north and south ends of the buildings—and the three approach tracks originally connected with the mainline at both ends of the yard. This brick structure measured 125 feet-8 inches x 60 feet, according to the D&RG plans for the building, printed on August 12, 1909. Thanks go to John W. Maxwell for the information regarding accurate dimensions.

Presumably, the flood of 1911 undermined at least part of the old enginehouse, knocking it off its foundation. The flood did other damage as well, and the wooden building probably was removed by 1912. During

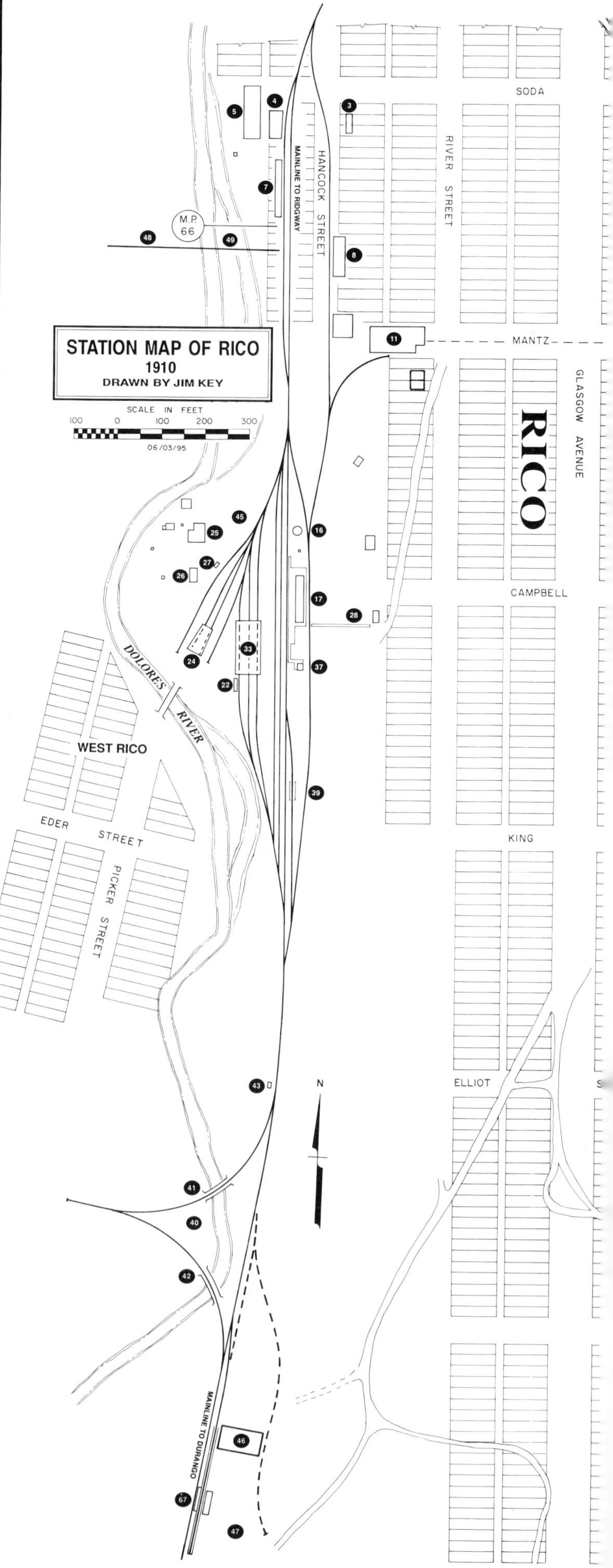

C. D. WOLFINGER PHOTO – JAN ROBERTS COLLECTION

C. D. WOLFINGER was the superintendent of the RGS between 1900 and 1910, and he shot a number of photographs while traveling over the line in the business car "Montezuma." For this trip, he had been accompanied by friends and was at a stopover in Rico. An oxide red Denver & Rio Grande refrigerator car had been spotted on the Rico house track, east of the depot. The white stuff beside the reefer no doubt was slush cleaned out of the car's ice bunkers. A boarding house and other dwellings are in view, above the embankment, where the town is located.

Adjacent to the small firehose shed—near the depot—there was another shed used by the Rico–Argentine Mining Company to store shovels, mallets and other tools used when ore was being dumped and loaded into boxcars at their loading ramp, east of the house track (next to the depot's access road.)

South of the depot's access road, there was a large building that housed an electric power plant for a short time. This plant was built in 1925, under the direction of the St. Louis Smelting & Refining Company, for the Rico Mining & Reduction Company. This was the structure with the three tall smokestacks, and it provided electricity for the company's processing mill. The Rico Mining & Reduction firm did some processing of metallic ores that came from the St. Louis Tunnel. However, the life of the power house was short—only about a year—be-

the same flood, two of the southern approach tracks to the brick enginehouse were partially washed out, undercut by the river, and only the first track (nearest the yard) remained connected to the yard tracks. This must have made providing sand for locomotives somewhat difficult since the sandhouse was just outside the southwest end of the enginehouse, on the western-most approach track. The fact that all locomotives had to go completely through the enginehouse and out the other end to use this facility did not make much sense.

C. D. WOLFINGER PHOTO – JAN ROBERTS COLLECTION

RGS SUPERINTENDENT Wolfinger had the use of the company's business car "Montezuma" (originally the "Rico") for his inspection trips over the line. While at the Rico depot, his porter took advantage of a free ride through the station grounds on one of the local burros.

CHARLES MILLER ENGEL COLLECTION – COURTESY RACHEL A. HARDWICK

WHILE ONE HOME OWNER in Rico was playing poker next door with friends, his house was carried downriver by the Silver Creek flood of 1911. In desperation, he drew his keys from his pants pocket and threw them into the raging flood water, shouting in so many words, "Here! You can have these, too!"

cause the Western Colorado Power Company ran a high-voltage line into Rico in 1926. Nevertheless, the building remained at this location for many years, perhaps used as a warehouse by the mining company.

Rico's colorful wooden depot retained its original appearance throughout most of its existence. However, the unusual tower of the depot was removed from the building after the second story was damaged by fire in 1936.

Next to the Rico water tank, to the north, was a section car and tool shed, first seen here ca. 1935. This building probably was moved from the wye area, west of the main track. And to the north of this shed there was

CHARLES MILLER ENGEL COLLECTION – COURTESY RACHEL A. HARDWICK

SILVER CREEK FLOODED through the town of Rico during October of 1911. On occasions, this happens after steady rain has come down for a period of a week or so. The mountains can absorb only so much moisture. Then, it runs off the surface to become a flood down the creeks and rivers. Silver Creek had burst a hole through the rock foundation of this brick building in its path.

CHARLES MILLER ENGEL COLLECTION – COURTESY RACHEL A. HARDWICK

SEVERAL HOUSES were destroyed during the October flood of Silver Creek in 1911. Remains of a bridge and buildings litter the creek banks after the high water had taken its grim toll. Some persons were left homeless, losing every possession they had. This picture was taken near the main street through town—Glasgow Avenue.

a coal bin and a place for kindling wood. Cabooses were spotted on the house track, near this little structure, as the caboose stoves obviously needed a supply of kindling (for starting fires) and coal. This shed was in place before the section-car shed arrived on the scene.

To the northeast of the water tank, there was a bulk-oil plant, located on the track leading to the Pro Patria ore-processing mill. This plant probably was a Conoco facility, because there definitely was at least one Conoco dealer in Rico during the 1920's and 1930's. And fuel oil was sold during this period to provide a supply for homes and businesses using oil-fired heating stoves.

The largest building in the yard area was the Pro Patria mill, sometimes called the Mantz Avenue mill. This ore-processing plant was built by David Swickhimer's Pro Patria mining firm in 1902. An aerial-tramway line brought ore down to the mill in buckets from the Pro Patria tramhouse, located above the grade of the Enterprise Branch, on Dolores Mountain. The aerial tramway was easy to see, located as it was in the middle of Mantz Avenue—going right past the Dolores County Courthouse. This mill operated on an on-and-off-again basis until 1948, when it was completely shut down. (It may have operated for a short period after 1948; however, the facts are unclear for this period.) Before that, in 1926, this facility was renovated by the International Smelting & Refining Company, which installed a selective-flotation mill. During 1941, the mill was taken over by the Rico–Argentine Mining Company, and it was operated through World War II, Rico's last big mining boom (as far as highly valued me-

CHARLES MILLER ENGEL COLLECTION – COURTESY RACHEL A. HARDWICK

AFTER THREE MONTHS of intermittent heavy rains and early snowfalls had soaked the mountain ranges surrounding Rico in 1911, a terrible flood rolled through the town in the Silver Creek channel. This normally small stream quickly swelled to a path 200 feet wide and roared down through the community and out into the Dolores River. This photograph shows the effect Silver Creek had after the flood waters had passed. The public school building is recognizable (at the left). This destructive flood also washed out a considerable amount of the Rio Grande Southern's freight yard and railroad buildings along the Dolores River.

GREAT DEVASTATION from the flood of October, 1911, moved almost every railroad tie in the Rico yard to some degree. At some locations, the railroad track was suspended in midair. This snapshot was taken as the camera was pointed toward the north, from the enginehouse roof. The enginehouse lead tracks, containing a three-way stub switch, are in the foreground. Several pieces of RGS rolling stock had already arrived in Rico, and they are visible on a hastily re-laid track at the right. One of the business cars is at the right, partly hidden by the water tank. Unfortunately, the sequence of pictures taken at this time did not include the enginehouse or depot.

COLORADO HISTORICAL SOCIETY COLLECTION

tallic ores were concerned).

Just north of the Pro Patria mill, there was an electric power plant, operated by the Rico Power & Light Company. This company provided electricity before the Western Colorado Power Company built its power line into town. For sometime, this plant was a hydroelectric operation, with a penstock coming down the hillside west of the Dolores River. This pipe crossed the river and went under the tracks to reach the power plant. A small building south of the plant evidently housed an alternative steam-generating facility (with a boiler) as a backup, in case the stream flow was low.

An eight-pocket Burnett & Clifton "Automatic Coal Chute" for engines was built at the southern end of the yard limits. For some reason, this trestle and tipple was not constructed until 1910. Coal cars (gondolas) were shoved up on the trestle by an engine to the delivery area, and the coal had to be hand-shoveled either directly into the pockets, or onto the storage platform if the coal was wet, or if there was an excess amount. From the storage platform, the coal had to be shoveled again to the opposite side of the track into the individual pockets. The spur for this facility was nearly 600 feet long, located a short distance south of the final wye.

CHARLES MILLER ENGEL COLLECTION –
COURTESY OF RACHEL A. HARDWICK

FROM THE EASTERN SLOPE of Expectation Mountain, Rico was captured on film as the photographer aimed his camera almost directly toward the northeast. This view probably was taken during the summer of 1912. The purpose for taking this picture was to present a record of the recovery work that had gone on after the bad flood during October of 1911. You can see that much of the damage has already been repaired.

ROLLING STOCK IN THIS VIEW includes a string of Bridge & Building Department work cars, lined up on the Pro Patria mill spur. A passenger car, believed to be the business car "Montezuma," was also on this track (partly hidden by a pine tree) near the RGS water tank. Notice damage done to the west wall of the Pro Patria mill, which took place in the fire of 1908. Flood water had cut a channel in front of the RGS section house and bunkhouse, and left the coal shed leaning at an angle. The old original enginehouse, which does not appear in this picture, may have received damage from the flood—after water undermined the foundation of the structure. It may have been removed late in 1911.

THE RGS DEPOT appears to have somewhat of a hump in the middle of the freight room, believed to have happened as flood water surged beneath the floor and shoved the freight room against the larger and heavier part of the building. Later photographs seem to document this assumption, although no records are available to tell of this damage.

THE ACCESS ROAD to the depot and yard suffered some damage because several sections of the log cribbing were washed out along the embankment by the heavy rainfall. Two D&RG boxcars appear in the south end of the yard, with an RGS drag flanger. The track to the right of these cars had been washed away, and two of the tracks leading to the southern end of the enginehouse were never reconnected to the mainline. Only the easternmost track was re-laid to its mainline turnout. You can see ties hanging out over the river bank to the right of the D&RG boxcar south of the flanger. What was known as West Rico by this time can be seen in the foreground. The houses in this area escaped the flood because they had been built above the lower portion of the Dolores River bottomland. The two-story Miners Union Hospital (with the front porch) is in this view, and the nurses' quarters were across Eder Street, in the house to the south.

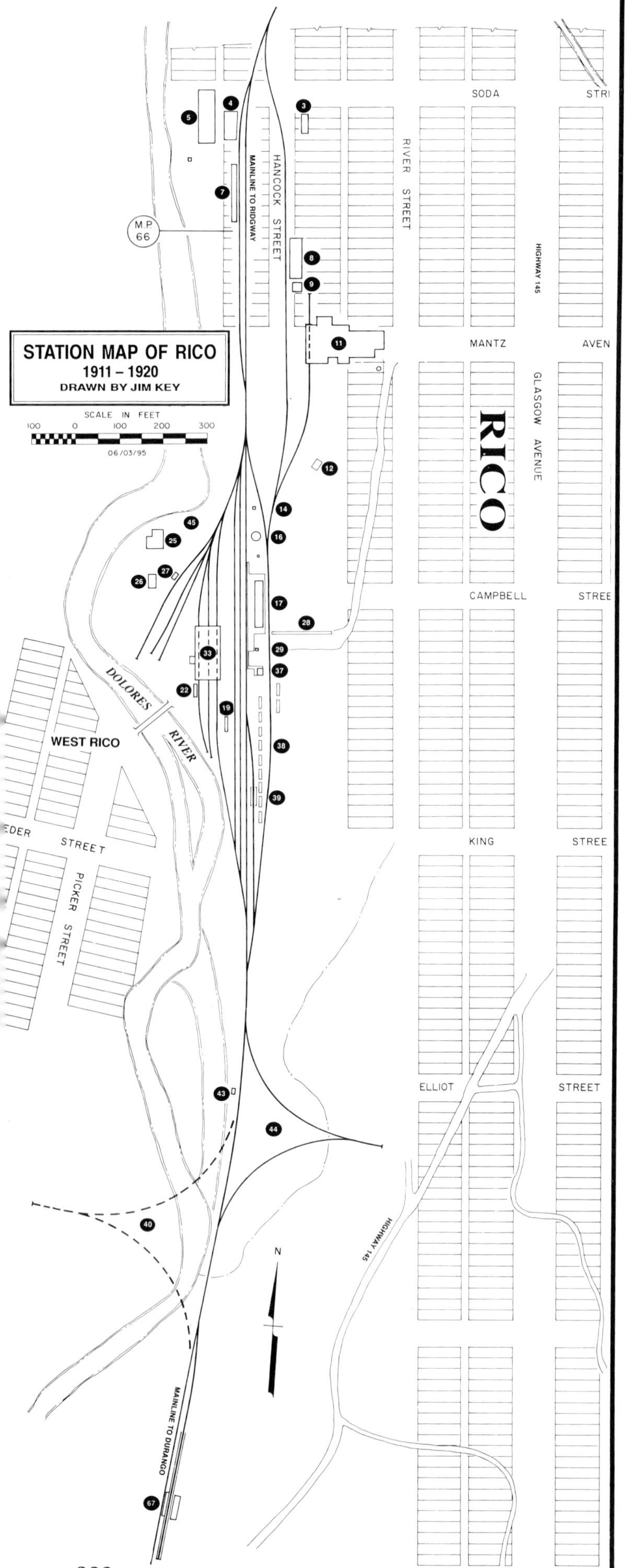

STATION MAP OF RICO
1911 – 1920
DRAWN BY JIM KEY

3. SAMPLING WORKS
4. RICO P&L WAREHOUSE
5. S. J. HOOD LUMBER YARD
7. COAL CHUTES
8. RICO P&L POWER PLANT
9. STEAM BOILER HOUSE
11. PRO PATRIA MILL
12. HOUSE
14. CABOOSE COAL & KINDLING BOX
16. RGS WATER TANK
17. RGS DEPOT
19. RPO CAR BODY FOR ROADMASTER
22. SAND HOUSE
25. SECTION HOUSE
26. BUNKHOUSE
27. COAL SHED
28. WOODEN BOARDWALK
29. FIREHOSE BOX AT DEPOT
33. SECOND ENGINE HOUSE (BRICK)
37. DEPOT OUTHOUSE
38. BRIDGE & TRESTLE TIMBERS
39. TRACK SCALE
40. SECOND WYE
43. SECTION-CAR SHED
44. THIRD (AND FINAL) WYE
45. LOCATION OF FIRST WYE
67. BURNETT & CLIFTON COALING FACILITY

Prior to the construction of this coaling facility, coal had to be hand-shoveled into tenders from a wooden platform (dock) adjacent to the original wooden enginehouse.

The stockyard at Rico was just south of the coaling pockets. It was similar to other livestock pens and loading chutes on the RGS, and it changed somewhat over the years. This was primarily because sheep came to dominate the type of livestock moved by the railroad. Originally, the stock pens were used for loading cattle into single-deck stock cars; however, it was later rebuilt to serve sheep ranchers, and a second loading chute was added. And both chutes were built to load sheep into double-deck stock cars, although cattle could still be shipped from the yards.

After the Rio Grande Southern finished laying track into Rico on September 30, 1891, there was still a considerable amount of work to do there—and from Rico south-

CHARLES MILLER ENGEL COLLECTION – COURTESY OF RACHEL A. HARDWICK

THE RGS YARD AT RICO was back in reasonably good condition when this photograph was taken, about 1913, two years after the terrible flood damage of 1911. An old harp switchstand was still in use at the stub-rail type of switch. The photographer was standing on one of the tracks leading into the enginehouse. The Pro Patria mill had been remodeled, and it remained in operation for a short time. By the time this picture was taken, the Rico Power & Light Company's plant had acquired more smokestacks to provide for additional coal-fired boilers. (Each boiler required its own individual smokestack.)

ward. Tracklaying continued before the formal celebration was held in Rico, and the last spike was driven into a rough-hewn tie on December 19, 1891. This was at a point about one mile beyond Milepost 84, about half way between Rico and Dolores. Unfortunately, there was still no connecting telegraph line between Rico and Dolores, and when the railroad's first head-on collision occurred during the early part of January, 1892, the lack of a telegraph line was blamed for the crash. (Tracklaying south of Rico will be covered in Volume VI of this series.)

RGS work trains began operating between Rico and Durango as soon as the connecting rails were laid. However, the rough track needed refurbishing in many places, including the line south of Coke Ovens into Rico. It was not until January 31, 1892, that through schedules went into effect between the northern terminus in Ridgway and the southern terminus in Durango. The train dispatcher's office in Rico was kept very busy during the early months of operations, and the records of this period were placed in pigeon holes (slots built in the walls of the Rico depot's freight room, adjacent to the telegraph office). In 1952, the Rico–Argentine Mining Company bought the RGS property in Rico, after the road was abandoned. Regrettably, the Rico–Argentine's manager ordered all of these historic papers burned.

From the beginning, Rico was intended to be both the road's only division point, as well as the station where helper engines were assigned to trains going northbound to Lizard Head Pass.

After regular service began in 1892, a southbound passenger train was scheduled to depart daily from Ridgway—complete with through D&RG Pullman cars from Denver. This train traveled over the railroad's First Division (later called the First District), and it tied-up at Rico overnight. The train continued on its run the following morning for the remainder of the scheduled service to Durango over the Second Division (later known as the Second District). Likewise, a northbound passenger train operated daily in similar fashion from Durango to Ridgway.

Freight traffic also was booming on the Rio Grande Southern. At Durango, the Denver & Rio Grande Railroad reported that

(Continued on Page 328)

CHARLES MILLER ENGEL COLLECTION – COURTESY OF RACHEL A. HARDWICK

THIS VIEW OF RICO was taken not long after the original wooden enginehouse had been removed after the flood of 1911. The brand-new brick enginehouse had been built about 1909 on the bottomland of the Dolores River, just west of the Rico yard and depot. Twelve smokejacks, plus a lengthy ventilator over the middle track, had been mounted on its roof to draw smoke and fumes from the building. An RGS water-service car, No. 01664, was standing in line with work car No. 04409, also being used in water service. Both had ladders on their roofs. Notice the D&RG lettering on the newer 3000-series boxcars, standing about the yard. This lettering style lasted until a "W" was added, after the Denver & Rio Grande Western Railroad was formed in 1921. The smaller boxcars are older 4000-series cars, which had smaller lettering. The town side of the Rico depot contained an entrance door, near the southeast corner, for patrons purchasing tickets. The RGS employees' door was standing open, farther north, also on the east wall.

NOTICE THE SAG in the roof line of the freight room, where it joined the two-story section of the depot. An ornamental tower was positioned above the second floor. This tower contained the telegraph cables, which entered the depot from both north and south. This depot originally contained the chief dispatcher's office, and extra lines were required for his use. Later, the dispatcher's office was moved to Ridgway and Durango.

A HELPER ENGINE stood on the throat track to the enginehouse lead tracks, beyond the water tank, at the far left. It would soon be pressed into service to help a northbound freight train up the steep grade to Lizard Head Pass. Notice the load inside D&RG boxcar No. 3176, spotted in the foreground, on the house track. It appeared to be loaded with bagged ore. Another carload of coal was in a gondola, to the right of the D&RG boxcar just mentioned, while yet another gondola load of coal was next to the depot. On across the river stands the Shamrock Mine.

CHARLES MILLER ENGEL COLLECTION – COURTESY OF RACHEL A. HARDWICK

A WOODEN FLUME that crossed the lower slope of Expectation Mountain and continued down the hillside as a penstock is the most interesting feature of this view. This flume tapped water from the Dolores River at Burns. Of course, the photographer also included the Shamrock Mine buildings, although his intention was to show the water lines that brought the hydraulic power source downgrade and across the Dolores River, via a suspension bridge, to the hydroelectric powerhouse. The penstock (water line) had to go under the RGS trackage to reach the Rico Power & Light Company's electric generating plant (out of sight behind the Pro Patria mill). The Pro Patria mill had a devastating fire in 1908, and some of the debris was still present from the damage done to the auxilary building during this fire, at the time this view was taken ca. 1915.

LOOKING BEYOND THE MILL and penstock, you can see a long wooden coal shed that had been erected beside a Rio Grande Southern spur. Coal was unloaded from solid-bottom gondolas (usually called "coal cars" by railroaders) by workers who had to hand-shovel the black nuggets into open chutes in the side of the bins. Teamsters pulled up to the coal bins to load coal into their wooden wagons. This coal was bought by coal merchants to sell to the residents of Rico, to keep fires going in their heating stoves during the cold winter months, as well as for use in their cooking stoves all year long. Farther to the south (left), you can see an RGS derail, located on the spur serving the coal bins. This derail was controlled by an old ex-D&RG harp switchstand.

DAN PRICE COLLECTION

A STRING of Bridge & Building Department work cars had been spotted on what was officially called "Track 72," across from the RGS depot in Rico, ca. 1915. The work car at the end of this lineup (at the right) appeared previously in this series—on Page 436 of Volume IV—with a mysterious box mounted on the roof. Notice the next two cars in this B&B train appear to be the same two cars photographed at Ophir, as printed in Volume IV, on Page 49. The northbound mainline to Lizard Head Pass begins its climb to the summit at the right-hand portion of this photograph, beyond the water tank.

during the last half of 1892, it received over 4,000 freight cars from the RGS, of which more than 1,000 were carloads of ore from Rico. Locomotives needed for helper service on the RGS mainline over Lizard Head Pass also were stationed at Rico. In addition to helper engines, maintenance-of-way cars and snow-fighting equipment—including the road's big rotary snowplow—and the personnel required for these services, were also stationed at Rico.

(Continued on Page 331)

BILL PLUNKETT COLLECTION

AN UNKNOWN PERSON had walked out onto one of the RGS yard tracks north of the Rico depot to get this snapshot of a locomotive, ca. 1915. The RGS engine appears to be No. 16, a Class 60 2-8-0. The United Rico Mines Company's Pro Patria mill appears at the right. Notice the section car standing at the left, on the three-way turnout.

CHARLES MILLER ENGEL COLLECTION – COURTESY OF RACHEL A. HARDWICK

A LONG STRING of pack burros was loaded with sacks of ore from one of the mines in the district. The burros were headed for the RGS depot grounds, where the ore would be loaded into boxcars. (Burros were nicknamed "Rocky Mountain canaries" for their braying). The Pro Patria mill had been remodeled into a single building, ca. 1912, with the lower portion built over the RGS spur used to load ore concentrates into boxcars. Also notice the wooden water tank, which provided water for the mill. The rear side of the Burley Block building shows at the right.

ONE BY ONE, the pack burros were relieved of their heavy loads of ore at the open boxcar door. A worker inside the car stacked ore bags over each truck, leaving the doorway area clear of the loads. This practice was used to keep undue stress off the middle of the freight car. (The center of this type of car was supported by four truss rods under the car's floor boards. The truss rods also helped keep the end sills from being pulled out.) Cars had been spotted on the house track, east of the depot. The depot was neatly painted in buff yellow and dark brown by this time.

CHARLES MILLER ENGEL COLLECTION – COURTESY OF RACHEL A. HARDWICK

CHARLES MILLER ENGEL COLLECTION – COURTESY OF RACHEL A. HARDWICK

FROM THE VICINITY of Charles Engel's cabin on Expectation Mountain, this scene was photographed about 1912. He was looking out over Rico, to the east. Nigger Baby Hill is at the left, and Silver Creek separates it from Dolores Mountain, at the right. Blackhawk Peak, at 12,681 feet elevation, has been renamed Blackhawk Mountain on the latest San Juan National Forest map. It is in view, rising back of Dolores Mountain on the far right side of this picture. Many of the old mine dumps appear in this scene, with the Enterprise Group Tunnel at the far right. Engineer Mountain appears on the distant skyline as a pyramid cone (near the center of the view), at an elevation of 12,968 feet. This peak is in the vicinity of Coal Bank Pass near Silverton. Whitehead Peak and the Continental Divide rise at the left. This is the most rugged alpine region in Colorado.

During the road's later years—from the 1920's and thereafter—the RGS leased engines from the D&RG, and K-27 2-8-2's were often used between Rico and Ridgway. However, between Dolores and Rico, lighter engines had to be used because of poor track conditions. RGS C-17 No. 42 and C-19's Nos. 40 and 41 were commonly used on runs between Dolores, Rico and Timber Spur during the 1940's. They hauled flatcars loaded with logs for the Montezuma Lumber Company during the summer and fall months.

The RGS had used a large quantity of secondhand 30-pound rail for its original construction, and the pounding freight traffic quickly wore out this light rail. The railroad decided to order heavier rail; however, Otto Mears placed a large order for 57-pound rail—an odd size—for the mainline between Vance Junction and Rico. Perhaps this unusual size was chosen because the Rio Grande Western (in Utah) had placed an order for the same size rail. In any case, the new 57-pound rail was laid during 1892, and this greatly eased the work load of the railroad's section hands. Patented Greer (often called "Jeffery") double-headed spikes also were ordered at this time for use with the new 57-pound rail. At the same time, special tie plates for the spikes also were ordered.

As of early September in 1892, the road had 462 kegs of Greer spikes on hand, enough for 12 miles of track. As of September 8, the new 57-pound rail had been laid nearly to Lizard Head, and later in the year the new rail was laid all the way into Rico.

RICO AND THE RIO GRANDE SOUTHERN

DURING ITS EARLY YEARS, Rico was a booming silver-mining town. However, despite ominous signs that the Nation was on the brink of an economic abyss, the mining town had almost a euphoric atmosphere. Of the estimated population of nearly 5,000 in Rico and its environs at the time the Rio Grande Southern reached this mountain town in 1891, nearly half the people were mine workers. To support this population, there were many establishments beyond the numerous saloons and dancehalls that went with early Western mining towns. Of Rico's hotels, three had electric lights from the beginning of electric service by Rico's local power plant, as well as hot-and-cold running water, improvements unknown in most small towns in the American West. Beyond that, there was a progressive newspaper in Rico, and as the seat of Dolores County, Rico, was a busy place because of this fact.

Unfortunately, Rico's big silver-mining boom lasted for less than two years after the Rio Grande Southern arrived, and for the rest of its mining life, Rico primarily depended on the fluctuations of the copper, lead and zinc markets, as well as depending on such minerals as iron pyrites, used to produce sulfuric acid. This meant that the mines and mills of the local Pioneer Mining District simply shut down during bad times, and the unfortunate miners were out of work. And, of course, Rico's local business establishments were hurt by this unfortunate circumstance, as was the RGS.

With high hopes, the masthead of one of the newspapers proudly boasted ***"Rico Means Rich."***

Incidentally, the people of Rico, as well as the RGS people, used the Spanish name "Río Dolores" for the Dolores River for a long time. Beyond that, there are many colorful Spanish names in this part of Colorado, and we have used them—including the Spanish word "cañon" (referring to a rocky gorge)—throughout this series because this is the way both the Rio Grande Southern and the Denver & Rio Grande spelled these words for many, many years. And the respected Colorado Railroad Museum followed this practice for their annuals... so, we have followed suit... with the reminder that much of southern Colorado was part of Spain's territory, Nueva España, and later, it was part of México before it became U.S. territory in 1848 (after the Mexican-American war). All the Spanish words bestowed on southern Colorado give this beautiful land an exotic, appealing flavor... including the Spanish name given to the little mining town of Rico... mindful of the fact that the name means "rich."

CHARLES MILLER ENGEL COLLECTION –
COURTESY OF RACHEL A. HARDWICK

THE ORIGINAL PICTURE of this scene was reproduced as a postcard view, which has been enlarged by 700 percent—and enhanced as much as possible with today's technology. This illustrates the entire town of Rico, as well as the railroad yard of the Rio Grande Southern. Most prominent in this picture is the Pro Patria mill, in view at the lower left center, with an aerial tramway cable line running up along Mantz Avenue to the Pro Patria Mine and tailings dump, visible near the top of the scene. Rico's public school is visible, toward the left, above and east of Glasgow Avenue, Rico's main street and the route of Colorado Highway 145 today. Silver Creek is the drainage stream between Telescope Mountain and Dolores Mountain, and it tumbled past the school, to the right (south). The RGS station grounds stretched from the left to the right (north to south), visible above the tree tops in the foreground. The Enterprise Mine and tailings dump appears at the far right, larger than the adjoining mine properties.

THE STORY OF RICO AND THE PIONEER MINING DISTRICT

By William A. Graves

(Continued From Page 264)

1911 The flood of October, 1911, was even more costly for the RGS than the one that occurred during 1909. Three months of heavy rain and a snowstorm that occurred during early autumn produced a disastrous flood along the Dolores River. Some 50 miles of RGS track north and south of Rico was washed out. Repairs cost the railroad $174,000, and the mainline north of Rico was out of service for 3 1/2 months. Silver Creek in Rico was particularly violent and destroyed everything in its path, to a width of 200 feet. Although the Dolores River flooded to some degree almost every year, the 1911 flood probably was the most disastrous flooding ever to hit the RGS.

1912 Many countries began to buildup their military forces, as the threat of war in Europe became stronger. The production of base metals—copper, lead and zinc—from the Rico mines experienced a big increase. Copper was the dominant metal, and undoubtedly it was used to produce brass for cartridge and shell casings. The mines of the Rico area produced 689,915 pounds of copper in 1912, and this amount was greater than Rico's total copper production for the preceding 10 years.

Much of this copper was mined at the Wellington group of mines, about two miles north of Rico, in an area known as the C.H.C. Hill, a location named for C. H. Carpenter. Fred Price of Salt Lake City formed the Rico-Argentine Company, which leased the old Black Hawk Mine. This company was active intermittently in the Rico area for over 60 years.

1913 With the threat of war coming ever closer in Europe, ore production in the Rico area took a very big boost, to over $500,000. The value of this production was greater than any year's production since 1894. The value of this ore was about evenly divided among silver, copper, lead and zinc. The Pro Patria mill was remodeled and operated for a short time.

(Continued on Page 337)

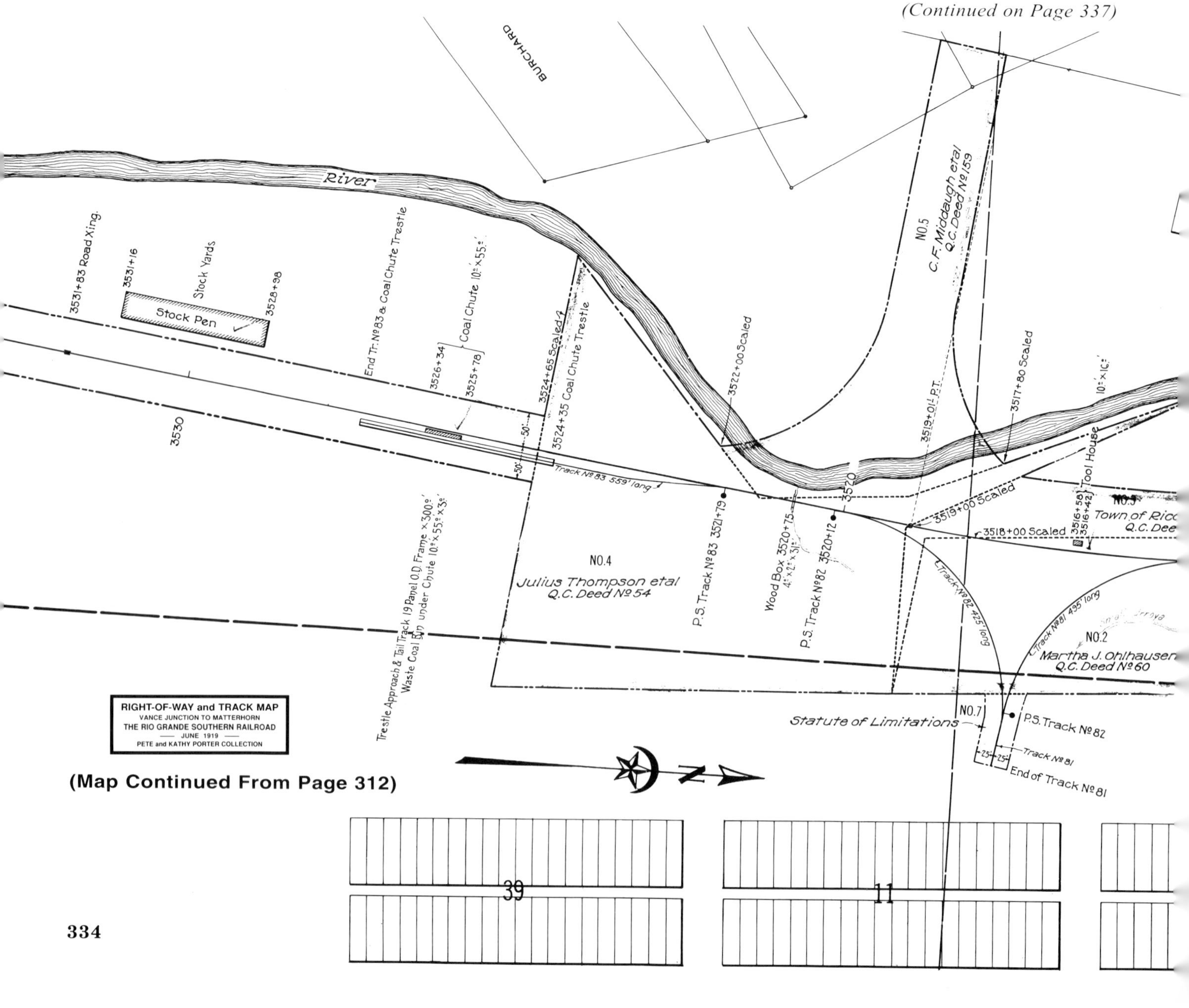

RIGHT-OF-WAY and TRACK MAP
VANCE JUNCTION TO MATTERHORN
THE RIO GRANDE SOUTHERN RAILROAD
—— JUNE 1919 ——
PETE and KATHY PORTER COLLECTION

(Map Continued From Page 312)

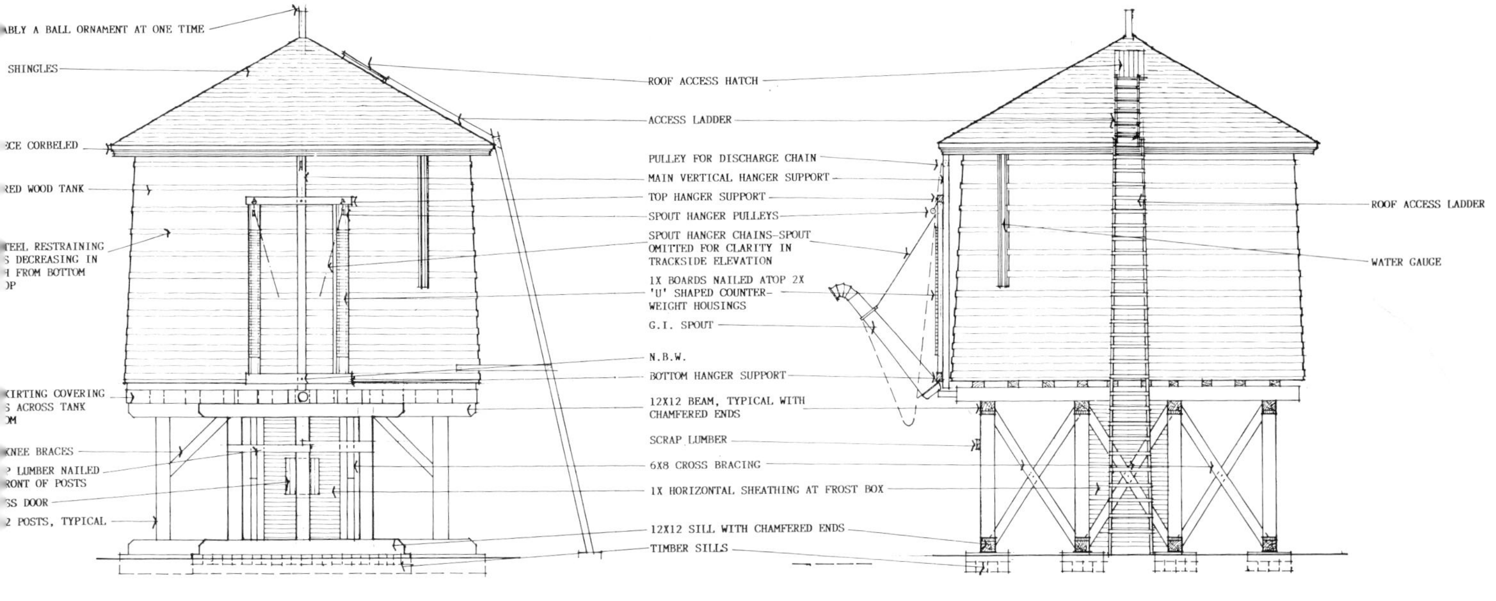

TRACKSIDE ELEVATION

SOUTH ELEVATION

WATER TANK AT RICO
ON WOOD BLOCKING

DRAWN BY MIKE BLAZEK
— Not to Common Scale —
ALL RIGHTS RESERVED
BY MIKE BLAZEK

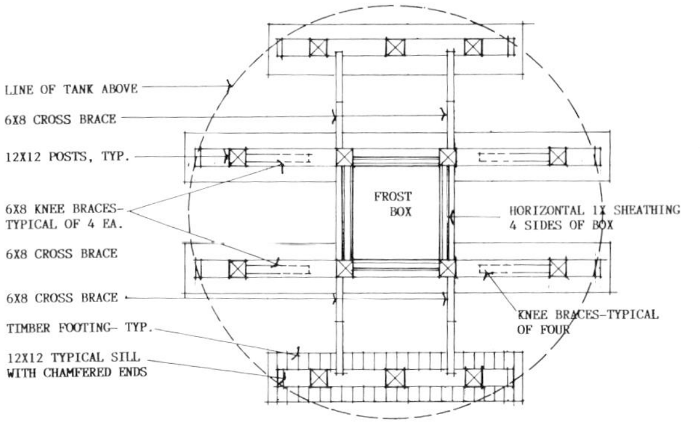

TANK PLAN
TRACKSIDE DOWN

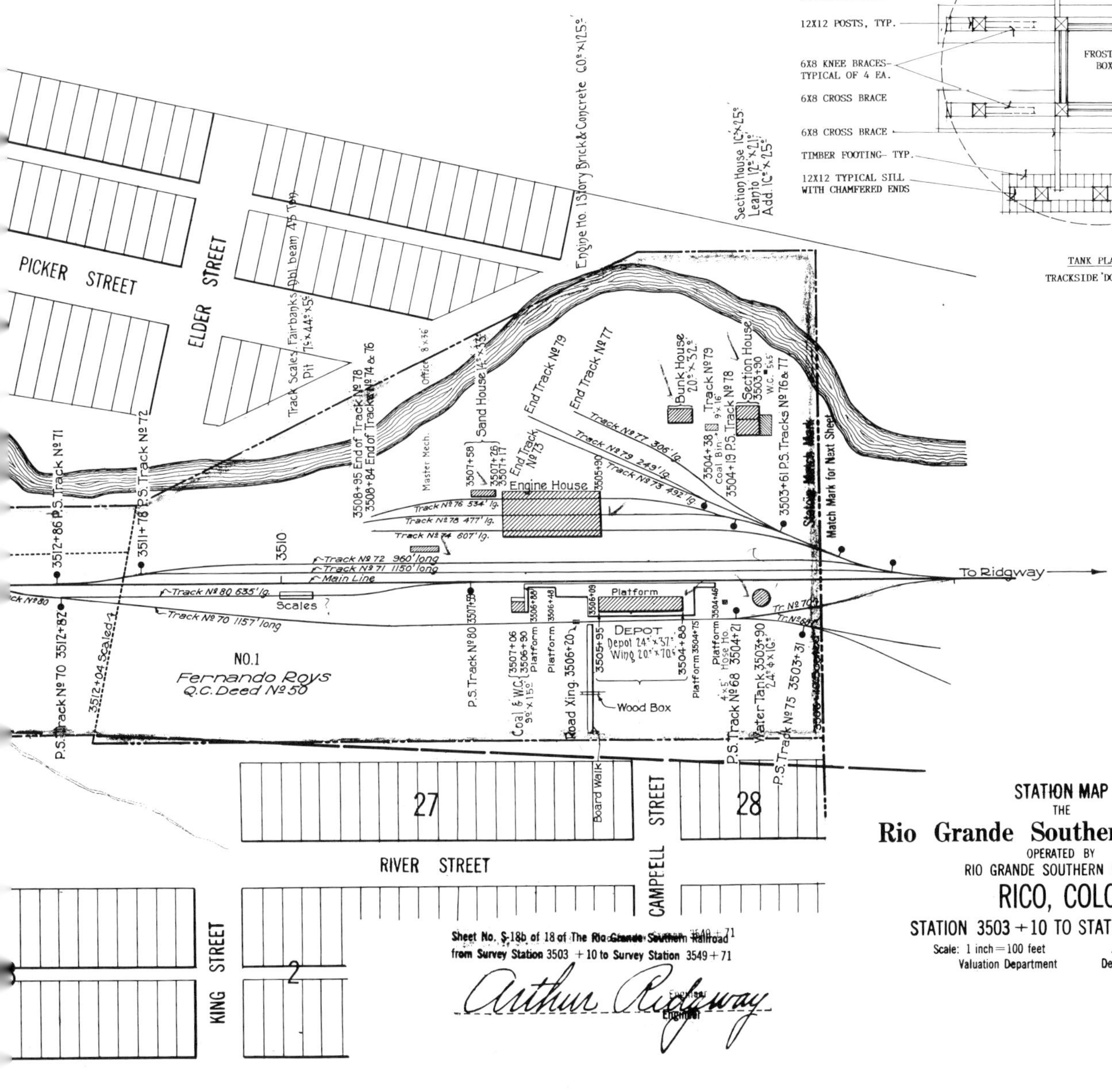

STATION MAP
THE
Rio Grande Southern Railroad
OPERATED BY
RIO GRANDE SOUTHERN RAILROAD
RICO, COLO.
STATION 3503 + 10 TO STATION 3549 + 71
Scale: 1 inch = 100 feet — June 30, 1919.
Valuation Department — Denver, Colo.

BILL PLUNKETT COLLECTION

RGS LOCOMOTIVE No. 25 had been spotted on an enginehouse lead at Rico one day ca. 1915–1920. A carbon-arc headlight had been placed on the extended smokebox of this ex-Florence & Cripple Creek 4-6-0, built in 1899 as the "Isabella." Two oil classification lamps were placed on the front of the engine, and the pilot beam had been outfitted with footboards on which brakemen would stand during their switching work. The Rico depot is nearly out of sight, in the far right-hand segment of this view. Nigger Baby Hill looms dimly behind the locomotive.

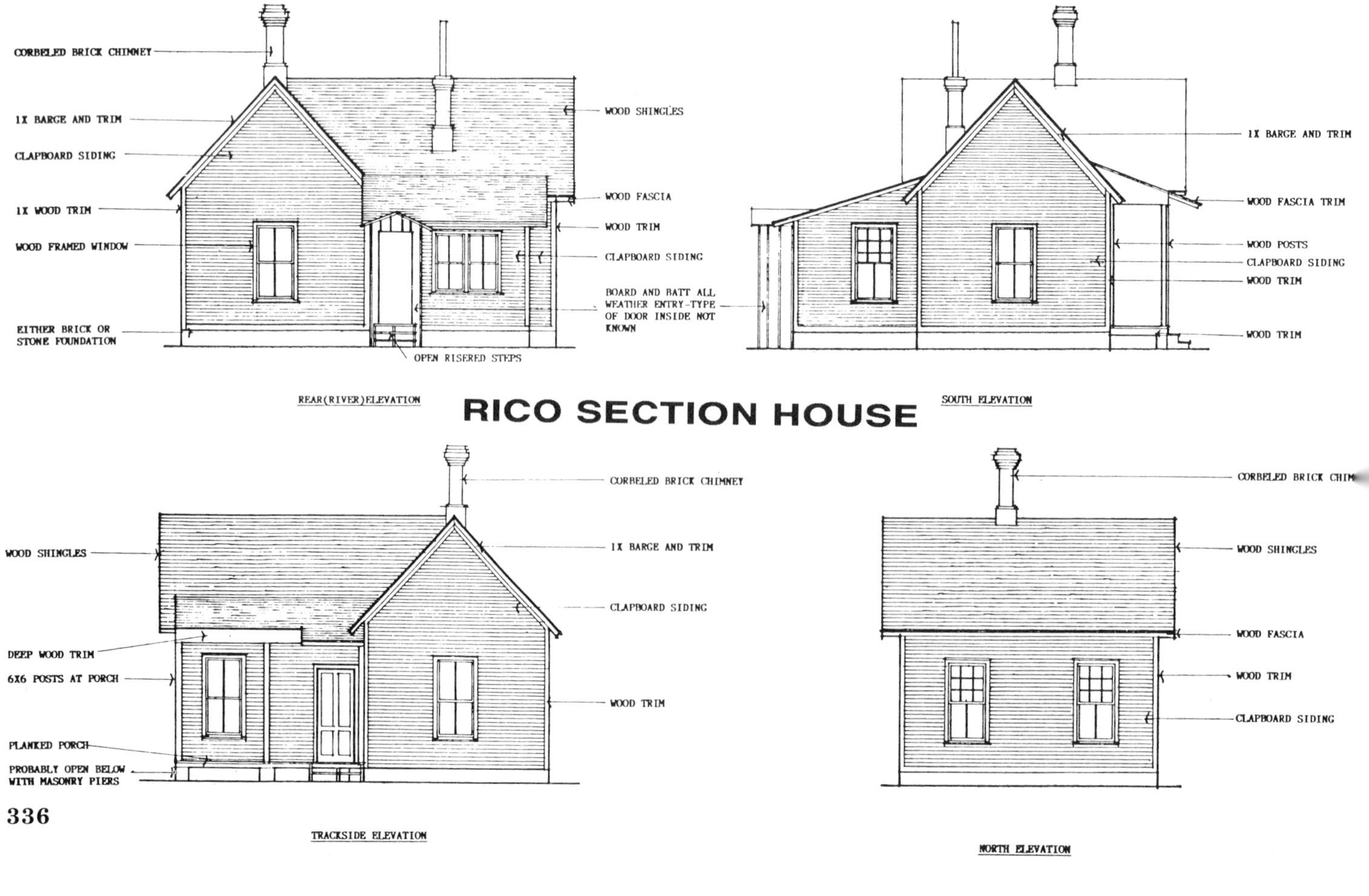

RICHARD H. KINDIG COLLECTION

"HOMEWARD BOUND" might well be the title of this photograph. The section hands, with dinner pails in hand, wearily left their train behind. They trudged through the snow toward the comfort of their homes for a hot meal and a soothing soak in a warm bath. They had been "up the hill," shoveling snow until their snowfighting train was able to break through the blockade. Now, it was time to coal and water the locomotives behind RGS Plow-Flanger 02. It looks as though the water gauge on the Rico tank was near the bottom, indicating the tank was nearly full. This picture may have been taken during the severe snow blockade of 1913–1914.

1914 For some unknown reason (or reasons), production of all metals in the Rico district took a big drop in 1914.

1915 During 1915, copper production in the Rico area reached an all-time high of over one-million pounds. This yearly production of copper was never mined before or since. However, lead and zinc shipments were down.

(Continued on Page 342)

DRAWN BY MIKE BLAZEK
— Not to Common Scale —
ALL RIGHTS RESERVED
BY MIKE BLAZEK

RICO BUNKHOUSE

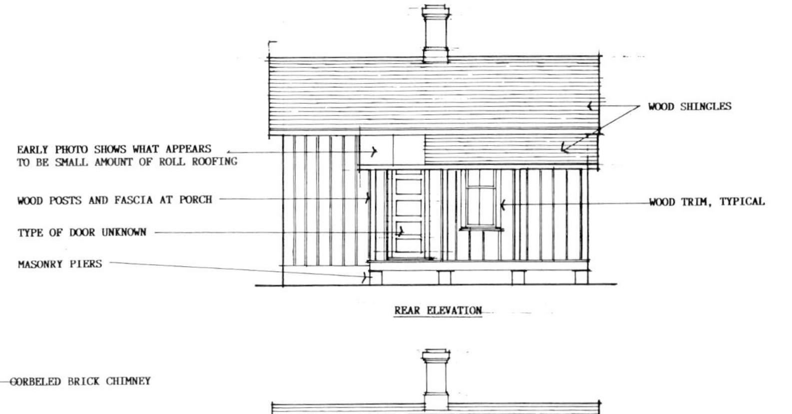

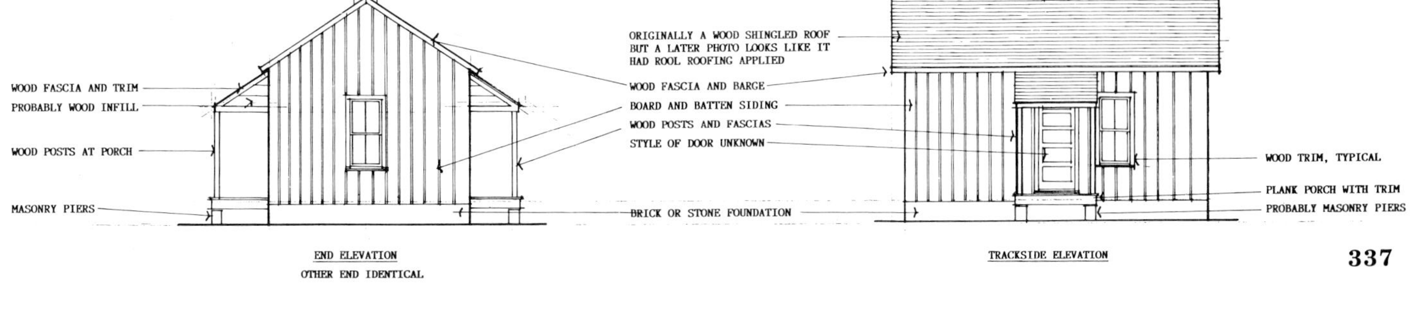

CHARLES J. LEE PHOTO – DAVID C. BUTLER COLLECTION

THE UNITED–RICO MINES company operation known as the Pro Patria (or Mantz Avenue) mill looked like this ca. 1916. Apparently, a new color scheme had been applied to the RGS depot sometime before this picture was taken—buff yellow with dark brown—used until the building was torn down. Of course, the paint faded during the final years of the RGS, since the building was not maintained properly (for lack of funds). However, the depot looked very attractive during this period, as did all the other railroad structures. Also of interest is the old wooden stairway that once gave access to the depot grounds from the foot of Campbell Street. The stairway had disappeared and taking its place was a pathway wandering down the face of the embankment. Notice

that the section house and bunkhouse also had been repainted in the new colors. Outbuildings previously located between the Dolores River and the section house are missing. A pile driver and B&B work car occupied tracks formerly located beside the original wooden enginehouse. The flood of 1911 had carved a new channel for the Dolores River, through the location where the rear wall of the old wooden enginehouse had been standing. Two boxcar bodies appear to be near the site of the original enginehouse inspection pits, with a third body turned about 45 degrees. Notice the old D&RG RPO car body used for the roadmaster's office, sitting on the ground at the right. It had been set up in the Rico yard to replace the older RPO car body (missing in this view). This body was later placed to the left of the section house.

COLORADO HISTORICAL SOCIETY COLLECTION

BY THE TIME this photograph was taken—between 1916 and 1918—the Syndicate Mining & Milling Company had acquired most of the patented mining claims in Rico's Pioneer Mining District. The Pro Patria mill was undergoing changes, and it was marked by an "X." Apparently, one of the mining company officials lived up Mantz Avenue, across the street from today's Presbyterian Church. This house is marked with a "✓." Interestingly, this home is now owned by Robert Metzler (as of 1995), who has purchased and moved the RGS freighthouse (warehouse) in Placerville to a different location. (This house was next door to Charles Engel's house, who was a photographer and reporter for the Rico newspaper, and

Robert W. Richardson stated that he stayed at his place at least once when shooting winter pictures of the RGS.) Notice the diminishing number of houses in Rico, as compared with earlier views of the town shot from this location. The RGS station grounds were very quiet on this summer day, although a string of boxcars had been lined up on the lead track to the Pro Patria mill. In addition, three boxcars had been spotted near the loading area, next to the main structure of the ore-processing plant.

DAVID AND FAYE WOLFORD COLLECTION

A RELATIVE OF HENRY WOLFORD, David Wolford, had the honor of trying out the front seat of Walter D. Lee's inspection car, lettered "R. G. S. No. 1." The Model "T" conversion was in Rico during either 1914 or 1915. The unnecessary steering wheel had been removed, and canvas side curtains had been installed. W. D. Lee was the superintendent of the RGS at this time. He had been photographed during 1913, near the Ophir Loop, featured in Volume IV of THE R • G • S STORY—as mentioned above—with the accompanying history concerning the happenings of this car. Notice the Rico depot in the background, which had been repainted in buff and brown by this time.

1916 Most of the patented mining claims in Rico's Pioneer Mining District were acquired by the Syndicate Mining & Milling Company.

1917 The price of silver, copper and lead increased as the United States entered World War I. The Rico district produced over $500,000 in silver and base metals.

1918 The price of silver rose to $1.00 per ounce for the first time since 1890.

1919 After the end of World War I, most of the world's attention was focussed on disarmament. Metal prices fell, and metal production in the Rico area dropped to one-third of the 1918 levels. Robert Pellet, a mining promoter, and his wife, Elizabeth Eyre Pellet, moved to Rico at this time. They would become very important residents for Rico, in their at-

HENRY WOLFORD stood beside Walter D. Lee's Model "T" Ford rail car, which had been converted from a secondhand automobile as RGS No. 1—officially called an inspection car. The car had undergone some modifications, which can be seen in this picture taken in 1915. The original headlights were replaced by one large headlight (with a curved lens), centered at the base of the windshield. This light may have been a carbide gas lamp judging from its domed top (with side vents), and the tube coming through the fire wall is of sufficient diameter to be a carbide-gas hose instead of an electric wire. Two smaller headlights had been mounted in the fire wall, along with a horn, at the right. (The two smaller lights may have been used as classification lights.) Another change was the removal of the foot boards over the front wheels, while the head end part of the roof had been filled-in with a curved down section, and some type of material closed the open spaces on each side of the windshield. The inspection car had no electric starter—it had to be hand-cranked. (See Pages 82, 83 and 123 of Volume IV in this series for other views of this car.)

DAVID AND FAYE WOLFORD COLLECTION

Ford

CHARLES MILLER ENGEL COLLECTION – COURTESY RACHEL A. HARDWICK

THE SILVER SWAN mining property, located about three-quarters of a mile south of Rico, had tunneled into the base of Expectation Mountain at the edge of the Dolores River, near its confluence with Sulphur Creek. The summit of Expectation Mountain is on the skyline (at the left), west of the river valley. Upstream, toward Rico, the engine-coaling facility used by the RGS is visible, at the south end of the yard. The southern end of Rico is at the far right.

tempts to keep some of the mines in operation. (See ***A Tenacious Woman... Elizabeth Eyre Pellet*** in Volume VI.)

1920 Many people left Rico for lack of employment. By the end of 1920, only about 150 people remained. Many houses were abandoned, and many were torn down to salvage their lumber. However, two general stores, a grocery store, a meat market and a pool hall kept their doors open—even though the ore-processing mills closed down.

1921 Ore production was the lowest in 40 years, and the remaining residents in town barely survived. Rico had neither a resident doctor nor a clergyman, and the community made do with occasional visits for their needs. Dances and social events helped hold the town together, and they helped to keep the people's spirits up.

(Continued on Page 415)

ACCIDENTS IN THE RICO AREA

By Robert W. Richardson

WHEN THE RGS LINE was built into Rico, the contractors were in a hurry, and they ignored the spring floods of the Dolores River. So, the yard, wye and wooden enginehouse were constructed in the only flat place available, next to the river. Worse yet, both of the approach tracks to the tail of the wye crossed the river! If several sunny days came together during the spring, with no clouds to keep the temperature down, the snow melt would flood the Dolores Valley. Such a flood came in 1911, which just about wiped out the entire yard of the RGS, structures and wye. The railroad had no choice but to rebuild in the same area; however, the wye was moved to the opposite side of the mainline. The new wye made use of a slight draw, where the tail ended at the steep embankment below Rico's business district.

Luckily, the railroad's newly completed brick enginehouse, with three run-through tracks, escaped destruction, but the old enginehouse was damaged so badly that it was summarily torn down. Only the old lead

(Continued on Page 354)

MARGARET MATZICK COLLECTION

THE PHOTOGRAPHER aimed his camera lens toward Rico to capture this distant view of the RGS station grounds from Newman Hill. His picture was taken on a cold winter day ca. 1920. At the far left, beside the RGS mainline leading toward Dolores, stood the Burnett & Clifton "Automatic Coal Chute." From this position, the trestlework was in view, along with support bracing against the coal bunker. The settlement of West Rico appears across the Dolores River, and the brick enginehouse and wooden depot can be seen to the right. The Shamrock Mine's buildings are visible above the station grounds. The dark line across the lower snow-covered Expectation Mountain was the flume of the Rico Power & Light Company, which carried water to their hydroelectric power plant in Rico. Snowclad Elliot Mountain is midway along the horizon.

BILL PLUNKETT COLLECTION

FOR AN HISTORICAL RECORD, it is fortunate that some unknown photographer walked the distance downgrade on the RGS track to visit the coaling facilities at Rico, in order to shoot this scene. He had walked up the ramp to take this picture from the top of the coal trestle, presumably to obtain a winter view of Rico and the surrounding snow-clad mountains. He was standing on the top level of the Burnett & Clifton automatic coal chutes that serviced engines at the south end of Rico. It is very clear in this picture that the mainline took the diverging route, while the coaling

track remained on the tangent to the coaling trestle. With one or two gondolas loaded with coal an engine had to take a good run to get up enough speed to make it to the top. The RGS station grounds were at the left, and the large ore-processing mill was located at the end of Mantz Avenue—constructed against the embankment at the western end of the business district. The houses and business buildings at the southern end of the town appear at the right. The snow-covered peaks at Lizard Head Pass rise in the far-distant background.

BILL PLUNKETT COLLECTION

A CLOSER LOOK at the previous photograph is illustrated here. The RGS mainline—southbound en route to Dolores—curved out of the Rico station grounds and followed the contour of the Dolores River Valley. It then curved again at the turnout for the coal chutes in the foreground. Notice the red-and-white target on the harp switchstand for the derail at the Burnett & Clifton coaling pockets. It showed a "red flag" signal for this derail (in the open position), because the red color was in the lower position. This so-called derail provided a safety measure should a coal car roll down the ramp toward the main track. Hopefully, such a runaway would grind to a stop on the ties just beyond the derail, short of the mainline switch; however, railroaders had their doubts about this particular derail! It looks as though the "finned" turret-type switchstands served both the switch at the coal chutes and the switch in the distance, at the southern leg of what was the third wye at Rico.

THE RICO WYE was primarily used for turning locomotives, although snowplows, Galloping Geese and passenger cars also were turned here. A combination section-car shed and tool shed occupied the river bank, and stacks of railroad ties are in view, both at the wye and next to the shed. The remains of the second wye at Rico, which crossed the Dolores River on two trestles, can be seen at the left. (As noted before, the first wye had been built where the enginehouse stands in this view.) Floodwaters of 1911 dispatched these trestles down the river, and it was decided to move the wye to its third location. Bridge 67-AS was nearer the depot, while Bridge 67-ASS is in the foreground of this scene.

JOHN W. MAXWELL PHOTO

THE TOWN OF RICO and the Rico–Argentine tailings pond—the former site of the Pasadena Smelter—are in this scene that was photographed after the abandonment of the RGS (in the left-hand portion, beyond the coaling trestle). The approach to the trestle, at the left, had about one foot of rise for every 16 feet of length. The trestle approach and tail track were built as a 19-panel open-deck frame that was a 300-foot long bridge. From the mainline switch, the track ran a distance of 559 feet. Four of the balanced aprons were in the down position in this view, and some coal had spilled out over the track.

BILL PLUNKETT COLLECTION

THE BURNETT & CLIFTON "Automatic Coal Chute" at Rico was photographed at the south end of the structure, showing a high-side gondola after it had been unloaded. Loaded gondolas were shoved up to the coal pockets from a ramp at the north end of the structure. The track in the foreground was part of the tail track that extended out of the coaling facility toward the south for about 33 feet. At least one loaded gondola could be stored there and coasted down to the coaling pockets to be unloaded. Additional detailed information regarding this type of coaling structure is on Pages 351–358 in Volume I of this series.

RICO COALING FACILITY

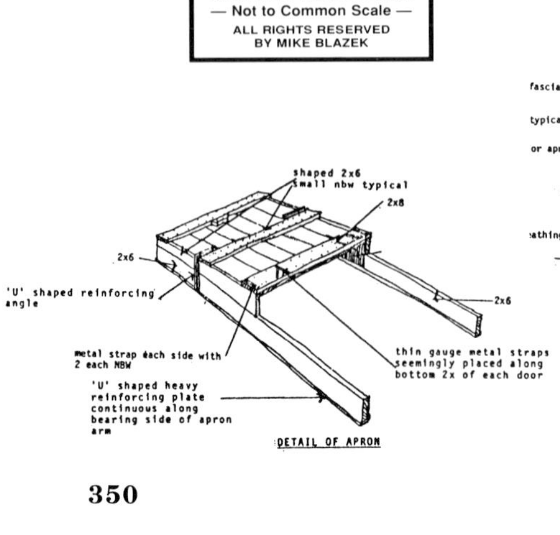

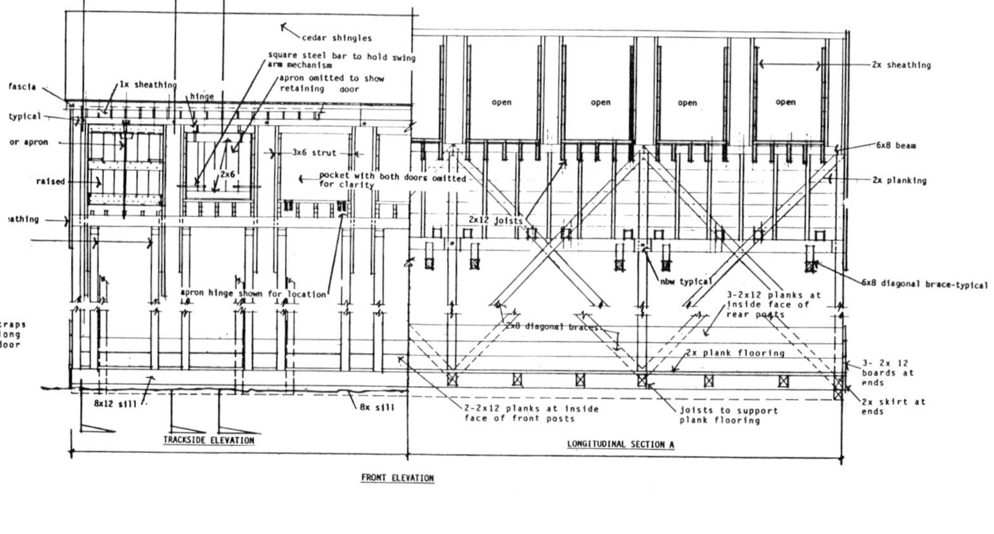

RICO COALING FACILITY

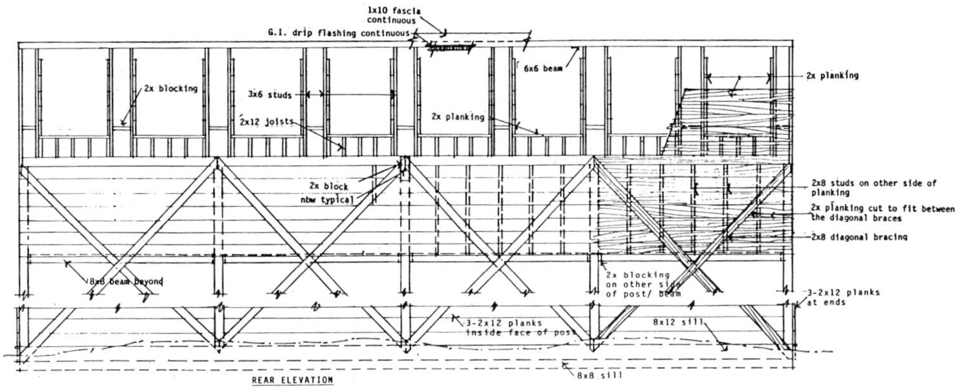

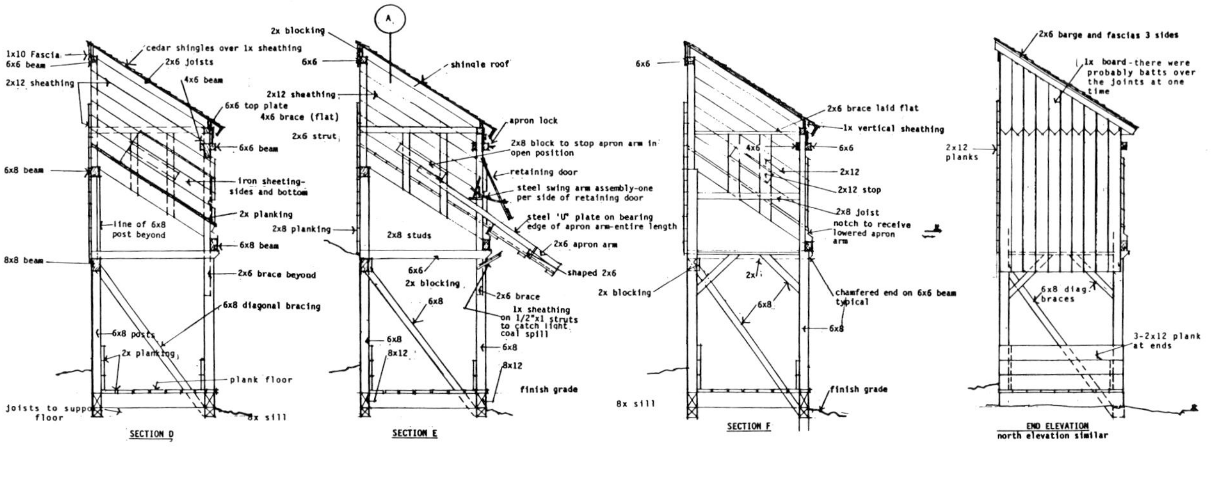

ROBERT W. RICHARDSON PHOTOS –
COLORADO RAILROAD MUSEUM COLLECTION

IT IS SURMISED that this grouping of three photographs were shot by Bob Richardson on September 16, 1946, when he took a trip "out West" to see the RGS before the start-up of his motel and railroad museum at South Alamosa, Colorado. In the upper view (at the left) are houses in west Rico, across the Dolores River, and in the center is the Pro Patria mill, while "uptown" Rico is to the right. The inclined timber-trestle approach to the Burnett & Clifton "Automatic Coal Chute" is in the foreground. Tailings from the Rico–Argentine's Pro Patria mill were piped to this location, and this waste material appears in the right foreground. The RGS wye was beyond this dump.

THE PICTURE at the lower left illustrates how sway braces were placed at the south end of the trestle for the engine-coaling facility. You can also see the row of supporting timbers propped against the east side of the structure. The RGS mainline between Rico and Dolores is at the left. The two heavy timbers propped against the end-of-track formed a barrier to keep cars from being pushed off the end of the coaling facility.

THE TRACKAGE in the scene at the lower right illustrates the derail and harp switchstand, with a flanger warning sign next to it. The derail was set in the through or clear position, indicating there were no cars on the coaling structure.

JOHN W. BARRIGER PHOTO, ST. LOUIS MERCANTILE LIBRARY

AS THE SPECIAL TRAIN carrying John W. Barriger passed northbound over the RGS during July of 1938, he shot this picture of the Burnett & Clifton "Automatic Coal Chute" at Rico. This coal-loading facility was almost identical to the one at Vance Junction, except that the ramp ascends from the opposite end. The old harp switchstand activated a derail, a safety device that was supposed to derail cars that might accidentally roll down from the coal pockets, preventing them from reaching the mainline. The target for this derail was painted yellow. This view was taken as John looked toward the south, and it shows the main track as it left Rico for Dolores (at the right).

tracks were left in place. However, only one of the new run-through tracks was left connected to the mainline at the south end of the yard. The other two tracks were partially destroyed and were never reconnected to the main track. The coaling facility was located nearby, perhaps out of reach of floods.

The line on southward from Rico to Dolores was built entirely too close to the river, seldom more than a few feet above normal high water. This was less expensive and quicker to build, as to construct the line up higher would have entailed much in the way of fills, cuts and bridges. Consequently, Rico occasionally was isolated from the south end of the line due to washouts. And this part of the line sometimes was out of service for months. Even during the final years of the railroad, such floods isolated Rico. When dismantling crews worked north and south of Rico, their task had to be accomplished by trucking stripped-down Galloping Goose vehicles and flatcars around the washed-out places. (The Rico yard tracks were sold to the Rico-Argentine mining firm.)

At one time, two hostlers took an engine down to the wye to turn it. Someone had been working on the coupling bar between the engine and tender, and it came loose. However, the tender continued to be pulled by the engine, as the safety chains held. Unfortunately, one hostler happened to be standing on the engine's apron when the tender separated. And he dropped through the gap, although he managed to grab some fittings and saved himself from falling completely down to the track. Nevertheless, his feet were touching the ties! The engineer did not dare apply the air brakes, for fear that the action might close the space and crush

BILL PLUNKETT COLLECTION

THIS SIMILAR VIEW of the Burnett & Clifton coaling facility at Rico was taken some years later than the previous one. Most noticeable, at first glance, are the four rails in the foreground, showing the setup of the derail switchstand and guardrails, with the derailer set in a position to put the wheels of a car coming down the ramp onto the ties between the guard rails (to the left) and the regular rails (to the right). A coal bunker for overflow storage had been constructed on top of the pilings, and it was braced against the hillside. Locomotive tenders were loaded with coal on the mainline, at the right. The mill tailings south of the coaling pockets had been left from the Rico–Argentine mining operation. Notice how the target on the switchstand had been bent.

the dangling hostler. So, he gradually let the engine slow to a halt, praying that the tender would not surge against the engine. It did not, and a badly bruised and frightened hostler survived the day!

Another time, Superintendent W. D. Lee had come in to town on a southbound freight train, and the engine had spotted his business car, No. B-21, in front of the depot. A brakeman elsewhere in the northern part of the yard released the brakes of two gondolas loaded with rock ore as he prepared to drop them down into the yard. However, as he went to slow them by turning the brake wheel, the chain broke! He had no time to reach the other car's brake wheel, and the cars coasted into the B-21 with a loud crash! Superintendent Lee was standing in the aisle at the time, looking southward out of the car. Unaware of the oncoming cars, the impending crash threw him on his face as he slid along the aisle. Noted for his fairness, the superintendent did not reprimand the brakeman, accepting it as a genuine accident, beyond anyone's control. Nonetheless, the superintendent had a badly scraped face for some time, and he walked rather stiffly for several days.

CHARLES J. LEE PHOTO – NORMA L. BUTLER COLLECTION

THE TRAIN-ORDER SIGNAL at the Rico depot was in the red (stop) position. The Ridgway-bound train was No. 372 on the RGS timetable, and it had arrived about on time this sunny afternoon, with passengers, mail and express for the bustling mining town of Rico. This scene was recorded on film during 1917, when all "ladies and gentlemen" wore hats when going outdoors. Notice the large mail sacks on the freight end of the depot platform. Somehow... maybe because of nostalgia... this scene looks as though it is straight out of a Hollywood silent film "Western" melodrama... complete with the tall hero wearing a felt hat with a satin band, jodhpurs and a vest... the maidens who will need to be rescued... and one must not forget to mention the depot dog standing in the doorway!

BEING IN NO RUSH to board their northbound train to Ridgway, two Rico women were chatting with a man at the Rico depot. RGS employees at the depot were busy with the baggage and mail at the head-end of the train. Notice the ramp leading up from the wooden platform for passengers, built to accommodate wagons moving express and L-C-L freight shipments in and out of the northern (freight) end of the depot. In this view taken during 1917, the original "RICO" depot sign was hanging on the wall nearby, on the western side of the building. The large boldface lettering was white, with a black background.

CHARLES J. LEE PHOTO – DAVID C. BUTLER COLLECTION

PASSENGER TRAINS INTO RICO

By Robert W. Richardson

PASSENGER-TRAIN SERVICE into Rico varied over the years, with the rise and fall of the mining industry. All the while, the RGS kept Rico as the dividing point between its two districts of operation. At times, a small force of car men, hostlers, etc., was stationed in Rico, while in poor times, perhaps only one car man and his helper were in Rico.

If a passenger rode on a train from the north, an overnight stay at Telluride was a necessity. If a passenger journeyed from the south, a similar stay overnight at Telluride was necessary. This was because RGS passenger trains through Rico terminated their runs at Telluride and Durango.

Mail service was very important for isolated places like Rico, and a result of this fact was that the volume of incoming mail was much greater than outgoing mail. By bulk, the mail was largely newspapers, and the postal people expedited the newspapers almost as fast as letter mail. After the discontinuation of the Railway Post Office car in 1917, mail was handled as "closed-pouch service" in the baggage cars by either the baggageman or by the conductors and trainmen.

L-C-L (less-than-carload) freight shipments for Rico constantly caused complaints, as the railroad had a monopoly, and small shipments were transferred from one freight car to another at Alamosa, Durango, Salida, Montrose and Ridgway—resulting in shipments taking a week or 10 days to travel from Denver or Pueblo, and that was when it was not further delayed. Then, in 1912, the U.S. Post Office Department instituted the Parcel Post, and many of the small shipments soon were handled by mail, as a package from Denver would reach Rico in no more time than a letter. The Denver & Rio

CHARLES J. LEE PHOTO – NORMA L. BUTLER COLLECTION

WHEN TRAVELING by train to a distant city during this period, the women always dressed in the latest fashionable attire. This picture was taken about 1917, when creative adornments were worn by the ladies, amidst items of jewelry and artificial flowers. The lady on the left was wearing a wrist watch, a novelty in 1917. This shot was taken on the south end of the depot platform at Rico, showing the electric power lines accessing the depot.

THIS GENTLEMAN was astride his burro, in front of one of Rico's older log cabins on River Street. This property was located in Block 12, Lot 11. Expectation Mountain is in view at the rear of this property, which had an assortment of storage sheds. A carriage house, which included a loft for hay, was at the rear of the log cabin. Notice that the house at the right had been decorated with fancy lace curtains, while a picket fence bordered the front yard.

MARGARET MATZICK COLLECTION

Grande—alarmed at the loss of revenue—immediately started speeding up its slow handling of L-C-L shipments, and more attention was paid to the transfer points. Where previously there had been only one L-C-L boxcar once a week in the RGS freight trains, now every freight was expected to handle some of the shipments.

THE EX-F&CC ENGINES ON THE RGS

By Robert W. Richardson

WHEN, IN EARLY 1916, the three former Florence & Cripple Creek locomotives came to the Rio Grande Southern, they were routed via Alamosa, being unloaded there from standard-gauge flatcars. At this time, the RGS passenger trains made the run from Durango through Rico to Telluride, and vice versa. Uncertain if these heavier engines should venture over the Ophir Loop trestles, they were used only as far as Rico, where a smaller engine was put on the train for the balance of the trip to Telluride.

Their performance was closely watched. In tonnage, they were rated the same as the Class 70 (C-19) 2-8-0's, which was 185 tons in both directions between Durango and Mancos, and 250 tons from Dolores to Rico. For years, the Ten-wheelers were restricted to this section of the RGS. One attempt to use them on the north end for passenger service was abruptly ended with a telegram issued by RGS Superintendent Carpenter to all concerned "...to keep those Class 85 [T-19] engines off our first district; they are knocking holes in our track." The 4-6-0's were finding soft places, and they kept section crews busy. The tendency of this class of engine to be hard on lightweight rail continued to plague the RGS. And the RGS more than once tried to trade them to the D&RGW for Class 125 (K-27) Mudhens, even offering three for two!

MARGARET MATZICK COLLECTION

THESE MODEST HOMES faced toward the east on River Street, ca. 1905. Residents were out in front, and one woman had her Sunday best on, including a fancy hat, perhaps ready to attend church. A boy was enjoying a ride about town on a burro, about midway on the block. The Colorado Light & Power Company's wooden flume from Burns—which provided water for their hydroelectric plant—crossed the east slope of Expectation Mountain behind the brick chimneys.

DELL A. McCOY PHOTOS

THIS PLEASANT LITTLE HOME, with a bay window, probably was located in Block 12, Lot 13, and it faced on River Street. The backyard view included the north portion of the RGS railroad yard, which would have provided excellent viewing enjoyment of trains coming from and going to Lizard Head pass. Indian summer was in the air when this view was taken, during September of 1966.

THIS OLD LOG HOME was occupied during September of 1966, located at the intersection of River and Soda streets. This is listed as Block 12, Lot 1. The next house, to the left, and a twin were probably constructed at some later date, after the log house was built. The next house to the left probably was on Lot 7, with the one at the far left being the little home on Lot 13. These houses faced on Silver Street.

CHARLES MILLER ENGEL COLLECTION – COURTESY RACHEL A. HARDWICK

THE UPPER END of Mantz Avenue was photographed during what likely was the early 1910's, showing the Dolores County Courthouse, at the left, a small residence, the Congregational Church, the Silver Street and Mantz Avenue intersection, a corner residence, the Charles M. Engel residence and one more house, at the right. Notice the square brick building to the right of the Congregational Church, on Silver Street. Also, the new brick schoolhouse stands in the distant background, at the right.

CHARLES MILLER ENGEL COLLECTION – COURTESY RACHEL A. HARDWICK

THIS WAS ONE of the more unusual residences in Rico. Built as an apartment building, it was made of brick and stood on Block 15, Lot 16, and it first appeared on the Sanborn Map of 1893. The photograph of this boarded-up dwelling shows the southern (left) and eastern (right) exposure, which fronts on Silver Street.

SILVER ST.
SODA ST.
COMMERCIAL ST.
GLASGOW AV.
RIVER ST.
MANTZ AV.
Dec.1899 RICO COLO.
CONG'L CHURCH
COUNTY COURT HO.
PUBLIC SCHOOL
Eng. Ho
Boardg
ENTERPRISE HOTEL
Dining Rm
Sample Rm
PO & Jewelry
UNITED STATES HOTEL
RICO TRANSFER CO AND CITY LIVERY
Baths
BRIDGE
Creek
Scale of Feet

RICO
1899

RICO

SANBORN

MAPPING AND GEOGRAPHIC INFORMATION SERVICE

— 1899 —

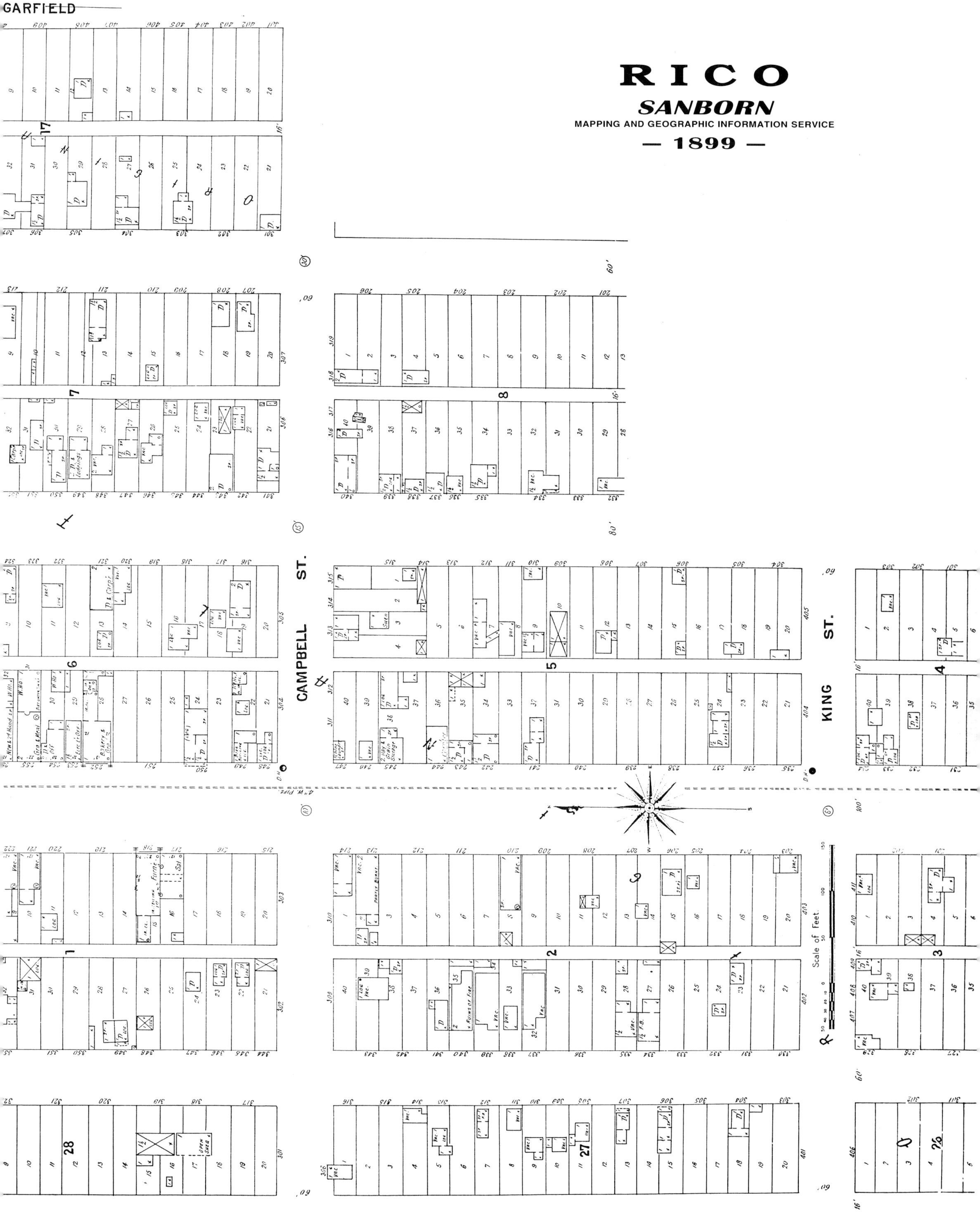

CHARLES J. LEE PHOTO – NORMA L. BUTLER COLLECTION

CHARLES LEE shot this picture near his front yard on Glasgow Avenue, at the south edge of Rico, during the 1920's. Sheep were being herded through town to be loaded at the stock pens for shipment on the Rio Grande Southern. At this end of town, the street was not much more than a couple of ruts through the grass at this time. However, it is a paved highway today.

CHARLES J. LEE PHOTO – NORMA L. BUTLER COLLECTION

DAVID BUTLER'S GRANDFATHER, Victor "Vic" Lee (Charles Butler's father), was taking care of his horse and colt on a vacant lot next to the Lee home in Rico. Victor ran the City Livery Stable in Rico at the time. A camera was hanging from his neck, and presumably he was shooting pictures of his horses. The old road to Dolores runs across the picture, just beyond the Lee house. Houses up on the base of Dolores Mountain would be located in Block 9.

THE FRONT PORCH of the Victor Lee home was the subject of this 1920's snapshot, taken during the winter. Skis were leaning on the front fence, as morning sunshine melted snow on the front porch. Notice the porch swing and the sled leaning against the wall. The house was equipped with electricity, and it may have had an indoor bathroom installed on the back porch. Unfortunately, this house was torn down approximately 15 years ago.

DAVID C. BUTLER COLLECTION

CHARLES J. LEE PHOTO – NORMA L. BUTLER COLLECTION

A GIRL NAMED EDNA, a friend of the family, posed out back of their home for Charles when he was trying out his new camera in the 1920's. Late afternoon sunshine warmed the west and south exposures of the house, facing on Glasgow Avenue.

THE HOUSE ON THE LEFT was taken out by a snowslide on February 8, 1932. The Victor Lee home was the second one from the left in this row, believed to have been located in Block 3, in Lot 2 or 3. The next house up Glasgow Avenue was sitting in Block 2, Lot 19, and next to it is a home that was probably on Lot 17, while the next (at the right) would be on Lot 15. The old-style Conoco gasoline pump, nearly buried in snow, was in the front yard of a home (out of view to the right). The pump was operated by Alfred Baer. The house at the far right was the Cal Trimble home.

CHARLES J. LEE PHOTO – DAVID C. BUTLER COLLECTION

FROM "THE RICO ITEM" IN 1932 – COURTESY WENDE P. STUART

NEW YEAR'S SLIDE DEMOLISHES HOUSE AND BURIES FAMILY

MONDAY NIGHT, February 8, about 11:00 o'clock p.m., the New Year's slide ran, striking the Butler house occupied by the H. E. Speer family, going through the house and pinning the entire family under wreckage and snow. Mr. Speer in some miraculous manner managed to release himself, and after digging through the debris, getting air to the others, rushing to the Torrence residence to summon help. Master Billy Torrence hurried to the siren and called for help and immediately sufficient of our townsmen were on the spot and quickly released Mrs. Speer, son Vincent and daughter Iva Belle, all unharmed, except for the severe shaking up.

FROM "THE RICO ITEM" – COURTESY WENDE P. STUART

THE AVALANCHE buried the Butler house on the east side of the highway. It then crossed over to what is now Sandy Payne's Mercantile store and on down the slope to the end of the RGS wye.

CHARLES J. LEE PHOTO – NORMA L. BUTLER COLLECTION

THOMAS WESTFALL, a relative living with the Lee family, was on the front porch of the Lee house during the late 1920's, practising his aim with a .22-caliber rifle. This home was in Block 3, Lot 2 or 3. The little house across Glasgow Avenue, at the right, with pine trees out front, still stands. From the front porch of the Lee house, you can catch a glimpse of uptown Rico.

OVERLEAF: Glasgow Avenue—Colorado Highway 145—curves into Rico's downtown toward the north in this picture, taken in 1994. Autumn leaves cover the slopes of Sandstone Mountain and Sockrider Peak. The last school building constructed in town is the third structure standing to the left. Beyond it is the old two-story Masonic Hall. The Rico Hotel (formerly called the Ricato), originally built to house miners, is beyond the tall spruce trees, at the right, on the east side of Glasgow Avenue.

DELL A. McCOY PHOTO

SPEED LIMIT
30

G. E. WILKINS PHOTO, BILL ELOE COLLECTION

A PHOTOGRAPHER from Wisconsin had come to Rico shortly after the mining boom had begun, probably in 1882, to shoot stereo views. In this case, G. E. Wilkins was fascinated by the houses in Block 23, with Nigger Baby Hill in the distance. The log cabin (still standing and now owned by David Butler) is located on Lot 26, about midway in this string of homes along Silver Street. Many of these houses still exist, although they have been remodeled in recent years. Neat wooden fences kept unwanted animals out of the yards. Home owners were proud of their houses and planned a long stay, evidenced by the planting of trees and shrubs. The home at the right sat on Lot 22. The house that is second from right is also still standing.

CHARLES J. LEE PHOTO – DAVID C. BUTLER COLLECTION

UPPER RIGHT: Charles J. Lee may have "gone a-courtin'" to visit Lenore Beaver, sister of Alma Beaver, seated on the front porch steps of this little cabin—now owned by David Butler. This cabin is in Block 23, Lot 26, and was once owned by Charles M. Engel. Victor Lee (David Butler's great grandfather) lived in this cabin when he first came to Colorado. It was called the "big house" by the family. This photograph was taken when David Butler's grandfather was courting Alma Beaver, he met Lenore afterwards and married her. This log cabin may well have been one of the first structures built in Rico. Records indicate that the cabin was deeded to two people by the names of Spurlock and Butler on August 10, 1881, but nothing is known about them.

CHARLES J. LEE was quite a photographer around the town of Rico. In this snapshot, he captured houses on the upper northeast side of town, located in Block 23. At the far left is Lot 30, and next to it was the Columbo house, on Lot 29. The bay-window home on Lot 28 housed the George Moore family, and the house at the right was on Lot 27.

CHARLES J. LEE PHOTO – DAVID C. BUTLER COLLECTION

DELL A. McCOY PHOTOS

PHOTOGRAPHED during September of 1966, the house located in Block 23, Lot 22 had undergone many changes since the photograph that was taken in 1882. The only recognizable thing left is the front of the building, facing on Silver Street. Additions have increased the size considerably. The ruins of the Van Winkle shaft house appear at the right. The scar across Nigger Baby Hill is that of the truck road to the Mountain Spring Mine, when ore was processed at the acid plant during the Korean War of 1950–1954.

AT THE INTERSECTION of Soda and Silver streets the camera lens was aimed toward the northeast for this autumn photograph of 1993. Nigger Baby Hill is resplendent in golden-colored aspen leaves. The Indian summers in this area may be brief, but they are spectacular. Many of these dwellings may be located on the Sanborn Map of 1893, situated in Block 23.

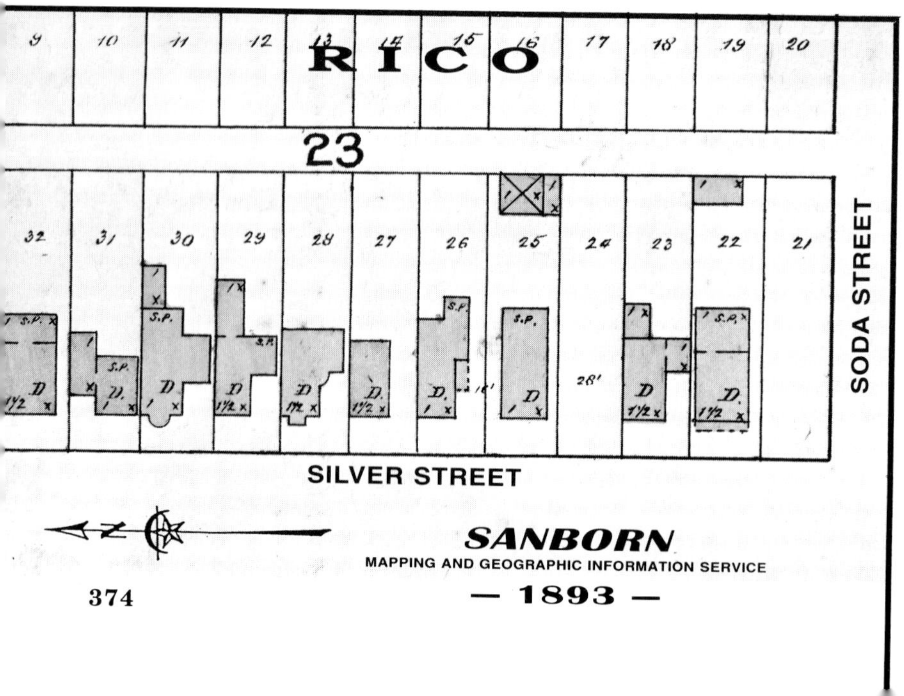

IT IS BELIEVED that this house, which was photographed during 1966, was in Block 15, Lot 33. Nigger Baby Hill is in the background, with the house facing on Silver Street, at the left.

DELL A. McCOY PHOTO

DELL A. McCOY PHOTO

IT WAS LATE SEPTEMBER of 1966, and the season for making outdoor improvements was just about to end. In this case, you are looking at the corner house at the intersection of Mantz Avenue and Silver Street. The upper view faces Mantz Avenue and the house sits in Block 16, Lots 23 and 24.

TAKEN ON THE SAME DAY, you can see that the former Albert Stampfel home had access to Silver Street from a side porch and door. It is quite possible that an indoor bathroom occupied the extension at the left. The Charles Miller Engel house is next door, at the left. Notice how the clothes lines have been mounted to porch posts.

DELL A. McCOY PHOTO

ORVAL L. JAHNKE COLLECTION

THE FRONT PORCH of this two-story house faced on Mantz Avenue, between Commercial and Silver streets. This is Block 7, Lot 3. Dolores Mountain forms the background.

THIS PICTURE was taken from Garfield Street, as Charles Lee looked down on Silver Street with twin houses, located in Block 7, Lots 5 and 6 on the Sanborn map. The house at the right is on Lot 3, facing on Mantz Avenue. Lower down Mantz Avenue, at the intersection with Glasgow Avenue, stands the original Rico State Bank building. A false-front building between the roofs of the twin houses sits in Block 6, Lot 7, formerly the I.O.O.F. Hall in 1893. The horses of Victor Lee were grazing in the foreground.

CHARLES J. LEE PHOTO – DAVID C BUTLER COLLECTION

DELL A. McCOY PHOTO

THE RICO TOWN HALL and Presbyterian Church were bathed in brilliant mid-morning sunlight during the winter of 1995. Commercial Street is in the foreground, where a residential home was in need of repairing and painting on the corner of Mantz Avenue, on Lot 40, in Block 7. Compare this winter view with the one on Page 375.

A RICO LADY was out for a stroll during the late 1920's. An ore bucket was hanging on the wire rope running up and down Mantz Avenue. The south and east walls of the Dolores County Courthouse are in view here, beside a small residential home. The Congregational Church, at the right, was badly in need of a coat of paint.

ORVAL L. JAHNKE COLLECTION

DELL A. McCOY PHOTO

A PORTION of Blackhawk Mountain is in sight on the skyline of this photograph, taken during the winter of 1990. Although nighttime temperatures drop below freezing in the winter—often below zero—the daytime sunshine melts the snow rapidly in this area. You are looking toward the east on Mantz Avenue.

WITH GENUINE PRIDE for our boys called up in World War I—the terrible conflict in which the United States was involved during 1917 and 1918—the people of Rico and many other citizens of Dolores County purchased U.S. war bonds beyond what was expected of them.

CHARLES MILLER ENGEL COLLECTION
– COURTESY RACHEL A. HARDWICK

DELL A. McCOY PHOTO

THE TWO-STORY HOUSE in this view is believed to have been at the corner of Commercial and Campbell streets, and it was photographed during September of 1966. It was in Block 7, Lot 21. An outhouse was up the hill, next to the barn. The barn was used to store coal and wood for winter heating. The trick was to visit the outhouse and come back with your arms full of wood or the coal scuttle full of "black diamonds." Surprisingly, this residential building was used as the local telephone-company office and operator's residence during the 1930's.

ANN AND JAMES FERANDO PHOTO

IVABELLE SPEER, at the left of "Babe" Heidt (Mrs. Ann Ferando), posed for this picture on Commercial Street in 1940. In this winter view, you are looking toward the north. Ann lives in Rico with her husband, James, a retired miner. The Dolores County Courthouse is on the site at the right. The homes on the left line the east side of Block 6, while Block 7 is across the street.

CHARLES J. LEE PHOTO – DAVID C BUTLER COLLECTION

ALMA NUTTER was a school teacher in Rico, and she was photographed during the late 1920's. The houses situated in Block 8 (on the Sanborn map) appear in the background of this view, located facing Commercial Street. The first log house was on Lot 38, and the rest were on Lots 37, 36, 35 and 34, and the shed was on Lot 31.

DELL A. McCOY PHOTO

SNOW-CROWNED ELLIOTT MOUNTAIN, and Sandstone Mountain form the skyline in this view. You are looking north on Commercial Street in this snowy scene. Homes line Block 6, facing the bright morning sun in this picture taken during the winter of 1995.

DELL A. McCOY PHOTO

FROM NEARLY the same photographic vantage point, houses lining the west side of Block 7 are featured in this view. Block 15, farther north, contains the Rico Town Hall—formerly the county courthouse. Distant Mount Wilson is visible at the left, above the base of Telescope Mountain. The crest of Nigger Baby Hill is at the right. Notice that the residential streets are still gravel-surfaced.

ANN AND JAMES FERANDO COLLECTION

DOLORES MOUNTAIN forms the background for this photograph, taken from Glasgow Avenue during the 1930's. The homes in the distance front on Commercial and Campbell streets (at the far left), in Block 8. The house in the foreground, nearest Campbell Street (at the left), is in Block 5, Lot 3. The log house was on Lot 4, while the board-and-batten house (at the right) was on Lot 7.

IT MAY HAVE BEEN that this photograph was taken ca. 1915, judging by the vacant lots beyond the Rico State Bank building. It looks as though there were charred ruins on the wall of the false-front building down the street from the bank, indicating that the intervening building had been lost in a fire. At the left, you can barely discern a Model "T" Ford sedan that had somehow rattled into town over the rocky, rough roads of this period. Pack burros had been left outside the business buildings, while provisions were purchased for some mining operation in the nearby hills. Obendorfer & Son, in the building at the right, carried canned goods and other staple items for the residents of the Rico area. This general store closed in 1936, during the Great Depression.

CHARLES MILLER ENGEL COLLECTION –
COURTESY RACHEL A. HARDWICK

SENSATIONAL Gold STRIKE!

RICO MEANS RICH

THE RICO ITEM

GOLD—$20.67 OZ.

VOLUME VIII — Entered as second-class matter, August 12, 1907, at the post office at Rico, Colorado, under the Act of Congress of March 3, 1879. — RICO, DOLORES COUNTY, COLO., FRIDAY, MARCH 12, 1915 — Forty-eight columns of interesting reading matter, including all the latest local, mining and general news, information, etc. — NUMBER 36

$10,000 ORE FOUND IN VIRGIN GROUND

Shelhamer and Batchellor Open up 8-inch Vein in Pro Patria

PROPERTY UNDER ARMED GUARD

Persevering Prospectors Discover What is Believed another Cresson

The startling news contained briefly in the above headlines is given out authentically this afternoon.

TUESDAY it was learned that Shelhamer & Batchellor, lessees on Jumbo No. 2 ground in the Pro Patria tunnel of the Rico Mining Co. had encountered what they believed to be the vein for which they were searching, but although rich ore was known to have been disclosed, the richness and probable extent of the find was not determined until today.

The developments now made are of the most sensational nature, the vein being from 8 to 12 inches in width and the mineral running, accord- [illegible] samples loosely taken [illegible]

Free gold is plainly seen in specimens that have been brought to town.

Superintendent George E. Hicks has been making frequent trips to the property during the past few days, and no attempt is now made to deny the importance of the strike to this camp.

The lucky lessees worked in territory adjacent to this ground some years ago, and at that time became convinced that they could find the Jumbo No. 2 vein, and that it contained rich ore in that part of the hill. Last fall they returned to the camp from outside points and went to work to prove their belief. After raising a distance of about 50 feet, they drove a crosscut east to the point where they had expected to encounter the vein. But no vein was found, and it was mighty hard sledding; besides, their grub-stake was about to give out. But their indomitable confidence and perseverance induced them to go a few feet further, with the result that their labors are to be rewarded to an extent that it is not now possible to compute. They are prospectors of the old type—men whose faith is exceeded only by their pluck, and their good fortune is a matter of pleasure to all who know them.

Not only management and lessees, but the entire town is greatly elated over the discovery, as the finding of gold ore to any considerable extent is a new thing to the camp, and holds marvellous possibilities for the whole district. The significant thing about the strike is that it was made in heretofore undeveloped territory, and may be the key to other like disclosures in Dolores mountain.

There seems to be no doubt but that, after twenty years of persevering loyalty on the part of those who believed in the wealth of her mineral resources, in the face of the most trying conditions, Rico, the brightest gem of the San Juan, is once more to come into her own.

We Recommend "Ball-Band" Goods Because We Want to See You Satisfied

Every time a customer goes out of this store we want to feel that he has got his money's worth.

BALL (Red) BAND

Every day that you wear a pair of rubber boots, you wear them out just so much. But with "Ball-Band" Boots the wear and tear is *less* because they are made of better stuff. You get more days' wear for dollars out of "Ball-Band" Footwear.

That makes "Ball-Band" cheapest in the long run. Every Rubber Boot with a Red Ball on the knee is a "Ball-Band" Boot. You'll see many of them because they have the quality.

We sell "BALL BAND" Rubbers, Overshoes and Boots because they are the best.

CHAS. ENGEL MERC. CO.

Report of the Financial Condition of

The First National Bank

of Mancos, Colorado

at the close of business Dec. 31, 1914.

RESOURCES		LIABILITIES	
Loans and Discounts	$189,157.37	Capital Stock	$ 50,000.00
Warrants, etc.	8,857.82	Surplus	10,000.00
U. S. Bonds	50,000.00	Undivided Profits	819.91
Bank Building, Furniture and Fixtures	19,500.00	Reserve for Taxes	2,928.20
Federal Reserve Bank Stock	566.67	Dividends Unpaid	5,000.00
U. S. Revenue Stamps	66.00	Deposits	224,460.32
Cash on Hand and in Banks	71,560.57	Circulation	50,000.00
Redemption Fund with U. S. Treasurer	2,500.00		
Total Resources	$342,208.43	Total Liabilities	$342,208.43

The FIRST NATIONAL BANK

KOSTER'S MARKET

M. E. STEWART

Dealer in

Tobaccos, Cigars, Soft Drinks, Confectionery, Fruit, Nuts, Stationery, Etc.

Also carry a choice line of Samples for Gents' Made-to-Order Clothing. Quality and fit guaranteed

THINGS I MIGHT DO DURING THIS LENTEN SEASON

MIGHT turn my thoughts upon myself; and awake to the sense of security that is born of self respect.

In so doing I might confess the need of care lest I confuse self respect with misleading self conceit.

And in my thinking I shall do well to remember the words of a far wiser man than I, who urged a man "not to think of himself more highly than he ought to think, but to think soberly, honestly."

To quicken my sense of self respect, must think of those elements of life that minister to it—grouping them thus—

Possession	Vocation
A lofty origin.	Worthiness of it.
A noble name.	To live up to it.
A great nature.	To develop it.
A dignified task.	To try to do it.
An exalted destiny.	Fitness for it.

—Rev. F. J. Bate, Pastor Calvary Episcopal Church, Columbia, Missouri.

For the "Kids"

Flexible Flyer Sleds and Coaster Wagons from $1.75 to $8.00 and say, they're DANDIES!

At H. Obendorfer's

The Metropole

THE Popular Resort

P. P. SCHIFFERER, Prop.

JOEL B. FRAZIER WHISKEY

GIBBS' CORNER

R H Gibbs Prop.

Our establishment is never without a large and complete stock of the finest brands of WINES, LIQUORS AND CIGARS both foreign and domestic. Drop in and sample them. Our aim is to please all.

Come and Enjoy a Sociable Hour With Us

PATRONIZE THE RICO PHARMACY

Read the Annual Report of the Rico-Wellington Mining Co. for 1914, on Last Page

DELL A. McCOY PHOTO

THE WEEKLY NEWSPAPER called THE RICO ITEM carried interesting advertisements on its front page on March 12, 1915. And the big news story obviously was the "SENSATIONAL Gold STRIKE!" heralded in the banner headline. Also notice that the price of gold was listed at $20.67 per ounce (to the right of the newspaper name).

AUTUMN COLORS were at their best on Newman Hill, southeast of Rico. Businesses still lined Glasgow Avenue during late September of 1993. The old Rico State Bank building and former Chas. Engel Merchantile store are at the left, in Block 6. The former Obendorfer store is at the right, beyond the restored bay-window building.

WILLIAM A. GRAVES COLLECTION

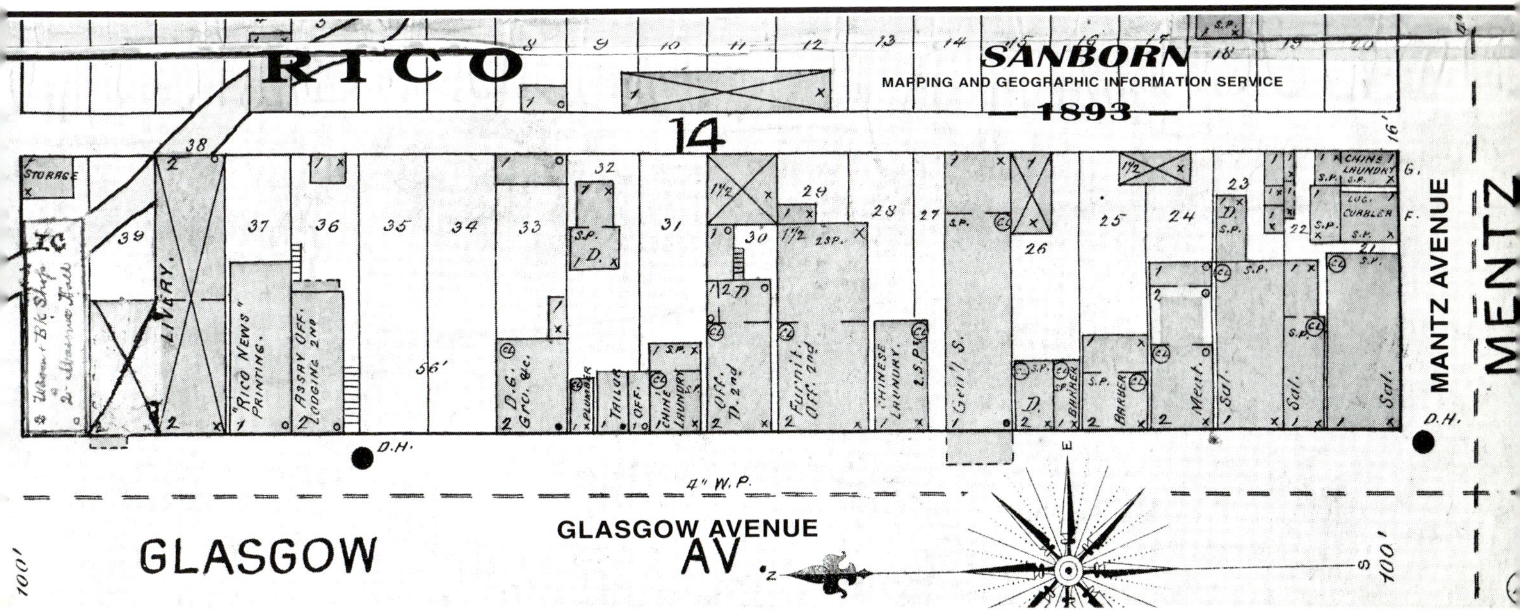

CHARLES MILLER ENGEL COLLECTION –
COURTESY RACHEL A. HARDWICK

THIS AERIAL-TRAMWAY TOWER was built in 1902, and the scene was photographed ca. 1915. The tram line carried the ore buckets from the Pro Patria Mine, up on the hillside above Rico, to the Pro Patria's ore-processing mill at the lower end of Mantz Avenue. The upper wire rope (cable line made of metal strands, twisted together) was stationary and supported the dual wheels of the ore buckets, while the smaller lower cable moved the loaded buckets down to the mill and the empties back up to the mine. The United States Hotel was located on the second floor of the building on the corner of Mantz and Glasgow avenues. This hostelry rented rooms for $35 and $50 per month, which was okay if you did not mind the noise from the saloon on the main floor! The next group of buildings had three entrances on the boardwalk. A confectionery shop, with a sign adorning its awning, was located next door to the stone Rico State Bank building. When this picture was taken, the sturdy flagpole was no longer surrounded by a bandstand.

RICO

SANBORN
MAPPING AND GEOGRAPHIC INFORMATION SERVICE
— 1893 —

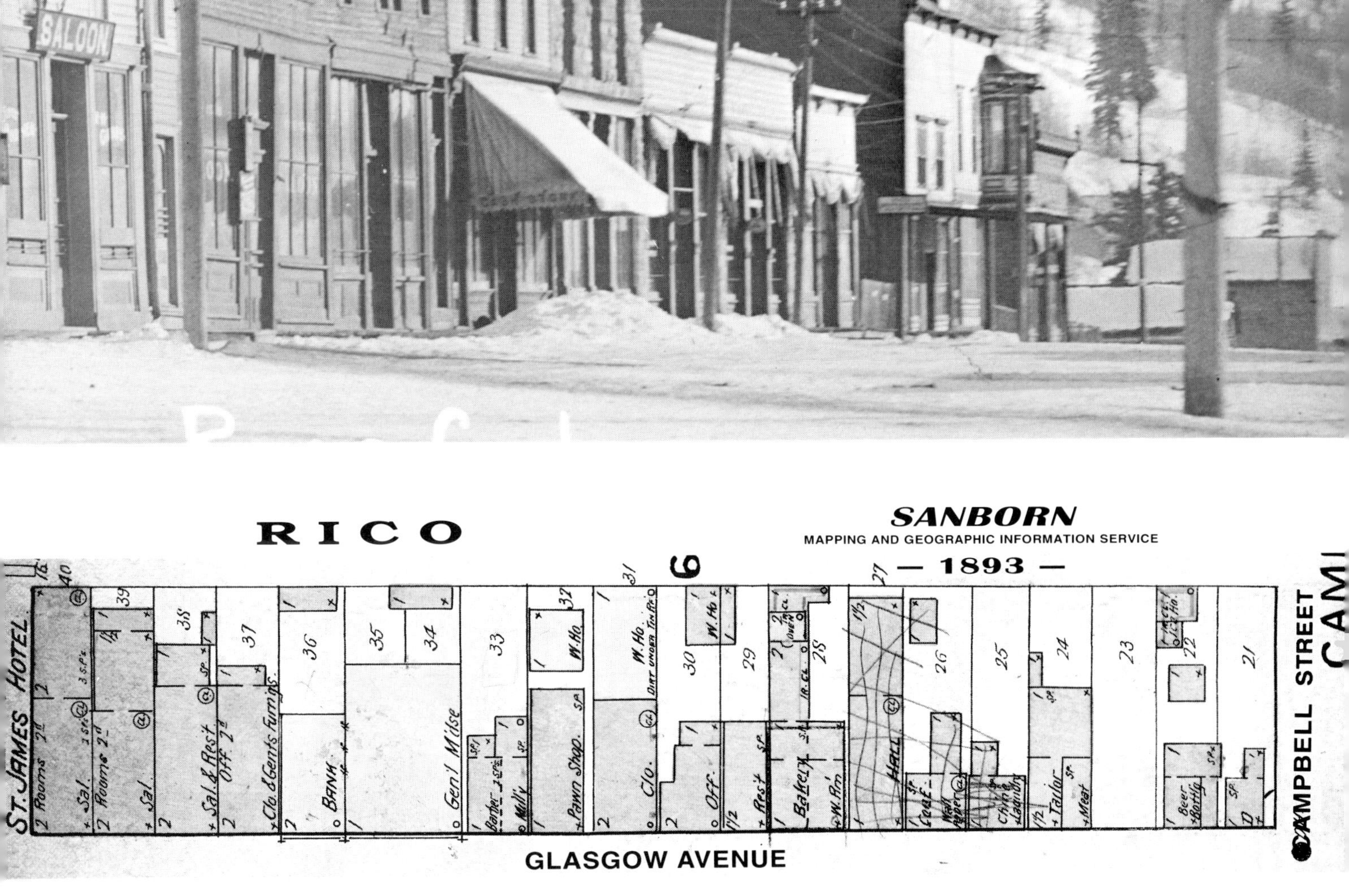

DELL A. McCOY PHOTO

THIS OLD BUILDING at 255 Glasgow Avenue barely survived until the present day. This is how the weathered wooden structure looked in 1983. Originally, the building was listed as a grocery-and-meat market on the Sanborn Map of 1899. The old frame building had an entrance off the front sidewalk that led to the second floor. The original paint colors were barely evident.

DELL A. McCOY PHOTO

IN THE AUTUMN of 1993, gold-colored aspen leaves were at their peak of glory. The old grocery-and-meat market had been remodeled into a bed-and-breakfast establishment for tourists. The former H. Obendorfer general store now houses the U.S. Post Office in Rico, to the south of this attractive bay-window building on Glasgow Avenue. A high-pressure weather pattern kept storm clouds out of the natural blue sky of southwestern Colorado's high country.

CHARLES MILLER ENGEL COLLECTION –
COURTESY RACHEL A. HARDWICK

A GLASS-PLATE NEGATIVE, which had been exposed around 1915, was used to produce this print. These old brick and wooden buildings were on the east side of Glasgow Avenue. The United States Hotel occupied the second floor of the white corner structure, while the Rico State Bank was in the stone building five lots this side of the hotel. The J. W. Burley Company previously occupied the two store fronts on this side of the bank, owned by the Charles Engel Mercantile Company. To the right (south), a tailor had set up shop next door, while a grocery-and-meat market utilized the building with the wooden roof extending out over the boardwalk. Another hotel, with a dining room, shared space in the building with the bay window overhanging the sidewalk. At the far

NEWSPAPER ADVERTISEMENTS FROM "THE RICO ITEM"
Dated March 29, 1912
MARGARET MATZICK COLLECTION

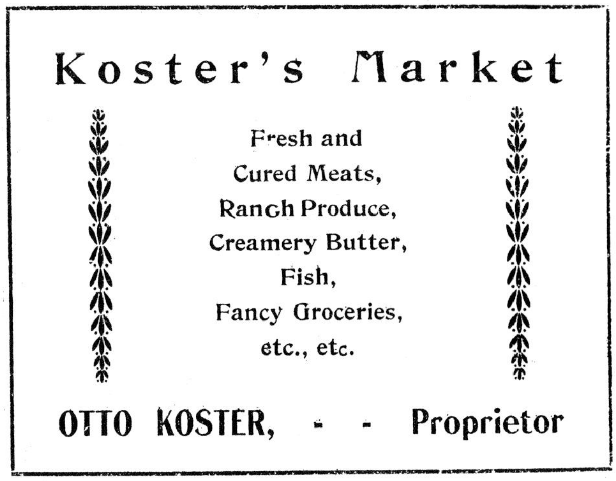

right, Obendorfer's establishment was offering competition for Charles Engel's store, only a short distance away. The Emporium sold canned goods and other staple food products in the structure taking up most of the scene on the right-hand page. Notice that there was enough snow in the area to warrant the use of a horse and sleigh.

OVERLEAF: Rico looked like this from what is known as Bemis Flats, on Expectation Mountain. Mantz Avenue is clearly visible, as it runs east and west, and Glasgow Avenue is the main street running north and south (horizontally). This 1995 view shows most of the buildings, including a glimpse of the old RGS water tank (visible in the lower part of the photograph on Page 395).

DELL A. McCOY PHOTO

Rico, Colo.

CHARLES MILLER ENGEL COLLECTION –
COURTESY RACHEL A. HARDWICK

IN THIS WINTER SCENE of ca. 1915, photographed facing toward the southeast across the town of Rico at Newman Hill and Dolores Mountain, important business buildings and residences that appeared in earlier views, are missing. They were completely destroyed in the severe flood of 1911, when Silver Creek overflowed its banks in a raging torrent. The Rico Schoolhouse was spared, but the flood destroyed the big barn housing the Masonic Hall on the second floor, as well as THE RICO NEWS building, located next door. The relatively small hoist house in the foreground belonged to the Atlantic Cable Mine. The big tailings dump that dominated the mountainside (in the center of the scene) marked the location of the Enterprise Mine. The Syndicate Tunnel appears at the far right.

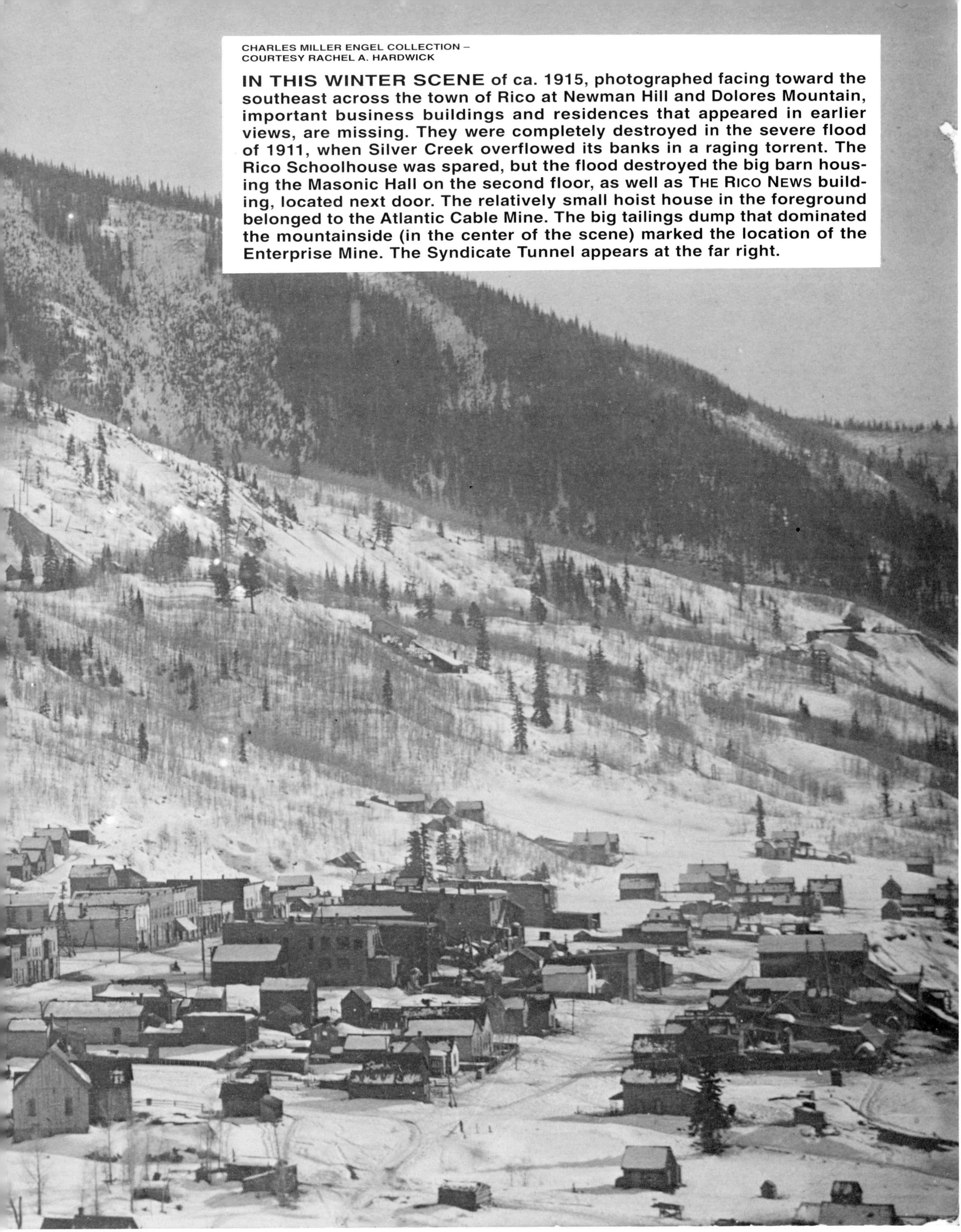

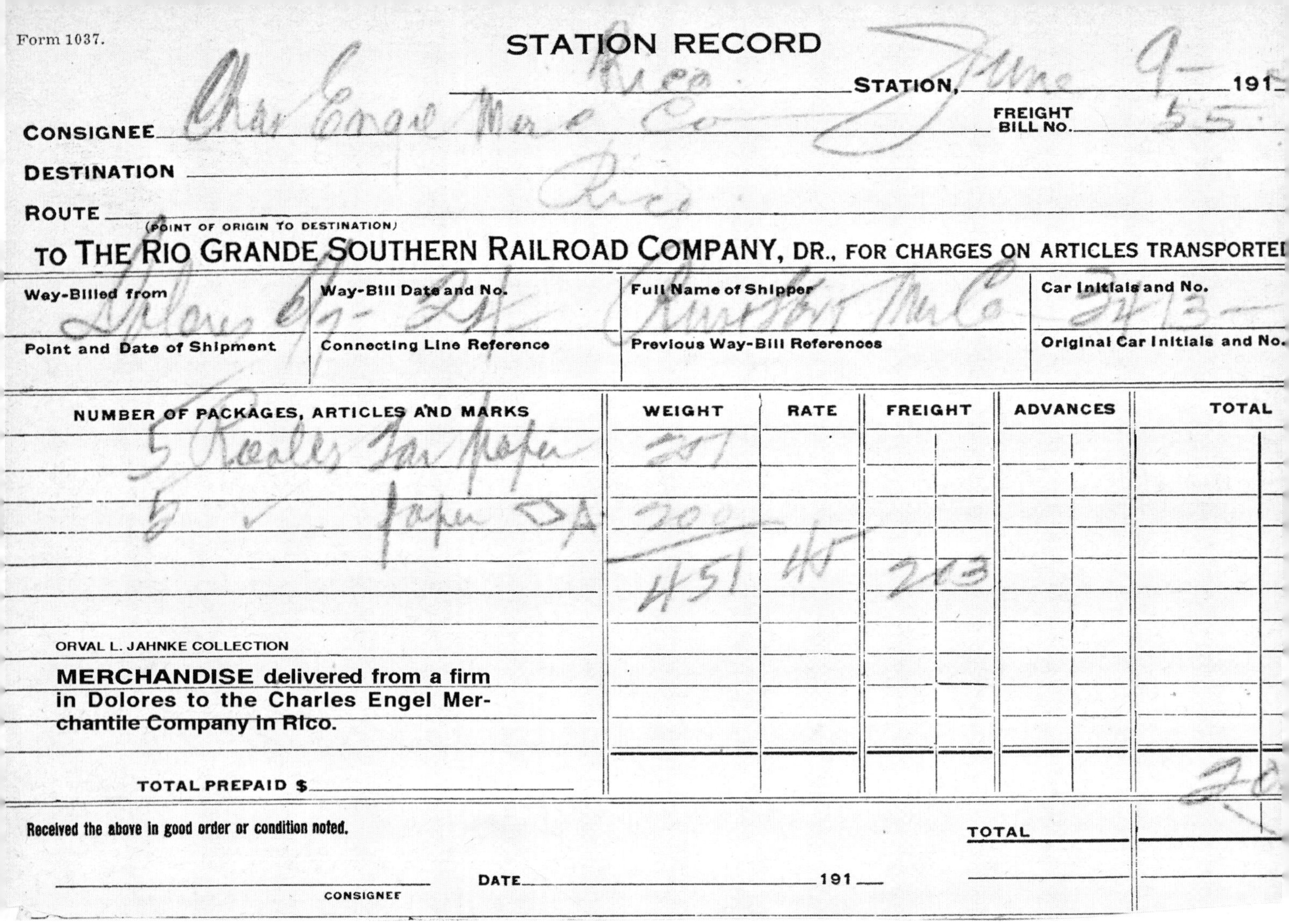

Form 1037.

STATION RECORD

Rico STATION, June 9- 191

CONSIGNEE Chas Engel Mer Co FREIGHT BILL No. 55

DESTINATION Rico

ROUTE (POINT OF ORIGIN TO DESTINATION)

TO THE RIO GRANDE SOUTHERN RAILROAD COMPANY, DR., FOR CHARGES ON ARTICLES TRANSPORTED

Way-Billed from	Way-Bill Date and No.	Full Name of Shipper	Car Initials and No.
Dolores 6/7 - 24		[illegible] Mer Co	2413
Point and Date of Shipment	**Connecting Line Reference**	**Previous Way-Bill References**	**Original Car Initials and No.**

NUMBER OF PACKAGES, ARTICLES AND MARKS	WEIGHT	RATE	FREIGHT	ADVANCES	TOTAL
5 Reales Tar paper	251				
5 ✓ paper DA	200				
	451	45	203		

TOTAL PREPAID $

Received the above in good order or condition noted.

TOTAL

CONSIGNEE DATE 191

ORVAL L. JAHNKE COLLECTION

MERCHANDISE delivered from a firm in Dolores to the Charles Engel Merchantile Company in Rico.

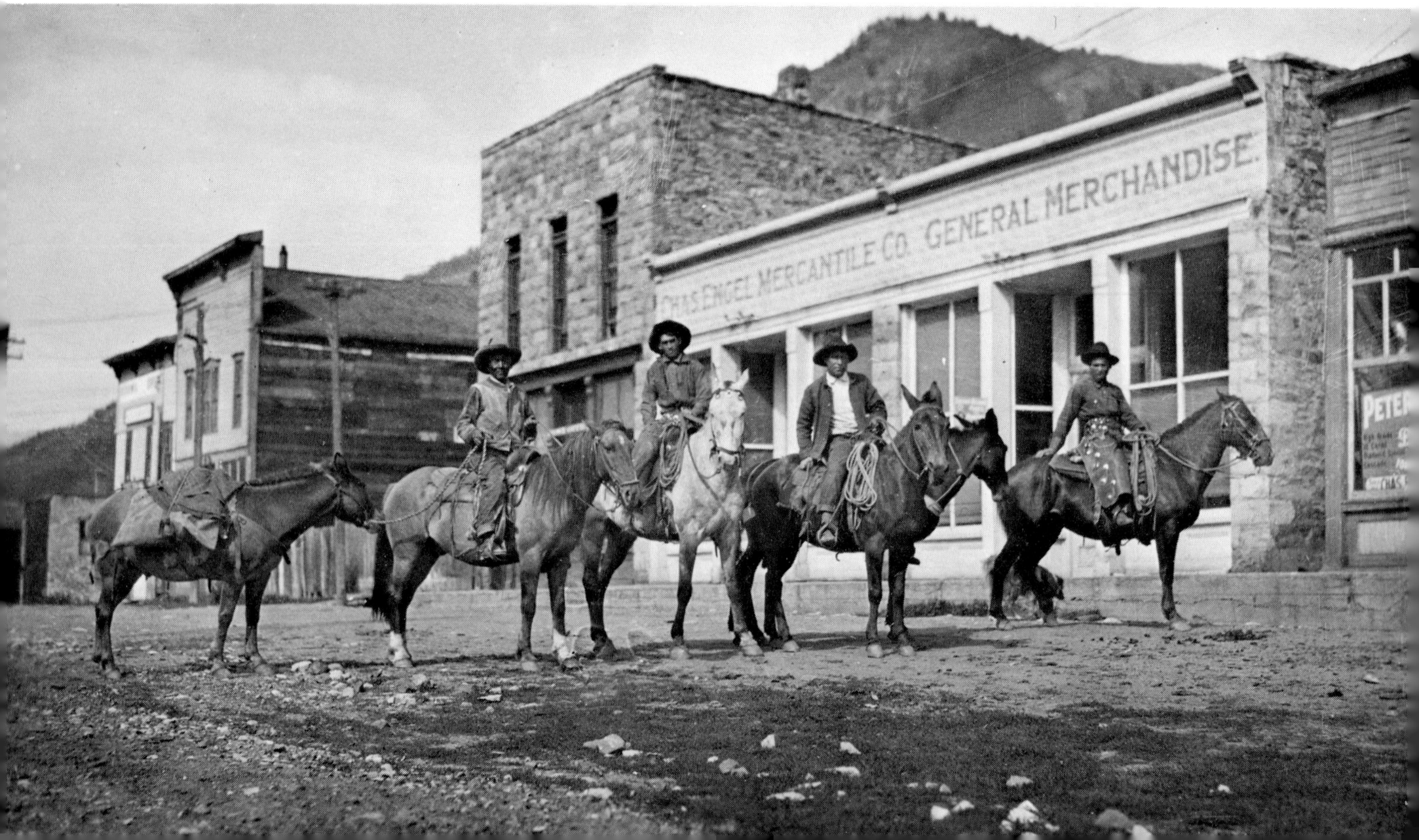

CHARLES J. LEE PHOTO – DAVID C BUTLER COLLECTION

CHARLES J. LEE must have had a job at the Engel Mercantile Company store ca. 1919, and he loaned his camera to Charles M. Engel for this portrait. Young Mr. Lee probably was about 18 years old at this time, and he was suitably dressed for work. He was wearing a dress shirt, tie and protective over-sleeves, along with bib overalls, to do his work in the store. The photographs of Charles Lee fill many pages in this volume. He became a pharmacist in Dolores and a mortician in Rico.

ENGEL'S GENERAL STORE and the Rico State Bank building—seen in many of these photographs—form the backdrop for this group of cowboys, in town from one of the so-called "cattle spreads" (ranches) down river from Rico. The fellow at the right was sitting on a horse, with a pack mule trailing behind him. The two "cowpokes" in the middle of the view were riding on mules, while the man at the left was astride a horse, with a pack mule following him. The "wranglers" appear to be American Indians, perhaps Navajos. The Engels were no longer using the building at the right as a warehouse, as seen in other views, but this picture shows a sign in the window advertising "Peters Shoes," with the Charles Engel name beneath the shoe sign. The son of the senior Charles Engel may have returned to Rico by this time. (He returned with his bride in 1924.)

CHARLES MILLER ENGEL COLLECTION –
COURTESY RACHEL A. HARDWICK

M E Stewart

Dealer in

Tobaccos, Cigars,
Soft Drinks, Confectionery, Fruit, Nuts,
Stationery, Etc.

Also carry a choice line of Samples fcr Gents'
Made-to-Order Clothing. Quality and fit
guaranteed

NEWSPAPER ADVERTISEMENT
FROM "THE RICO ITEM"
Dated March 29, 1912
MARGARET MATZICK COLLECTION

CHARLES J. LEE PHOTO – NORMA L. BUTLER COLLECTION

HARRIET LEE was the matronly woman in this view, according to one of her grandchildren. Mrs. Lee had been shopping in Charles Engel's general store, possibly during 1917, according to the dress style. It looks as though Mr. Engel needed to do some relettering on his store windows. Notice the tall fellow standing beside the store, the same one who appeared at the depot on Page 257.

CHARLES J. LEE PHOTO – NORMA L. BUTLER COLLECTION

YOUNG CHARLES LEE posed for his friend, Charles M. Engel, outside the Chas. Engel Mercantile Company, where Lee was working. This shot was taken on Glasgow Avenue, ca. 1917, with the Burley Block across the street. By this date, the Rico Fire Department's wooden building for Hose Team No. 1, at the left, was badly in need of a paint job.

CHARLES J. LEE PHOTO – NORMA L. BUTLER COLLECTION

CHARLES MILLER ENGEL was posing for his buddy, Charles J. Lee, about 1917, facing his father's store in morning sunlight. Across the avenue, at the right, you can see that the Burley Block had an open entrance for the stairs that went up to the second story. You can also see the rather faded sign advertising "GROCERIES" over the store, to the left of the stairway.

CHARLES J. LEE PHOTO – DAVID C. BUTLER COLLECTION

THIS YOUNG LADY probably was Mary Nixon, a local Rico lass. She was all dressed up in her Sunday best one day during 1917.

CHARLES J. LEE COLLECTION – COURTESY NORMA L. BUTLER

CHARLES M. ENGEL was busy shooting photographs of Rico residents, as they paid a visit to the Engel general store. The man posing at the right is unidentified. The young woman was feeding snacks to a friendly dog, and Charles J. Lee was holding a box camera, farther up the street. This was the way life was in Rico during 1917.

RON RUHOFF PHOTO

OVERLEAF: From a location on Nigger Baby Hill, Ron Ruhoff captured this fall scene on film. The Presbyterian Church, Burley Block and Enterprise Cafe & Lounge stand out prominently. The distinctive home with the porch and fireplace chimney had belonged to Elizabeth Pellet. Myron Jones, local historian, lives next door, to the right. Notice the timber platform beyond the Dolores River, at the south end of the settlement of West Rico. These ruins are the remains of a mining venture named the "Iron Clad," begun in the 1950's with an ore bin, jig mill and trestle, owned and operated by Myron Jones.

ORVAL L. JAHNKE COLLECTION

THE BURLEY BLOCK first appears on the Sanborn Map of 1893. According to lettering on the windows, clothing was being sold there, although the word "groceries" was lettered on the sign above. The Rico Drug Company was located on the other side of the building as of this date, ca. 1917. The vacant lot once housed a confectionery store, and the board-and-batten wall was part of an old 1880 building that housed a restaurant called "Mack's Place."

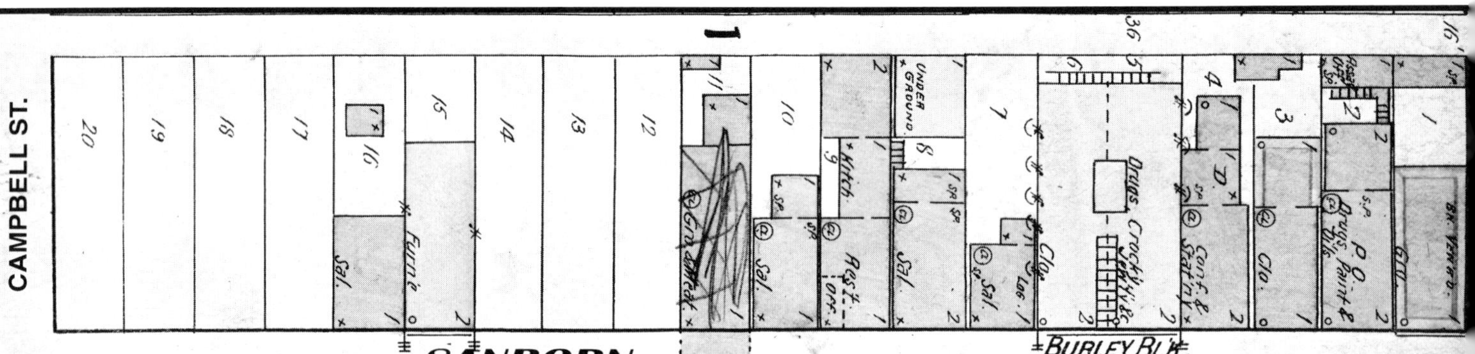

RICO

SANBORN
MAPPING AND GEOGRAPHIC INFORMATION SERVICE
— 1893 —

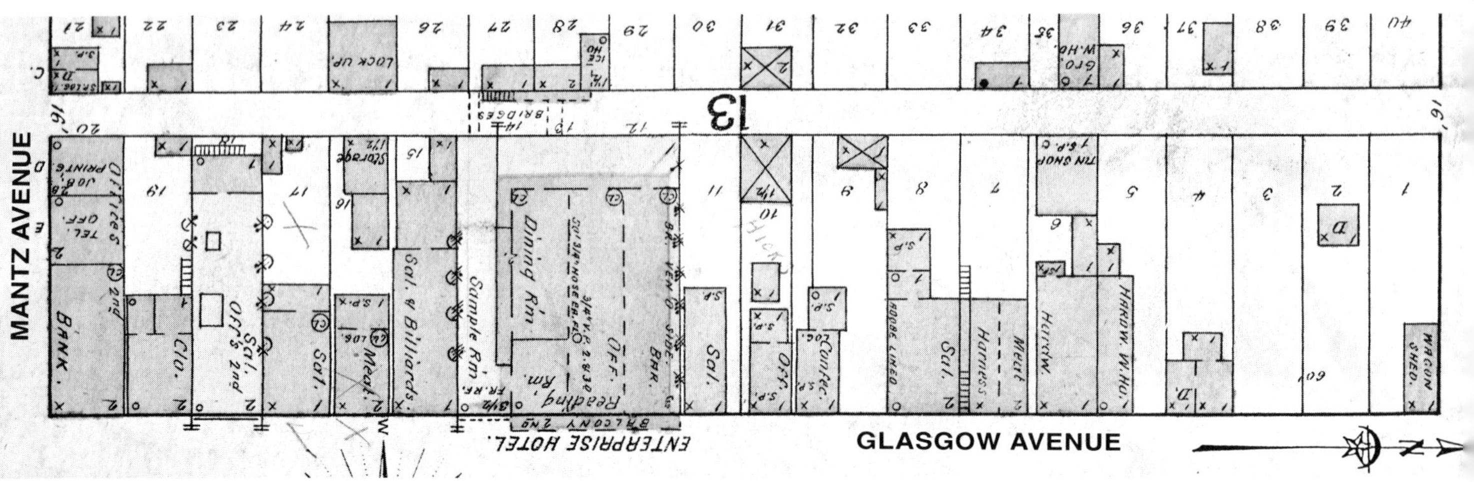

CHARLES MILLER ENGEL COLLECTION – COURTESY RACHEL A. HARDWICK

RICO'S FIRST BRICK BUILDING, erected on the southwest corner of Mantz and Glasgow avenues in 1889, reportedly was this single-story structure shown in this 1917 picture. The restaurant, at the left, named "Mack's Place," appears in a photograph dated 1880, and it is still standing today (1996). It is interesting to see the house plants growing in containers placed just inside the front windows of these business buildings. Also of interest is the timber tower at the right, supporting part of the aerial tramway leading to the Pro Patria mill at the end of Mantz Avenue.

CHARLES MILLER ENGEL COLLECTION – COURTESY RACHEL A. HARDWICK

TODAY'S ENTERPRISE Bar and Restaurant occupies the stone building, which appears on the Sanborn Map of 1893, located in Block 13, Lot 18. The building next door, to the left, housed a clothing store during that year, while the corner building originally was constructed for the First National Bank. Harry Kauffman was standing in his characteristic pose next to the corner. Notice the ore bucket hanging from the wire rope, as well as the flagpole that had been erected in the center of the intersection of Mantz and Glasgow avenues. The view was taken ca. 1915.

RUSS COLLMAN PHOTOS

THE WEST SIDE of Rico's business block containing the Burley Block was photographed during late September of 1965. The original Fire Department building in Rico, at the left, had been sheathed in metal siding. Next door, at the right, is the new annex, with the fire truck inside. Notice the siren atop the old firehouse. A Shell gasoline station and garage occupied Rico's first brick building.

THE TWO VIEWS below illustrate the Atlantic Cable Mine's shaft and supporting structure. These pictures were photographed on the same day in 1965 as the above view.

Going South before
snow Comes

CHARLES J. LEE PHOTO – DAVID C. BUTLER COLLECTION

THIS DISTANT VIEW of "Block No. 1" (according to the Sanborn map) was taken by Charles Lee as he looked toward the north on Glasgow Avenue during 1917. The old saloon building was standing next to a brick structure that had been the E. Carey & Company's furniture store—facing toward the east on Glasgow Avenue, strewn with rocks in this scene. The old furniture store had become the home of the Masonic Hall when Charles shot this picture. Notice the wooden-frame bell tower, which had been built out over the sidewalk at the firehouse.

CHARLES MILLER ENGEL COLLECTION – COURTESY RACHEL A. HARDWICK

THE LAST DAYS of September had arrived when this scene was photographed. Charles M. Engel's father had leaned out of a bay window in a building on Glasgow Avenue to get this shot. A sheepherder was moving his flock of wooly animals from summer pasture in the high country to the loading pens along the RGS, at the south end of town. The sheep were passing the Masonic Lodge on Glasgow Avenue, formerly occupied by a furniture dealer.

CHARLES ENGEL, SR. moved on down to the sidewalk to shoot this picture of sheepherders moving south on Glasgow Avenue. Mules were used to haul the covered wagons that carried all the belongings of the sheepherders, including their bedrolls, as well as their food supplies and cooking utensils. The building on the other side of the rear wagon was used by the Masonic Lodge in Rico, and they had rented the main floor to Charles M. Engel to use as a theater.

CHARLES MILLER ENGEL COLLECTION – COURTESY RACHEL A. HARDWICK

OVERLEAF: The south end of Rico was photographed at the peak of Indian-summer colors in late September. Block 3, in the left foreground, contains one residence at the intersection of Glasgow Avenue and King Street. In the next block, No. 2, is a group of homes and sheds, with the 1950's brick schoolhouse (now the Rico Visitors Center) at the intersection at Campbell Street and Glasgow Avenue. The Masonic Lodge and Burley Block are farther up in Block 1. What appears to be the ruins of a former house is in this view at the lower center. It probably was the snowslide casualty remarked about in the caption on Page 369.

RON RUHOFF PHOTO

CHARLES J. LEE PHOTO – NORMA L. BUTLER COLLECTION

CHARLES J. LEE lived in Rico during 1928, when he ventured out for a hike up Nigger Baby Hill, for a view of the town. The big bend in the Dolores River, at the far left, was the location of charcoal kilns used to produce fuel to fire furnaces at Rico smelters during the early days. The railroad's wooden Burnett & Clifton "Automatic Coal Chute" was just beyond the mill tailings, which appear like an arrow of white sand. There were quite a number of vacant lots in the south end of Rico at this time, either due to fire or the removal of unoccupied houses. Today's Presbyterian Church and the former Dolores County Courthouse are in view, in the lower left-hand portion of this scene. The top part of the Pro Patria processing mill appears at the lower end of Mantz Avenue, and you can see one of the aerial-tramway towers in the middle of the avenue, to the right of the courthouse. Stacks of lumber are visible in the Rio Grande Southern's freight yard, near the brick enginehouse. At the far right, in the northern end of the RGS station grounds, two warehouses and a coal shed can be seen next to the tracks.

ANN AND JAMES FERANDO PHOTO

A RICO LOVELY by the name of Valera Winkfield stood atop the very summit of Telescope Mountain, at 12,210 feet in elevation. This was during the 1930's, when hikers wore riding breeches known as "jodhpurs." They fit closely from the knee to just above the ankle.

CHARLES MILLER ENGEL COLLECTION –
COURTESY RACHEL A. HARDWICK

ON THIS OCCASION, a northbound locomotive, equipped with a pilot plow, was breaking open the snowdrifts south of Rico, in the vicinity of the locomotive coaling facility. Section men and laborers were on hand in case they were needed to hand-shovel snow or re-rail the engine. This winter scene must have been photographed after Charles Engel, the newspaper photographer, returned to Rico in 1924.

AN RGS LOCOMOTIVE had come to a sudden halt while blasting her way through deep snow a short distance south of Rico. Now, the manual laborers were about to earn their pay as they attempted to trench out enough snow to enable the little engine to make another run at plowing through the white stuff.

CHARLES MILLER ENGEL COLLECTION
COURTESY RACHEL A. HARDWICK

Mile by Mile to...

The Rico Division Point

The End of the First District and the Beginning of the Second District

The St. Louis Tunnel Period at Rico

No. 9

THE STORY OF RICO AND THE PIONEER MINING DISTRICT

By William A. Graves

(Continued From Page 344 in Chapter 8)

1922 A young mining engineer by the name of Duane Kline visited Rico. He represented the St. Louis Smelting & Refining Company (the StLS&R) of St. Louis, Missouri. The StLS&R used sulfuric acid to process lead from their Missouri mines, and Kline's job in Rico was to evaluate the iron-pyrite deposits on Telescope Mountain. He found the pyrite deposits to be rich in iron and sulfur. Silver, lead and zinc values also were present, and the St. Louis Smelting & Refining Company negotiated a contract with the owner, the Rico Mining & Reduction Company, to develop the property. During the following year, development began on the St. Louis Tunnel. The portal was located about one mile north of Rico, on the east side of the Dolores River. A narrow-gauge electric tramway (i.e., a mine railroad) was to be built to "tram" (haul) the pyrite out of the tunnel.

1923 Robert Pellet hired a few men and reopened the Princeton Mine on C.H.C. Hill. This was the beginning of Bob Pellet's efforts to save the old mining town from becoming just another deserted Colorado ghost town.

Bob Pellet did not have the capital to work any of the Rico mines on a large scale. However, after several rather unsuccessful attempts, Bob negotiated a contract with the International Smelting & Refining Company, a subsidiary of the great Anaconda Company. With this

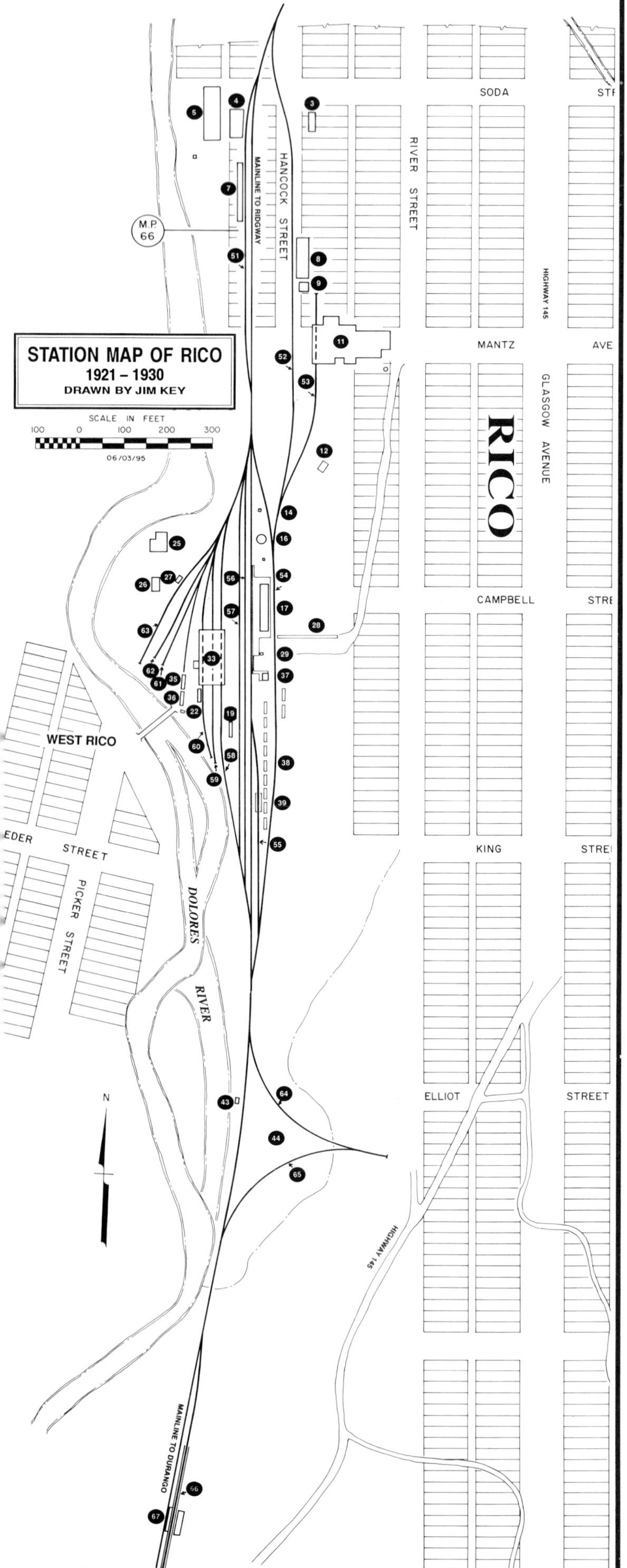

STATION MAP OF RICO
1921 – 1930
DRAWN BY JIM KEY

3. SAMPLING WORKS
4. RICO P&L WAREHOUSE
5. S. J. HOOD LUMBER YARD
7. COAL SHED
8. RICO P&L POWER PLANT
9. STEAM BOILER HOUSE
11. PRO PATRIA MILL
12. HOUSE (OFFICE)
14. CABOOSE COAL & TINDER BOX
16. RGS WATER TANK
17. RGS DEPOT
19. RPO CAR BODY FOR ROADMASTER
22. SAND HOUSE
25. SECTION HOUSE
26. BUNKHOUSE
27. COAL SHED
28. WOODEN BOARDWALK
29. FIREHOSE BOX AT DEPOT
33. SECOND ENGINE HOUSE (BRICK)
35. OLD BOXCAR BODY
36. OLD BOXCAR BODY
37. DEPOT OUTHOUSE
38. BRIDGE & TRESTLE TIMBERS
39. TRACK SCALE
43. SECTION-CAR SHED
44. THIRD (AND FINAL) WYE
51. TRACK No. 69 – 789 Ft. Long
52 .TRACK No. 68 – 1,257 Ft. Long
53. TRACK No. 75 – 515 Ft. Long
54. TRACK No. 70 – 1,157 Ft. Long (HOUSE TRACK)
55. TRACK No. 80 – 635 Ft. Long
56. TRACK No. 71 – 1,150 Ft. Long
57. TRACK No. 72 – 960 Ft. Long
58. TRACK No. 74 – 607 Ft. Long
59. TRACK No. 78 – 477 Ft. Long
60. TRACK No. 76 – 534 Ft. Long
61. TRACK No. 73 – 492 Ft. Long
62. TRACK No. 79 – 249 Ft. Long
63. TRACK No. 77 – 306 Ft. Long
64. TRACK No. 81 – 495 Ft. Long
65. TRACK No. 82 – 425 Ft. Long
66. TRACK No. 83 – 559 Ft. Long
67. BURNETT & CLIFTON COALING FACILITY

CHARLES MILLER ENGEL COLLECTION –
COURTESY RACHEL A. HARDWICK

THE SOUTHBOUND passenger train of the RGS had reached Rico on a snowy Easter Sunday during 1922. The engineer was at the right, holding his oil can, while the fireman was standing next to him. The conductor had climbed up on the snow-covered locomotive, hanging onto a grab iron. Their trip must have been an unnerving experience, as the little engine had to buck through snowdrifts on Lizard Head Pass. Chances are, if the locomotive did not have a pilot plow, the train would have stalled somewhere on the northern side of the pass. The young man in front of the engine probably had been hired by the roadmaster to keep the yard's switches in operating condition during freezing weather, as well as to work at other tasks in the Rico yard, including the removal of snow from the locomotive before the train continued on its run. Notice the railing along the top of the Railway Post Office car, used by trainmen to hold onto while walking on the roof while going to and from the engine, since passage through the RPO car was not possible.

powerful backing, Bob formed the Falcon Mining Company and reopened and reworked the Falcon and Yellow Jacket mines on Nigger Baby Hill, under a lease-and-option arrangement with the Syndicate Mining & Milling Company.

1924 Bob Pellet was elected as the mayor of Rico—for the first of three consecutive terms. He continued to expand his mining operations, and he formed a second mining firm, the Pell-Eyre Mining Company. Bob also secured options to work the Atlantic Cable and Shamrock mines, as well as the Pro Patria mill. Over $350,000 worth of ore was shipped. Lead and zinc predominated.

(Continued on Page 436)

CHARLES MILLER ENGEL COLLECTION – COURTESY RACHEL A. HARDWICK

THE NORTHBOUND passenger train was late in arriving at the Rio Grande Southern's depot in Rico on this snowy winter day. Passengers were lined-up on the depot platform, waiting as a lone engine plowed through the snow at the southern end of the yard. Since no other buildings were in sight near the depot, other than the combination outhouse and coal shed, this picture must have

PASSENGER TRAFFIC TO RICO
— Continued —
By Robert W. Richardson

During 1923, when the RGS was made an operating part of the Third Division of the D&RGW (headquartered at Gunnison), only a few hardy people chose to go to the outside world via Dolores and Durango. The D&RGW claimed that their schedule change was made to ensure better mail and passenger connections at Durango. However, the resulting schedule seemed to have been designed to discourage rail travel. The RGS train left Durango at 6:00 p.m., after the arrival of the Alamosa train. Then, it would take nearly six hours to reach Rico.

Following arrival at Rico, the train would be turned and serviced, and almost immediately head back to Durango around midnight, arriving there in time to catch that morning's train for Alamosa. As there were only regular station stops at three points—all other places being flag stops—this meant that passengers had to wait out in the

been taken ca. 1925. Beyond the engine you can see a section-car shed, and farther yet is a glimpse of the large Cliffton & Burnet locomotive-coaling facility.

weather for a train not noted for being on time. The records did not show many Rico passengers in either direction, for what was a long roundabout way to anywhere. When the RGS was again its own master, with headquarters at Ridgway—after the Third Division was discontinued during March of 1925—this schedule was changed back to a daytime train again.

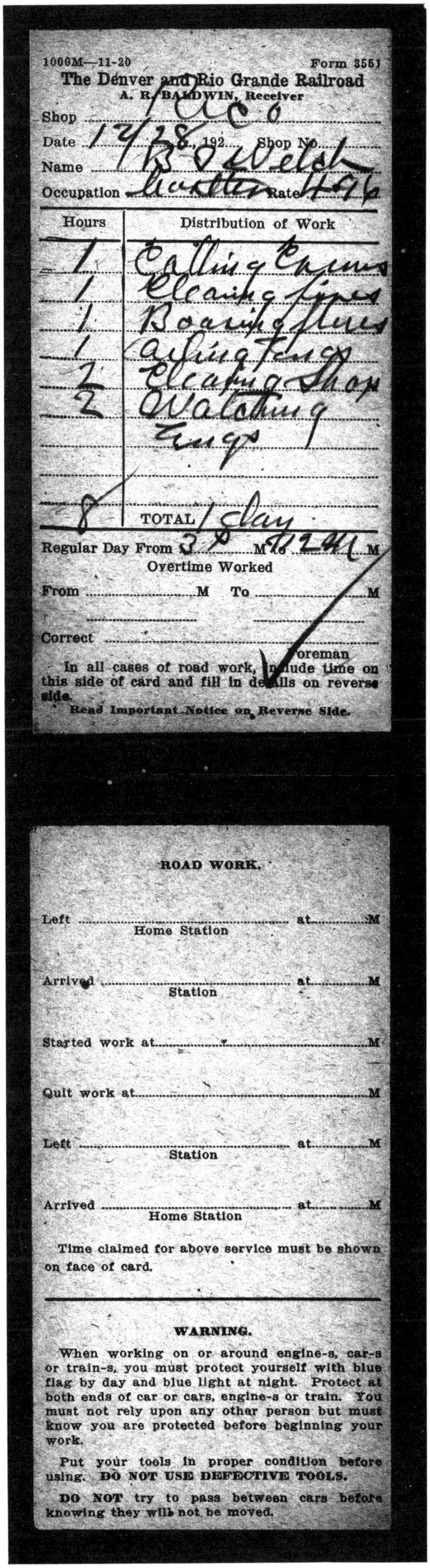

1000M—11-20 Form 3551

The Denver and Rio Grande Railroad

A. R. BALDWIN, Receiver

Shop Rico

Date 12/28, 192 Shop No.

Name

Occupation Hostler Rate

Hours	Distribution of Work
1	Calling Crews
1	Cleaning fires
1	
1	
2	Cleaning Shop
2	Watching Engs
8	TOTAL 1 day

Regular Day From M to M

Overtime Worked

From M To M

Correct Foreman

In all cases of road work, include time on this side of card and fill in details on reverse side.

Read Important Notice on Reverse Side.

ROAD WORK.

Left at M
Home Station

Arrived at M
Station

Started work at M

Quit work at M

Left at M
Station

Arrived at M
Home Station

Time claimed for above service must be shown on face of card.

WARNING.

When working on or around engine-s, car-s or train-s, you must protect yourself with blue flag by day and blue light at night. Protect at both ends of car or cars, engine-s or train. You must not rely upon any other person but must know you are protected before beginning your work.

Put your tools in proper condition before using. DO NOT USE DEFECTIVE TOOLS.

DO NOT try to pass between cars before knowing they will not be moved.

E. J. HALEY COLLECTION

THE RGS HOSTLER PRIMARILY USED THIS FORM DURING THE 1920'S

P. DAVID SMITH PHOTO

WINTER CONDITIONS at the station grounds in Rico were anything but pleasant during this snowstorm in 1925. The enginehouse and the north end of the depot appear in the left-hand portion of this snowy scene. Rio Grande Southern Ten-wheeler No. 22 and another locomotive were steamed up, and they were on the enginehouse lead west of the water tank. The larger buildings beyond the two locomotives were the bunkhouse and the section house. An assortment of car bodies were used in the Rico yard for housing workers and for other purposes. Notice that a string of empty gondolas had been spotted on the yard track not far from the depot.

Ex-Crystal River 2-8-0 No. 361 on the RGS

By Gabe Bradford

FOR AN INTERESTING and somewhat unusual story about an RGS train dispatched northbound from Rico to Ridgway, we are indebted to Gabe Bradford of the Rio Grande Southern Historical and Technical Society. In their newsletter, THE GALLOPING GOOSE, published in 1990 as their Volume III, No. 3, they document the run of D&RGW locomotive No. 361, which ran northward from Rico.

What is especially interesting about this story is that one of the D&RGW's ex-Crystal River 2-8-0's was assigned to an RGS train, and there was no extraordinary mention about using this engine. Gabe discovered a letter documenting an inspection trip made on the RGS over a four-day period from Ridgway to Durango and return. The person who made this trip typed his report

(No. 361 Continued on Page 423)

CHARLES MILLER ENGEL COLLECTION –
COURTESY OF RACHEL A. HARDWICK

THE BIG ROTARY SNOWPLOW of the RGS, No. 2, had been steamed up, and the tender had just been filled at the Rico water tank, north of the depot. The snowplow was nearly ready to clear wet snow from the tracks in the station grounds. The "D&RGW" lettering on the gondola to the left of the plow indicates this scene was photographed after the reorganization of the D&RG system, which took place in 1921. And since the newspaper photographer had been absent from Rico for awhile, and did not return until late in 1924, it is assumed that this view was taken during the late winter (or early spring) of 1925. The snow appears to be heavy and rather wet, not the type that falls during the first half of the winter in the San Juan Region.

ANN AND JAMES FERANDO COLLECTION

RESIDENTS OF WEST RICO living along Picker Street, west of the Dolores River, had an excellent vantage point to watch action in the RGS yard—if one was so inclined! The date for this view must have been about 1922, as you can see the freshly painted "W" that had been added to the D&RG boxcars. The D&RG became the D&RGW following the reorganization of August 1, 1921, and their freight cars received a "W" in the appropriate places as money and time allowed. At least two of the boxcars had not yet received the new lettering. Bridge timbers had been stacked on platforms next to the rail yard—at or above the level of flatcars. This permitted faster transfer of the timbers when trestle repair was required and kept the wood off the damp ground. The shed covering the track scale mechanism is in view beyond the flatcar.

RGS LOCOMOTIVES were painted and lettered in this livery about 1922, when local photographer Charles J. Lee took an interest in taking pictures of the railroad action. Engine 15 was a Class 60 2-8-0 (a year or so before becoming a Class C-16 engine). She had derailed and was firmly on the ground, including the lead truck and all eight drivers. This was next to the switchstand (draped with a man's white shirt) that was located at the northern turnout of the Rico wye—the third and last wye at Rico! The engine sported a shiny dark-green boiler jacket and graphite-gray smokebox at this time, with a black cab and tender. No. 15 was built in 1881 as the D&RG's No. 253 by the Baldwin Locomotive Works.

CHARLES J. LEE PHOTO – DAVID C. BUTLER COLLECTION

on the morning of Thursday, February 25, 1926. The original copy was signed and sent to Superintendent Bowen at Alamosa, while the unsigned copy was put in the file at Ridgway. It was this carbon copy that Gabe located and then shared the information contained in it. We have slightly paraphrased Gabe's article, but he invites you to get a map of the RGS, use your imagination and settle back…

Now, to the story…

RGS Train 392, a scheduled freight, had their regular crew, and Conductor Houk was called for the train at 3:30 p.m., while both helper crews were called at 4:30 p.m. After

(No. 361 Continued on Page 433)

JOHN W. MAXWELL COLLECTION

THE LIVESTOCK PENS AT RICO
— AS REBUILT IN 1923 FOR SHEEP —

The old livestock pens were retained, but were altered as required for sheep

One-half mile north to Milepost 66

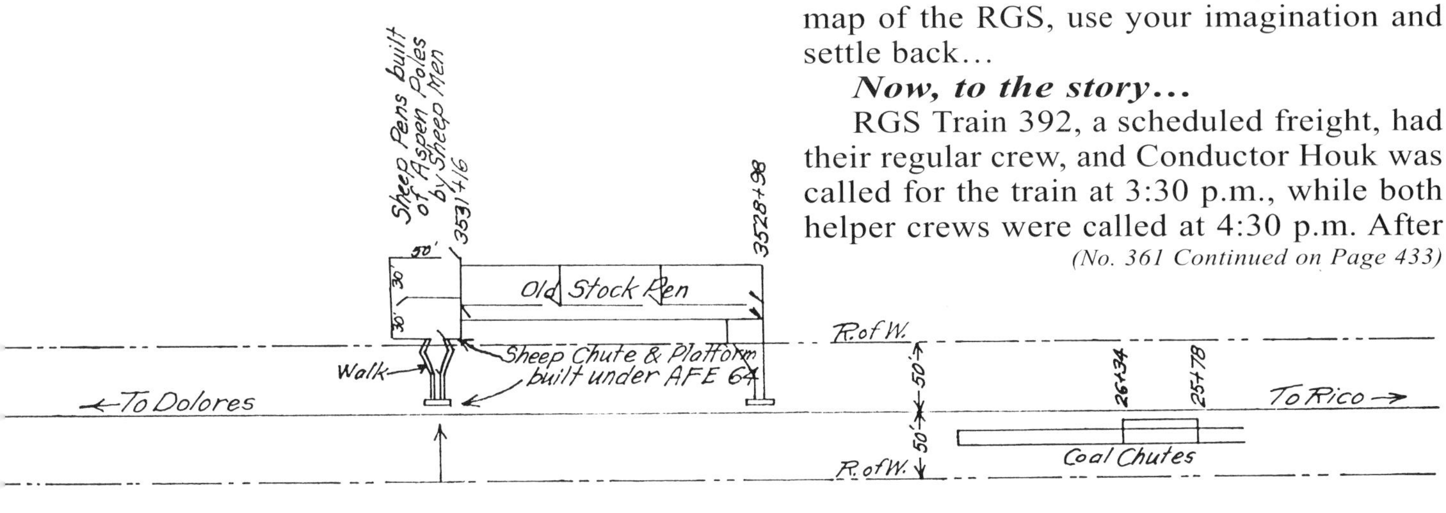

D&RG
D&RGW
D&RGW

CHARLES MILLER ENGEL COLLECTION – COURTESY RACHEL A. HARDWICK

RICO'S LIVESTOCK PENS were at the south end of the RGS yard. In this view, several head of cattle were in the pens, waiting for the arrival of stock cars. (One stray would miss the train if no one noticed his presence outside the pens.) The livestock movements on the RGS were planned for a tight schedule, due to the fact that the animals required feed and water after a fixed amount of time. One car at a time was moved into position at the loading ramp. The hinged doors at the end of the chute were moved out toward the side of the stock car being loaded, preventing the escape of animals.

RGS LOCOMOTIVE No. 10 was moving stock cars at the south end of the Rico station grounds. Engine 10 was a Baldwin-built Class 60 2-8-0 received by the D&RG as their No. 250 in 1881, and acquired by the RGS on April 9, 1891. The Consolidation was retired in the 1920's and dismantled in 1942. The Rico stockyard was near Milepost 67, where cattlemen had built pens and a loading chute, mostly out of aspen logs. This scene was recorded on film ca. 1922, when the D&RG became the D&RGW. It probably was about the first week of October, when the ranchers moved their livestock off the high mountain rangeland to avoid freezing weather.

CHARLES MILLER ENGEL COLLECTION – COURTESY RACHEL A. HARDWICK

DAN PRICE COLLECTION

THE DEPOT TOWER was retained until a fire destroyed part of the upper floor in 1936. Its original purpose was to channel telegraph wires down to the main-floor offices, where the dispatcher was stationed during the early years. The tower windows offered a good view of the yard and the distant southern mainline. The station agent and his family had living quarters upstairs, and you can see lace curtains at the upper windows. Notice the fire hose lying on the station parking area, no doubt placed there to dry. Also notice the depot's extended boardwalk in this picture taken beside the main track.

A GOOD PLACE to find action in Rico during the 1920's was down at the RGS depot. Here, you could see passengers coming and going, and people were picking up express and L-C-L shipments. Notice that a boxcar had been spotted on the house track at the freight end of the depot. Since it was summertime, school was out, and boys liked to lounge around the depot when trains were due. The morning passenger train may have been late in arriving, and Rico folk were enjoying the summer sunshine on the depot's platform at the south end of the building.

CHARLES J. LEE PHOTO – NORMA L. BUTLER COLLECTION

GABE BRADFORD COLLECTION

THIS YARD VIEW shows a two-dome Texaco tankcar that was sitting on the siding next to the mainline at Rico. It may have either been waiting to be added to an outbound train, or waiting to be spotted on a siding at Rico. The picture was taken ca. 1924. This picture also provides you with a view of the old First National Bank building's south wall, at the left.

CHARLES J. LEE PHOTO – DAVID C. BUTLER COLLECTION

SOME MOUNTAIN CLIMBING was done by Charles Lee to get this panoramic picture of Rico. Looking down from the east slope of Expectation Mountain, he photographed the RGS yard and facilities, as well as Rico, with Nigger Baby Hill, Silver Creek Valley, Blackhawk Peak and Dolores Mountain forming the magnificent backdrop. This view probably was taken shortly after the St. Louis Smelting & Refining Company had taken over and started to remodel the old Pro Patria (Mantz Avenue) mill in 1925. A new structure had been built at the foot of the bluff, on the south side of the access road to the depot. It is believed this was the new coal-fired electric-generating plant (the building with three tall smokestacks emitting a plume of steam toward the rear). There were a large number of boxcars and gondolas in the railroad yard, indicating that mining was active. Notice the tents for miners at the southern end of town.

DURING THE BOOM YEAR of 1925, the National Lead Company solved the housing shortage in Rico by erecting tents at the south end of town for their employees. A new coal-fired electric-generating plant had been constructed in 1926 for this company's needs, and it was emitting steam. The third and final RGS wye and tail track, which was built to turn locomotives and other equipment (in place of a turntable) is in the foreground, in an area that was rather marshy. Sandstone Mountain and Sockrider Mountain rise high into the southwestern Colorado sky, beyond the town and station grounds.

FORT LEWIS COLLEGE, CENTER OF SOUTHWEST STUDIES COLLECTION

ORVAL L. JAHNKE COLLECTION

HORSE TEAMS AND WAGONS were still the main type of freight transportation on the roads and streets of the Rico district during 1924. The men in the flatbed wagon were at the intersection of Mantz Avenue and Silver Street, while a wagon load of ore had just passed them as it headed down toward the railroad yard. The Silver Street bridge over Silver Creek was half a block north (to the right) of the Presbyterian Church, and quite often this was a busy intersection. The overhead wire cables of the Pro Patria tramway are in the upper left-hand corner of this view.

A STEAM BOILER had been in use at the small building with a cupola, adjacent to the Pro Patria mill. Rico's hydroelectric power plant was to the left (north) of this little building. (Perhaps this small structure was used to generate electricity when water was in short supply during a dry fall.) This snapshot was taken during the late 1920's.

CHARLES J. LEE PHOTO – NORMA L. BUTLER COLLECTION

CHANGES THAT TOOK PLACE on the west side of the Pro Patria mill complex appear in this photograph, taken in 1926. Most of the changes took place along the upper section, constructed to allow the mill to accept ore from dump trucks. The section of the mill at the left probably was used to sort and move the ore up to the top part of the mill to start the milling process. The rear wall of the Burley Block building on Glasgow Avenue appears at the right.

ORVAL L. JAHNKE COLLECTION

BILL PLUNKETT COLLECTION

THE LOWER NORTH SIDE of the Pro Patria mill was photographed ca. 1924. A leased D&RGW slide-valve K-27 2-8-2 (possibly No. 457) was heading north, leaving the Rico yard. Part of the ruins of the 1908 Pro Patria fire were still lying on the ground. From this rare view, you can see where the covered RGS ore-loading track entered the large ore-processing mill.

ORVAL L. JAHNKE COLLECTION

THE OVERHEAD WIRE ROPES (cables) of the Pro Patria's aerial tramway were still in place at the ore-processing mill, located at the west end of Mantz Avenue. The latest addition to the Pro Patria mill can be seen to the right of the main structure. At this point, Mantz Avenue ended, and a 90-degree turn led off onto the narrow depot-access road. This narrow road continued down the face of the bluff to the Rio Grande Southern's depot and freight yard. Albert Stampfel's small dray wagon was parked by the side of Mantz Avenue. The scale shed at the right is now a private residence. Notice the Model "T" Ford coupe standing next to the scale house.

one hour and 25 minutes of switching, as well as 25 minutes taken for air tests, the 517-ton train finally departed from Rico for Ridgway at 6:20 p.m.

It was a cold Wednesday afternoon in Rico, during late February of 1926, and it was starting to get dark. Two K-27's, Nos. 463 and 464, and a C-21, No. 361, were called for a northbound freight train to Ridgway. After the 14 cars and the caboose that made up this 517-ton train were all tied-on to the locomotives, we finally whistle out of Rico, almost two hours late. As we blast up the ruling 3.0-percent grade toward Lizard Head Pass, we stop at the Coke Ovens water tank to top off the tender tanks on all three engines. Between Murphy and Gallagher, we stop to blow up steam on the ex-Crystal River engine, No. 361. You can be sure that at 9,800 feet above sea level in late February, around 7:00 p.m., the temperature probably was around 10 degrees (F.), if the crews were lucky! Once the 361 was steaming again—which took about 20 minutes—the short climb to the summit of Lizard Head was continued.

At Lizard Head Pass—10,250 feet above sea level—the helpers were cut out of the train, and the road engine continued down the grade, while the helpers drifted downgrade by themselves, running light. The train drifted down and around Trout Lake, through "Otto's Puzzle," past Matterhorn, over the Highline bridges, Nos. 46-E, 46-D, 46-C, 46-B and 46-A, as well as over the short Bridge 45-B, to a short supper break at the Ophir depot. Once again—after their repast, which included hot coffee—the crews climbed aboard the engines and caboose, and the entourage resumed their journey over Bridge 45-A, the "Big Bridge" at the Ophir Loop, and on over the Butterfly trestle, Bridge 44-A, and continued downgrade to Ames and Vance Junction. Here, cars were set out for Telluride and two loaded cars were picked up. Following that, the fires were cleaned and coal was replenished in the tenders before the train left the junction for the final 3.0-percent grade to Dallas Divide.

The trip continued to Vanadium, where one car was added to the consist. At Placerville, the helpers were coupled back into the consist before the climb was made from Placerville to Dallas Divide (Peake). Before reaching the summit, another car was picked up at Brown—for 630 tons from Brown to Ridgway. At the 8,989-foot summit, the helpers once again were uncoupled from the train, and the K-27's and C-21 rolled on downgrade to Ridgway. Both of the Mudhens and the outside-frame Consolidation were needed to get the 630 tons up to Dallas Divide.

After setting retainers and making the required airbrake test, the train descended the 4.0-percent grade the eight miles to Hagen. And the train finally arrived in Ridgway at 3:15 a.m. Following another 40 minutes of weighing the cars on the scale, the train and engines were put away at five minutes to 4:00 in the morning, a little over 12 hours since the first crew was called in Rico.

Gabe Bradford also located an entry written in the Dolores train register on February 24, 1926, showing that C-21 No. 361 ran light from Dolores to Rico earlier on the same day. She joined the K-27's, Nos. 463 and 464, on the scheduled Rico–Ridgway freight train, No. 392. Gabe and Herb Kelsey believe that the C-21's saw a fair amount of service on the RGS during the 1920's. However, finding photographic evidence is slim at best, as Gabe says.

THE PRO PATRIA MILL is shown here (at the left) as it appeared in 1926, after a fire and remodeling had taken place. During 1926, the Rico–Argentine Mining Company gained control of the Pro Patria mill, and the International Smelting & Refining Company was given the task of renovating the old mill into a flotation operation. This picture portrays the southern exposure of the mill structures, and it shows the addition at the upper end of the mill. The structure in the distance was the new scale house, where truck loads of ore were weighed. This ore-reduction plant included a relatively small water tank (typical for mills of this type). Originally, the ore came into the mill only via ore buckets transported into the top of the structure by an aerial tramway, as you can see from the wire ropes (cables) entering the mill. The narrow access road that led to the RGS depot and yard is in the foreground of this scene.

ORVAL L. JAHNKE COLLECTION

FORM 1028

Uniform Bill of Lading—Standard Form of Straight Bill of Lading approved by the Interstate Commerce Commission by Order No. 787 of June 27, 190

The Rio Grande Southern Railroad Company

THIS SHIPPING ORDER must be legibly filled in, in Ink, in Indelible Pencil, or in Carbon, and retained by the Agent.

Shippers No. ______

Agents No. ______

RECEIVED, subject to the classifications and tariffs in effect on the date of issue of this Shipping Order,

at Wakefield Switch July 13 191

from Rico Wellington Mng Co

the property described below, in apparent good order, except as noted (contents and condition of contents of packages unknown marked, consigned, and destined as indicated below, which said Company agrees to carry to its usual place of delivery at said dest nation, if on its road, otherwise to deliver to another carrier on the route to said destination. It is mutually agreed, as to eac carrier of all or any of said property over all or any portion of said route to destination, and as to each party at any time intereste in all or any of said property, that every service to be performed hereunder shall be subject to all the conditions, whether printe or written, herein contained (including conditions on back hereof) and which are agreed to by the shipper and accepted for himse and his assigns.

The Rate of Freight from ______

to ______ *is in Cents per 100 Lbs.*

IF ___ Times 1st	IF 1st Class	IF 2d Class	IF 3d Class	IF 4th Class	IF 5th Class	IF Class A	IF Class B	IF Class C	IF Class D	IF Class E	IF Special per ___	IF Special per ___

(Mail Address—Not for purposes of Delivery.)

Consigned to United States Zinc Co

Destination, Blende State of Colo County of

Route, ______ Car Initial D. & R. G. Car No. 305

No. Packages	DESCRIPTION OF ARTICLES AND SPECIAL MARKS	WEIGHT (Subject to Correction)	Class or Rate	Check Column
	Zinc Ore	30000		
	Value $4000			
	Weigh & Seal	768 / 2 / 5.5 / 5		
	at Rico			

If charges are to b prepaid, write or stam here, "To be Prepaid."

Received, $ ______ to apply in prepayment c the charges on the prop erty described hereon.

Agent or Cashier.

Per ______

(The signature here acknowledge only the amount prepaid.)

Charges Advanced,

$ ______

Rico Wellington Mng Co Shipper. ______ Agen

Per H R Frenholm Per ______

(This Bill of Lading is to be signed by the shipper and the agent of the carrier issuing same.)

ORVAL L. JAHNKE COLLECTION

RGS UNIFORM BILL OF LADING
Winkfield Shipper – July 13, 1912

CONDITIONS

Sec. 1. The carrier or party in possession of any of the property herein described shall be liable for any loss thereof or damage thereto, except as hereinafter provided.

No carrier or party in possession of any of the property herein described shall be liable for any loss or damage thereto or delay caused by the act of God, the public enemy, quarantine, the authority of law, or the act or default of the shipper or owner, or for differences in the weight of grain, seed, or other commodities caused by natural shrinkage or discrepancies in elevator weights. For loss, damage, or delay caused by fire occurring after forty-eight hours (exclusive of legal holidays) after notice of the arrival of the property at destination or at port of export (if intended for export) has been duly sent or given the carrier's liability shall be that of warehouseman only. Except in case of negligence of the carrier of party in possession (and the burden to prove freedom from such negligence shall be on the carrier or party in possession), the carrier or party in possession shall not be liable for loss, damage, or delay occurring while the property is stopped and held in transit upon request of a shipper, owner, or party entitled to make such request; or resulting from a defect or vice in the property or from riots or strikes. When in accordance with general custom, on account of the nature of the property, or when at the request of the shipper the property is transported in open cars, the carrier or party in possession (except in case of loss or damage by fire, in which case the liability shall be the same as though the property had been carried in closed cars) shall be liable only for negligence, and the burden to prove freedom from such negligence shall be on the carrier or party in possession.

Sec. 2. In issuing this bill of lading the company agrees to transport only over its own line, and except as otherwise provided by law acts only as agent with respect to the portion of the route beyond its own line.

No carrier shall be liable for loss, damage, or injury not occurring on its own road or its portion of the through route, nor after said property has been delivered to the next carrier, except as such liability is or may be imposed by law, but nothing contained in this bill of lading shall be deemed to exempt the initial carrier from any such liability so imposed.

Sec. 3. No carrier is bound to transport said property by any particular train or vessel, or in time for any particular market or otherwise than with reasonable dispatch, unless by specific agreement indorsed hereon. Every carrier shall have the right in case of physical necessity to forward said property by any railroad or route between the point of shipment and the point of destination; but if such diversion shall be from a rail to a water route the liability of the carrier shall be the same as though the entire carriage were by rail.

The amount of any loss or damage for which any carrier is liable shall be computed on the basis of the value of the property (being a bona-fide invoice price, if any, to the consignee, including the freight charges, if prepaid) at the place and time of shipment under this bill of lading, unless a lower value has been represented in writing by the shipper or has been agreed upon or is determined by the classification or tariffs upon which the rate is based, in any of which events such lower value shall be the maximum amount to govern such computation, whether or not such loss or damage occurs from negligence.

Claims for loss, damage, or delay must be made in writing to the carrier at the point of delivery or at the point of origin within four months after delivery of the property, or, in case of failure to make delivery, then within four months after a reasonable time for delivery has elapsed. Unless claims are so made the carrier shall not be liable.

Any carrier or party liable on account of loss or damage to any of said property shall have the full benefit of any insurance that may have been effected upon or on account of said property, so far as this shall not avoid the policies or contracts of insurance.

Sec. 4. All property shall be subject to necessary cooperage and baling at owner's cost. Each carrier over whose route cotton is to be transported hereunder shall have the privilege, at its own cost and risk, of compressing the same for greater convenience in handling or forwarding, and shall not be held responsible for deviation or unavoidable delays in procuring such compression. Grain in bulk consigned to a point where there is a railroad, public, or licensed elevator, may (unless otherwise expressly noted herein, and then if it is not promptly unloaded) be there delivered and placed with other grain of the same kind and grade without respect to ownership, and if so delivered shall be subject to a lien for elevator charges in addition to all other charges hereunder.

Sec. 5. Property not removed by the party entitled to receive it within forty-eight hours (exclusive of legal holidays) after notice of its arrival has been duly sent or given may be kept in car, depot, or place of delivery of the carrier, or warehouse, subject to a reasonable charge for storage and to carrier's responsibility as warehouseman only, or may be, at the option of the carrier, removed to and stored, in a public or licensed warehouse at the cost of the owner and there held at the owner's risk and without liability on the part of the carrier, and subject to a lien for all freight and other lawful charges, including a reasonable charge for storage.

The carrier may make a reasonable charge for the detention of any vessel or car, or for the use of tracks after the car has been held forty-eight hours (exclusive of legal holidays), for loading or unloading, and may add such charge to all other charges hereunder and hold such property subject to a lien therefor. Nothing in this section shall be construed as lessening the time allowed by law or as setting aside any local rule affecting car service or storage.

Property destined to or taken from a station, wharf, or landing at which there is no regularly appointed agent shall be entirely at risk of owner after unloading from cars or vessels or until loaded into cars or vessels, and when received from or delivered on private or other sidings, wharves, or landings shall be at owner's risk until the cars are attached to and after they are detached from trains.

Sec. 6. No carrier will carry or be liable in any way for any documents, specie, or for any articles of extraordinary value not specifically rated in the published classification or tariffs, unless a special agreement to do so and a stipulated value of the articles are indorsed hereon.

Sec. 7. Every party, whether principal or agent, shipping explosives or dangerous goods, without previous full written disclosure to the carrier of their nature, shall be liable for all loss or damage caused thereby, and such goods may be warehoused at owner's risk and expense or destroyed without compensation.

Sec. 8. The owner or consignee shall pay the freight and all other lawful charges accruing on said property, and, if required, shall pay the same before delivery. If upon inspection it is ascertained that the articles shipped are not those described in this bill of lading, the freight charges must be paid upon the articles actually shipped.

Sec. 9. Except in case of diversion from rail to water route, which is provided for in section 3 hereof, if all or any part of said property is carried by water over any part of said route, such water carriage shall be performed subject to the liabilities, limitations and exemptions provided by statute and to the conditions contained in this bill of lading not inconsistent with such statutes or this section, and subject also to the condition that no carrier or party in possession shall be liable for any loss or damage resulting from the perils of the lakes, sea, or other waters; or from explosion, bursting of boilers, breakage of shafts, or any latent defect in hull, machinery, or appurtenances; or from collision, stranding, or other accidents of navigation, or from prolongation of the voyage. And any vessel carrying any or all of the property herein described shall have the liberty to call at intermediate ports, to tow and be towed, and assist vessels in distress, and to deviate for the purpose of saving life or property.

The term "water carriage" is this section shall not be construed as including lighterage across rivers or in lake or other harbors, and the liability for such lighterage shall be governed by the other sections of this instrument.

Sec. 10. Any alteration, addition or erasure in this bill of lading which shall be made without an indorsement thereof hereon, signed by the agent of the carrier issuing this bill of lading, shall be without effect, and this bill of lading shall be enforceable according to its original tenor.

1925 Pellet's mining operations continued to prosper, and his success attracted other smelting companies. The National Lead Company came to town and invested in mining operations that employed 125 men. The International Smelting Company built a 40-room bunkhouse on Glasgow Avenue for their men—originally named the Ricato Hotel. This building recently was renovated, and it has been operating as the Rico Hotel (as of 1995).

1926 Lengthy litigation between the Rico-Argentine Mining Company and the Syndicate Mining & Milling Company was resolved in the Rico-Argentine's favor. And the Rico-Argentine company acquired control of the old Pro Patria mill. The Western Colorado Power Company

(Continued on Page 439)

ORVAL L. JANKE COLLECTION

DURING THE 1925 BOOM the Pro Patria mill had undergone extensive improvements. You can see the exterior changes in this scene, including the extensive trestlework east of the landmark mill, used for refining ore from the St. Louis Smelting & Refining Company or the International Smelting Company. Mining was booming when this photograph was taken, and the price of metals was rising, which might account for the number of boxcars spotted on the RGS yard tracks. Notice the business car "Montezuma" at the north end of the depot, spotted on an added spur. Also notice the small Conoco bulk-oil depot, to the left of the RGS water tank. It was painted silver and reflected brightly in the brilliant summer sunshine. A new building was constructed on the west side of Glasgow Avenue (to the right), between King Street and Elliott Avenue, in Block 3. It was a boarding house for mine-and-mill workers during this great boom period. The old D&RG RPO car body had been moved near the RGS section house.

ALLEN LEE PHOTOS – DAVID C. BUTLER COLLECTION

THE TIPPLE AND TRAMHOUSE at Winkfield was owned by the Mountain Spring Mine, and it was producing carloads of ore, as these views show, taken sometime during the 1910's. Ore chutes had been set up here, which looked rather crude, but they got the job done. The RGS listed this spur as "Winkfield," and it was 331 feet in length, located at Milepost 64.7, at an elevation of 8,905 feet. An aerial tramway carried ore down the mountain in buckets from an adit some distance uphill. The buckets dumped the ore into the tipple, and two delivery chutes loaded the ore into gondolas on the short spur. A large box, filled with rock, hung in the side framework to stretch the wire cable of the tramway. This operation closed and the track was removed by 1920.

built a high-voltage electric line over Lizard Head Pass to serve the Rico area.

In its turbulent history of "boom-and-bust" cycles as a mining town, Rico was well into its third "boom cycle." Rico's population multiplied to over 1,000 residents; however, so many houses had been torn down during the slow days after World War I, that a severe housing shortage developed. Old businesses were revived, and new businesses opened their doors. A new dancehall was built, and social action got a lift. The "good old days" had returned in Rico.

1927 Prosperity continued in the Rico area. The average wage for miners rose to $6.00 per day, an increase of $2.00 from 1923. Bob Pellet's payroll increased to 300 workers. The old Pro Patria mill, at the foot of Mantz Avenue, was totally rebuilt as a modern selective-flotation mill, with a capacity of 250 tons per day. Ore worth almost $1.5 million was produced in the district during the year.

All of this success was achieved in spite of a bad flood on June 28. The Dolores River and Silver Creek both ran wild, and the town and the RGS suffered heavy damage, particularly to bridges. A mile of track on the RGS was washed out, and no trains made it in or out Rico for about three weeks.

1928 Eleven active mines produced almost $1.5 million in ore, mostly lead and zinc, as prosperity continued in the Rico area. The town's population continued to increase, and about 20 new houses were constructed.

(Continued on Page 482)

FROM "THE RICO ITEM," FEBRUARY 1932 –
COURTESY WENDE P. STUART

NINE MEN started from town Wednesday morning, on foot for Coke Ovens to assist in bringing the two engines and plow in to Rico after which the Railroad Company will open the line between here and Vance Junction with the rotary.

THE NIGGER BABY SLIDE was cleared Friday of last week. The county machinery, with five men, and Mr. Kline of the St. Louis S. & R. Co. donating three men, shoveled out 120 feet of snow four feet deep. This again enables the men to drive their cars to work at the tunnel.

A Report From...

"The Galloping Goose"

A Quarterly Publication of the Rio Grande Southern Historical & Technical Society

TWENTY YEARS IN RICO

By William F. Horky, Jr.

Published in Volume II, No. 3

THIS RGS TALE has its beginnings a long way from the Silver San Juan Country, in St. Louis, Missouri. In the early 1920's, ... [officials of a] firm called the St. Louis Smelting & Refining Company came to the decision to increase their production of sulfuric acid. This fairly routine business decision was to have a significant economic impact on the Rio Grande Southern and the country [it] served....

ORVAL L. JAHNKE COLLECTION

THE TOP OF TELESCOPE MOUNTAIN was the highest point on the horizon, near the center of this panoramic view, taken during the 1910's. The Mountain Spring ore chute is in view, standing beside the RGS grade, at the lower left. The removal of trees formed a line up the lower part of the mountain, which cleared an opening for the aerial tramway to reach the mine entrance. A second mine portal, above the Mountain Spring, was known as the Wellington. Two unknown mines produced large dumps near the center of this picture. The Nora Lilly Mine was above the Nutmeg Mine, at the right, near the erosion marks in the snowslide chute on Nigger Baby Hill. The two summits looming above the snowslide chutes on Nigger Baby Hill form Blackhawk Mountain, with the right summit reaching 12,677 feet. On the right edge of the photograph is the 12,350-foot summit of White Cap Mountain, partly hidden behind Dolores Mountain.

WORK HAD BEGUN in earnest on the Mountain Spring Mine's portal, located on the lower slopes of Telescope Mountain. A miner by the name of Harry Barnes stood just outside the mine portal. Huge aspen logs had been cut to be used as support and roof timbers in this mine—no easy task for this old timer. This was a renovation project at this mine.

ORVAL L. JAHNKE COLLECTION

The company used sulfuric acid at their St. Louis plant, to process lead from their mines in southern Missouri. They also sold the acid to rubber processors in the automobile industry.

The automobile industry was in a boom following World War I and needed the lead for batteries and rubber for tires.

In the early part of this century, most sulfuric acid was produced from iron pyrites imported from Spain. This was before the large sulfur fields of Texas came into production. The St.L.S.&R.Co. [officials] decided to find a source of iron pyrites closer to home that they could control. By some odd twist of fate, they learned of a whole mountain of pyrites near Rico, Colorado, owned by the Rico Mining & Reduction Company.

In early 1922, they sent Duane Kline, a young mining engineer from the Missouri School of Mines at Rolla, to investigate this possible source of pyrites. He was to evaluate the deposit, and if it proved suitable, to start work toward getting it into production. Mr. Kline traveled from St. Louis to Rico by train, the last part of the journey on the Rio Grande Southern.

One of his first duties upon arrival in Rico was to acquire office space for his survey work. He took over and renovated a large vacant, rundown stone building on the

CHARLES MILLER ENGEL COLLECTION – COURTESY RACHEL A. HARDWICK

THE NORTHBOUND passenger train rumbled along the RGS mainline above the adlt to the St. Louis Tunnel during February of 1931. Engine 22 was chuffing along as it began its long climb up the southern slope of Lizard Head Pass with a baggage-mail car and a coach, operating as the daily passenger train.

THREE EMPLOYEES of the St. Louis Smelting & Refining Company were watching RGS 4-6-0 No. 22 with a mixed train, as it passed directly above the entrance to the St. Louis Tunnel in January of 1923. At least two passenger cars were attached to the rear of the train. The company's main source of high-grade iron pyrites was at this mine, and the ore was processed into sulfuric acid. This mine was located near Milepost 65 on the RGS. Notice the diamond-shaped warning placard on the door of the boxcar behind No. 22.

CHARLES MILLER ENGEL COLLECTION – COURTESY RACHEL A. HARDWICK

west side of Glasgow Street. This had been the Rico Drug Company and the A.S. Ditchfield Grocery Store. This building still stands and is used by the Anaconda Company for offices.

Survey work was commenced on the Rico Mining & Reduction Company's iron-pyrite deposit, which proved it indeed was a large deposit of exceptionally pure iron pyrites. It was also determined that there were recoverable amounts of lead, zinc and silver. The St.L.S.& R.Co. entered into an agreement with the Rico Mining & Reduction Company, whereby the St.L.S.& R.Co. would provide capital to develop and mine the Rico Mining & Reduction Company's ore body, in return for the iron pyrites.

After careful study, it was decided to dig a new haulage tunnel at the base of Telescope Mountain and stope-mine the ore deposit by tapping the old workings above it. "Stoping" is digging from the bottom of an ore body, using gravity to bring down the ore. It is easier to mine up than to mine down. The ore was to be trammed to the tunnel mouth [portal] by a narrow-gauge electric mining railway [i.e., a tramway].

In 1923, work was commenced on what was to be called the "St. Louis Tunnel." (What else would they name it?) The St. Louis Tunnel was located about a mile north of Rico, on the east side of the Dolores River, at the base of Telescope Mountain. [The tunnel ran] almost due east, under the track of the RGS into the heart of the mountain.

The St. Louis Tunnel was to operate until the early 1970's, under various owners,

CHARLES MILLER ENGEL COLLECTION – COURTESY RACHEL A. HARDWICK

THE ST. LOUIS TUNNEL had been bored into the base of Telescope Mountain by the St. Louis Smelting & Refining Company. The passenger coach was directly above the entrance to their tunnel. The previous mine owner had driven a shaft from above (and to the east), and the St. Louis firm dug their tunnel to connect with this shaft. Telescope Mountain tops out at 12,208 feet above sea level, in view at the upper center of this scene. The northbound mixed train of the RGS was headed by a 4-6-0 Ten-wheeler, en route to Ridgway in this afternoon view. The train consisted of a boxcar, tankcar, baggage-mail car and a coach.

CHARLES MILLER ENGEL COLLECTION –
COURTESY RACHEL A. HARDWICK

AS THE WORK PROCEEDED at the St. Louis Smelting & Refining Company's property, two additional supporting structures were placed north of the tunnel. A short passenger train, bound for Ridgway, was heading upgrade for Lizard Head Pass, at this time—probably near the end of the prosperous 1920's.

AN RGS MIXED TRAIN, consisting of Engine 20, four freight cars, an RPO-baggage car and a coach, was traveling northbound—about to cross over the top of the portal of the St. Louis Tunnel. This view was shot during the early 1930's, and you can see small ore cars standing on the dump trackage.

DUANE KLINE PHOTO – W. F. HORKY COLLECTION
COURTESY "THE GALLOPING GOOSE" – VOLUME II, NO. 2

STATION MAP OF RICO
1931
DRAWN BY JIM KEY

3. SAMPLING WORKS
4. RICO P&L WAREHOUSE
4a. ORE-LOADING RAMPS
6. ST. LOUIS MINE LOADING PLATFORM
8. RICO P&L POWER PLANT (ABANDONED)
9. STEAM BOILER HOUSE
10. TRACK SCALE
11. PRO PATRIA MILL
12. HOUSES (OFFICES)
13. CONOCO BULK-OIL DEPOT
14. CABOOSE COAL & KINDLING BOX
16. RGS WATER TANK
17. **RGS DEPOT**
18. BUNK-CAR BODY
19. R.P.O.-CAR BODY
22. SAND HOUSE
25. SECTION HOUSE
26. BUNKHOUSE
27. COAL SHED
29. FIREHOSE BOX AT DEPOT
32. FORMER POWER PLANT OF THE ST. LOUIS MINING COMPANY
33. SECOND ENGINE HOUSE (BRICK)
35. OLD BOXCAR BODY
36. OLD BOXCAR BODY
37. DEPOT OUTHOUSE
38. BRIDGE & TRESTLE TIMBERS
39. TRACK SCALE
43. SECTION-CAR SHED
44. THIRD (AND FINAL) WYE
67. BURNETT & CLIFTON COALING FACILITY

and it reached a length of several miles into the mountain.

The St. Louis Smelting & Refining Company originally dug and operated the tunnel. Later, they changed the name to the National Lead Company. The property was acquired by the Rico-Argentine Mining Company in 1942. Later, it was owned by the Anaconda Company, which changed their name to the AMAX Company. At the present time, it is for sale by AMAX, at a reported price of $3 million.

The surface facilities for the tunnel were built on the spoil [waste material] from the tunnel excavation, just west of the RGS track. These included repair shops, change house, air compressors and ventilation-fan buildings.

Originally, the pyrite ore was hauled by truck to a loading dock on the far north end of the Rico yard. Later, a new loading dock was built north of the old [Pro Patria] mill building on the town side of the track. One of the truck drivers once backed his truck off this dock into a gondola. I have a photograph of this accident. [The original view is included with this volume, from the Charles Miller Engel collection in Rico.]

The St.L.S.&R.Co. generally shipped their ore out via Ridgway and Montrose in gondolas. Who would want to steal iron pyrites, also known as "fool's gold"?

During 1925, Mr. Kline directed construction of a power-plant building for the

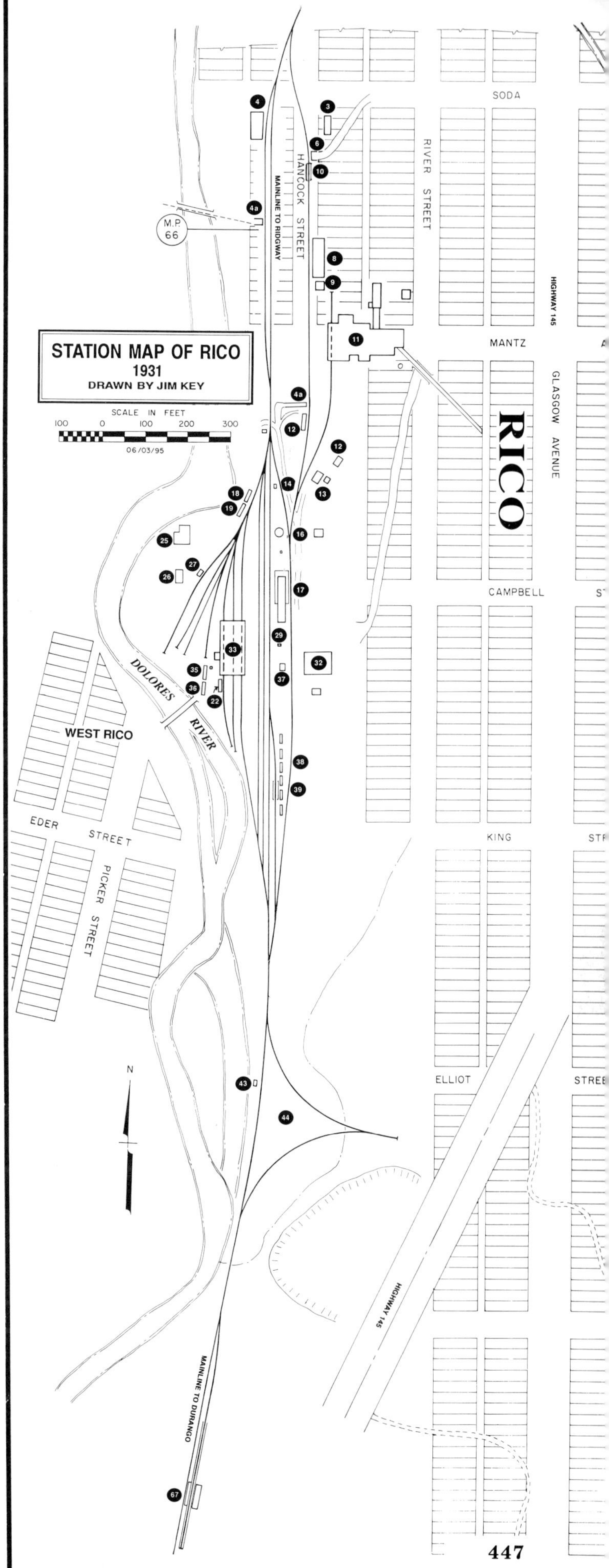

D&RGW
1562
W-A-BRAKE

CHARLES MILLER ENGEL COLLECTION – COURTESY RACHEL A. HARDWICK

THIS HALF-TRACK TRUCK needed to be rescued from its sudden stop inside a D&RGW gondola. The first order of business was to find someone with a strong back to hand-shovel the ore from the bed of the truck. Unfortunately, pictures do not exist showing how the truck was pulled out of the gondola, but it probably was extracted by a winch, or something of the sort, using cables to pull the vehicle back up onto the dock. Notice that one of the gondola's trucks had been derailed after breaking the rail from the force of the impact—no doubt upsetting the section foreman!

CIRCA 1930, a truckload of ore from the St. Louis Tunnel was backed up to unload at the dock, which had been built for this purpose near the north end of the railroad yard in Rico. However, the brake did not hold, and over the truck went, ore and all, dumped into the D&RGW gondola. According to researchers who pictured this accident in the California-based publication, *The Galloping Goose,* Volume III, No. 3, of 1990, the vehicle is a 1927 Model 60 Linn Republic half-track tractor truck, with a 2.5-ton capacity and a six-cylinder 100-horsepower engine. The St. Louis Smelting & Refining Company was trucking this ore to the railroad for shipment to a processing plant, where it was converted into sulfuric acid.

CHARLES MILLER ENGEL COLLECTION – COURTESY RACHEL A. HARDWICK

Rico Mining & Reduction Company in the RGS yard area. This is the building south and east of the RGS depot, with the three large smokestacks, seen in many photographs. The power plant was built to provide electricity for the refurbished RM&RCo. mill. The mill was to process some of the metal ores from the St. Louis Tunnel. This is the mill [that was] located east of the RGS water tank. The power plant was only used for about a year, as in 1926, the... [Western

GREG LePAK COLLECTION

THE TRUCK DOCK in this view—to the left of the Pro Patria mill—was used by the St. Louis Smelting & Refining Company (along with independent miners) to unload ore from their trucks into waiting gondolas. It looks as though a scale house had been set up to the right of the gondola, in this view taken during 1946. Perhaps the loaded gondolas were weighed on the spot. The Pro Patria mill appeared to be in need of repairs.

AS 1936 ROLLED AROUND, the miners at the St. Louis Tunnel drove their cars to the work site. Two miners were pushing a tramway ore car to the edge of the mine dump, at the left, not an easy job. The so-called "highway" into Rico is visible on the opposite bank of the Dolores River, little more than a dirt road over Lizard Head Pass to Ridgway. Horse Creek runs down through the ravine above the highway, and at the top center is a glimpse of the basin below Elliot Mountain, an area of gold and silver mining during earlier years. The RGS mainline between Rico and Ridgway is in the foreground.

CHARLES MILLER ENGEL COLLECTION – COURTESY RACHEL A. HARDWICK

Colorado Power Company] built a [high-voltage] power line over Lizard Head Pass into Rico. After that, both the mill and the St. Louis Tunnel used... [WCPCo.] power.

When the Ames slide came down in 1929, it cut off the St.L.S.&R.Co.'s favored shipping route out of Rico. Mr. Kline, on behalf of his employer, along with other shippers, made a formal protest to the ICC and PUC, when the D&RGW procrastinated in clearing the slide. Mr. Kline said you might as well talk to the wall as talk to the D&RGW about removing the slide.

While the Rio Grande Southern was in the process of being forced into receivership, Mr. Kline attended many formal and informal meetings to find out how this would affect his company's ore transportation. After all, a mountain of iron pyrites in Colorado was no good if they could not get it to St. Louis.

WILLIAM A. GRAVES COLLECTION

SOMETIME AFTER the 1930's, the property at the St. Louis Tunnel was cleaned up and rearranged for a more efficient operation at the surface. At the same time, additional buildings were erected for a blacksmith shop and office. The camera had been pointed toward the west for this picture. (Colorado Highway 145 is across the Dolores River Valley.)

DURING THE LATE 1930's, the portal at the St. Louis Mine had an extended shed built over the tramway track exiting the tunnel. At the left, one of the miners was on a battery-powered "trammer" (a mine engine), with a string of ore cars about to enter the mine. Tramway ties, wornout rail and used support timbers were stacked on the dump.

ORVAL L. JAHNKE COLLECTION

DAVID C. BUTLER COLLECTION

DAVID BUTLER'S GRANDDAD worked at the St. Louis Tunnel during the 1940's, when this view was photographed. The mine workers were carried inside the mine by an electric-tramway motor (locomotive) to work on the lode. The tramway had been electrified by this time, and you can see the overhead trolley line. The engines and cars of the tramway passed under the RGS railroad grade to reach the work site underground. The large contraption at the left was a "mucking" machine, which did a lot of the backbreaking work of hand-shovelers.

Remember, at this time, there were no paved roads in the area, and the RGS was the only reliable transportation into or out of the area. There may be folks who will dispute the use of the word "reliable" and the RGS in the same sentence though! The road over Lizard Head Pass was not paved until the 1960's [Colorado Highway 145].

After Victor A. Miller was appointed receiver for the RGS, Mr. Kline accompanied him, along with the superintendent, Forrest White, on an inspection of the Ames slide. The purpose of this trip was to assess the time and cost of removing the slide. During this trip, Mr. Kline took many photographs of this slide to send to the home office of his employer. He also wrote many letters keeping them apprised of the progress in getting the railroad open again.

Until the slide was cleared, the company shipped its ore out via Durango and Alamosa [also over the D&RGW], but it was more expensive and took longer.

The following incident happened during this time. Mr. Kline related a story about how they once loaded some rather wet ore in a boxcar. As it was winter and a long way to Alamosa, the ore froze into a solid mass during the trip. This presented a rather tough problem to the Alamosa transfer crew. After a futile exercise with picks, one of them suggested they try a little dynamite to loosen the ore. The dynamite was duly produced and applied to the frozen ore and the fuse lit. The amount of explosive was in excess of need, and it blew the sides out of the boxcar, scattering the ore all over the yard area.

Much of the ore was not recoverable, so Mr. Kline dutifully filed a damage claim for the lost ore. After a long wait, the claim was finally settled to his satisfaction. Not so fortunate was the poor transfer-crew foreman. He was given a few days off, without pay, to ponder the error of his judgment....

ORVAL L. JAHNKE COLLECTION

THIS GROUP OF MINERS is believed to be at work inside the St. Louis Tunnel, ca. 1932. The men have been identified by Orval Jahnke as follows (left to right): In the top row were Herb Lines, Vic Lee Torrence, Roy Conkling and an unidentified man, while in the lower row were Bill Conkling and Frank O'Driscoll. Al Edwards and Clyde Overman were sitting underneath the drilling platform. The temperatures underground usually remained at close to 60 degrees (F.).

Another time, a new truck driver caused the RGS a few problems, as if they did not have enough already! When pyrite ore was shipped in gondolas, it was loaded in two rather small piles, one over each truck of the car. Iron pyrites [ore] is quite heavy, and it did not take much to make the 25-ton limit. The new driver was not told this, and he filled the gondola full! The RGS switch crew came to move the car down to the scale track, and in the process, broke several rails. Some of the load had to be transferred to another car. Mr. Kline's company received a bill for the damages this time. The truck driver received some more instructions as to what constituted a "carload" of pyrites!

Mr. and Mrs. Kline frequently made shopping trips to Montrose over the RGS, first on steam trains and later, on the Galloping Geese [railbuses]. [Mrs. Kline]... said they were cramped and generally dirty, but she rode them anyway. A dirty bird was not

(Continued on Page 459)

CHARLES MILLER ENGEL COLLECTION –
COURTESY RACHEL A. HARDWICK

SEVERAL NEW BUILDINGS had been built at the St. Louis Tunnel by 1929, and you can see the additional shop buildings at the lower right. The tailings dump had been pushed out in all directions. Looking to the left, there was an old ore bin, located at Milepost 64.7, known as Winkfield on the RGS. It was the trackside tipple and lower tramhouse, connected by aerial tramway to the Mountain Spring Mine's tunnel, located higher up the mountain, but not in view in this photograph.

CHARLES MILLER ENGEL COLLECTION – COURTESY RACHEL A. HARDWICK

THE MINING INDUSTRY was booming during the mid-1920's, when this picture was taken, and numerous carloads of ore concentrates were being transported by the RGS. The house track east of the depot appeared to be filled with boxcars awaiting ore loading. The demand for metals had made mining profitable again. The International Smelting Company had utilized the Pro Patria mill, and it was now greatly expanded. The St. Louis Smelting & Refining Company graded an access

road down to the Pro Patria mill spur, where a truck dock is visible this side of the power plant. Empty gondolas had been spotted on the Pro Patria mill spur, and several had been filled (beyond the truck dock). A second truck ramp was on the west side of the mill-loading spur, north of the RGS water tank. Another spur, west of the RGS mainline, held three boxcars and eight loaded gondolas. The lower RGS yard tracks were occupied with additional gondolas, filled with ore.

PETERS SHOES
FOR EVER
CHAS. ENGE
ROOMS

ANN AND JAMES FERANDO PHOTO

ACCORDING TO RELATED VIEWS of buildings still in existence, this picture was taken ca. 1925. The old saloon building is still standing, next to the furniture store at the left, constructed of brick on Block 1 (according to the Sanborn map). The Engel's general store is visible across the street, next to the stone building that had been the Rico State Bank. The backsides of buildings located on Block 6 include a large carpenter shop, and two rustic dwellings are at the right.

THIS "OLD BOY" had loaded up his string of pack burros at the stores in Rico for another go at prospecting in the hills above town. A store building advertised "rooms" on the second floor, where the bay window extended out over the sidewalk. This multi-sided bay window once housed a dentist's chair—the five windows providing sunlight for dental work.

CHARLES MILLER ENGEL COLLECTION – COURTESY RACHEL A. HARDWICK

about to keep her from a shopping trip!

The Klines also rode the trains and Geese to Dolores to see movies on the weekends. Dolores had the nearest movie theater at that time. The Goose schedules made for a nice weekend trip! [The railbus left]... Rico Friday afternoon and [returned on]... Sunday morning, if Mother Nature and the Rio Grande Southern cooperated!

Mr. Kline worked in Rico until 1942, when the National Lead Company sold their interest to the Rico-Argentine company.

I hope this article sheds a little light on the "lineside life" in RGS country. I thank Mr. Duane Kline and his wife for sharing it with me, so I could share it with all my fellow RGS fans.

CHARLES MILLER ENGEL COLLECTION – COURTESY RACHEL A. HARDWICK

THE GOOD TIMES had returned to Rico, as the summer of 1925 passed. Mining was booming again, and the businesses in this little mountain community were making money again. Traffic regulations even had to be implemented, such as the sign on the box covering the hole for the placement of the town's flagpole! This read: "Keep to the right–drive slow." The United States Hotel no longer sported a sign, and the saloon that had been on the ground floor had suffered the "ax" of Prohibition (in effect between 1920–1933). Two of Rico's ladies may have been coming from the Rico Cafe, a family restaurant. At least one of them was wearing a shorter skirt, typical of the

"Roaring Twenties," the era of the "flappers." The bank building was now listed as having a hotel upstairs. The fire department's bell tower, placed over the sidewalk at the right, was used to notify the volunteer firemen when they were needed to fight a fire. And a restaurant signboard (in the shadow above the Model "T" pickup truck) was hanging over the boardwalk, at the right, offering residents and visitors a second choice, at the intersection of Mantz and Glasgow avenues. A new building in town—the Ricato Hotel—was under construction in the distance, on the east side of Glasgow Avenue. This structure originally was built to house mining workers.

RICHARD A. RONZIO COLLECTION

WHILE VISITING RICO on a bright-and-sunny day, an unknown photographer shot this overview of the town. The Rico schoolhouse stands out, near the center, and the downtown business district is at the right. This same photograph has been greatly enlarged and reproduced on the following three pages, making details of the town much easier to see. By examining the Sanborn Map of 1899, printed on Pages 364–365 in this volume, many of the buildings can be identified.

PAGES 463-465: The public schoolhouse built in 1903 was located at the corner of Commercial Avenue and Soda Street. Directly beyond the school is the Dolores County Courthouse. Robert Pellet, the mining tycoon, and his wife, Betty, later moved into the large house in the middle of the block to the south (right) of the school. Below the school was the narrow dirt road that led onto Glasgow Avenue. Later, this rough-and-rocky road that came in from Lizard Head Pass became Colorado Highway 145, and it passed through the business district and went on south toward Dolores. The Atlantic Cable Mine's hoist house and supporting structure was in the right foreground, built on the south bank of Silver Creek. Notice the 1920's automobiles on Glasgow Avenue. And you can also see the 40-room boarding house, with the many dormers on its roof, at the end of Glasgow Avenue. This structure was built about 1925, and it was originally named the Ricato Hotel. Currently, it is known as the Rico Hotel. The sprawling Enterprise Mine properties can be seen on Newman Hill above town. These mines were now dormant. The newly completed trestle work extending from the Pro Patria mill ended at the alley behind the Burley Block, at the far right.

RICHARD A. RONZIO COLLECTION

SANBORN
MAPPING AND GEOGRAPHIC INFORMATION SERVIC

— 1893 —

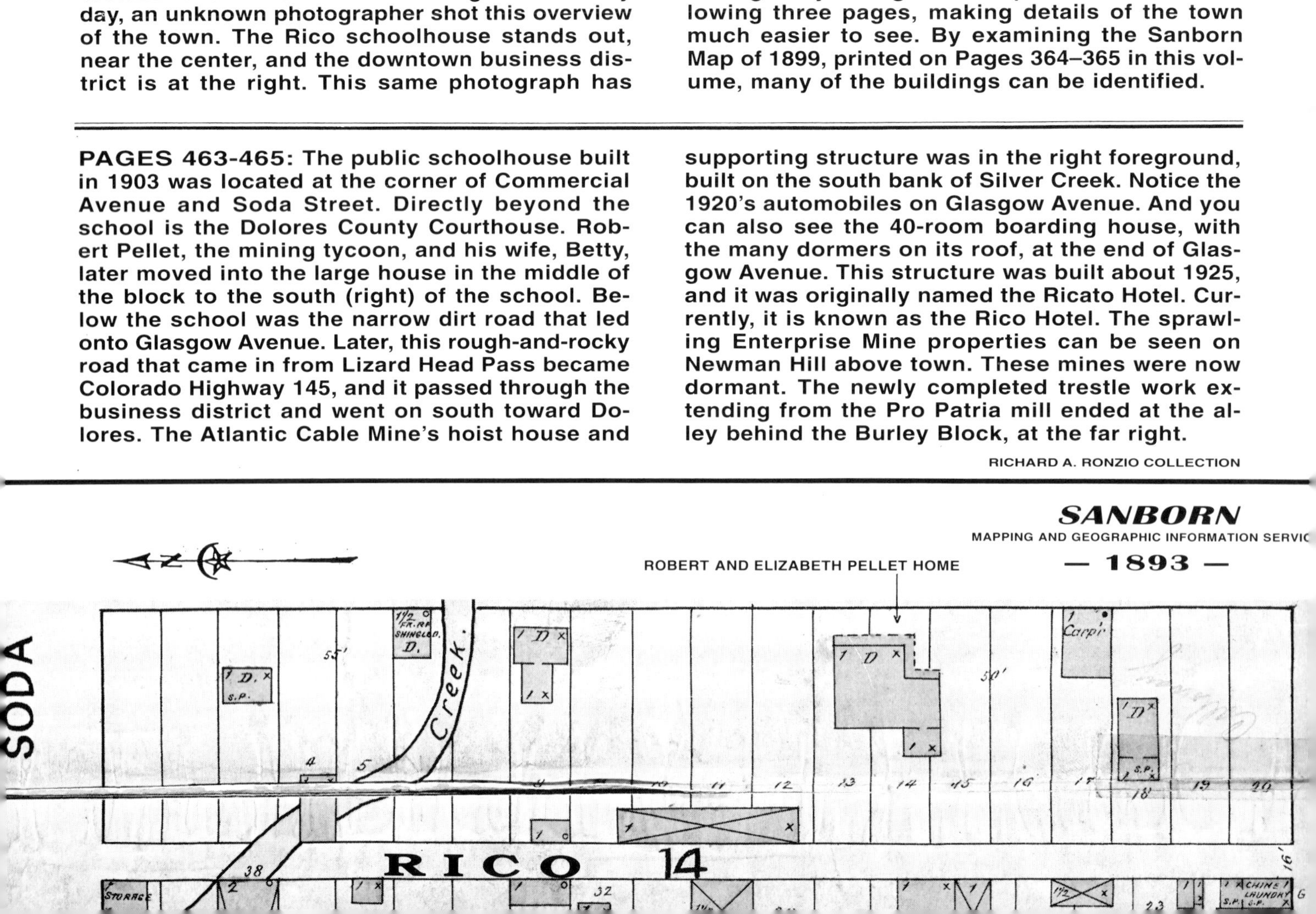

DELL A. McCOY PHOTO

THE ENTERPRISE Cafe & Lounge was the only active business on this block when this picture was taken in 1995. This block once boasted of the large Hotel Enterprise, as well as six saloons. Farther to the north on Glasgow Avenue, is the photographer's trusty old blue Ford four-wheel-drive "Bronco," and there is a glimpse of the Atlantic Cable hoist structure behind it.

THE ROOF of Rico's fire-department building was being repaired on October 7, 1995. The crew was (left to right): Todd Jones, fire chief; Dennis Swank, assistant chief; Rachel Hardwick; Kim Jones, emergency medical technician; Jonathan Hay, fireman; and Ken Hazen, ambulance driver. The firetruck is a 1953 Ford pumper, originally bought by the Rico–Argentine mining company.

DELL A. McCOY PHOTO

WILLIAM A. GRAVES COLLECTION

THE BURLEY BLOCK continued to look impressive in this 1995 winter view. Stamped-metal trim added old-time character to this building, and it outlasted wooden products on other structures in Rico. The locally quarried stone has also held up well during the 104 years since this building was constructed. This was the third location for Burley's once-lucrative department store, located across the street from the general store Charles Engel, Sr., operated for so many years. His locally well-known son, Charles M. Engel, succeeded him in the Charles Engel Mercantile Company.

CHARLES MILLER ENGEL PHOTO – LARRY PLEASANT COLLECTION

CHARLES M. ENGEL found a clearing back of Rico (to the east of town) during the late 1920's, to try out his panoramic camera. This was taken during the spring melt, when dandelions cover meadowlands with yellow in green grassy areas. Expectation Mountain is the high point at the left.

CHARLES MILLER ENGEL PHOTO – LARRY PLEASANT COLLECTION

A FRESH SNOWFALL had come to Rico, with a lot of moisture in it, making it stick to branches. The water flume that had been placed across the base of Expectation Mountain is in sight in the background. The snowstorm helps to bring out the detail of the distant buildings, including the trestle going off to the left, near the top of the Pro Patria mill (near the center). This would place the photograph in the time period just past 1925.

Horse Creek drains the area at the right, where snowcapped Elliot Mountain rises above Sockrider Peak and Sandstone Mountain. The front entrance to the schoolhouse in Rico is visible, at the right of center in this scene. Before long, it will be springtime in the Rockies once again.

CHAS. ENGEL MERCAN
HOTEL

CHARLES J. LEE PHOTO – NORMA L. BUTLER COLLECTION

DURING THE LATE 1920's, one of the local cowboys from down along the Dolores River had ridden into Rico. He posed on his prize horse, astride a fine hand-tooled Western saddle. This snapshot had been taken near the photographer's home, at the south end of Glasgow Avenue, and the photographer was looking toward the north.

A PROUD AMERICAN INDIAN sat astride his spunky, lean cattle pony. The heavy leather chaps protected his legs against thorny bushes as he rode on his horse. (The word "chaps" comes from the colloquial Mexican Spanish word "chaparejos"—which, in turn, is from "chaparro," referring to a thicket of shrubs. "Chaparral" is the Southwestern U.S. word for the same thing.) This cowboy had deeply tanned skin, from being out-of-doors much of the year, and you can see his lariat (rope) just in front of his right hand. The old wooden boardwalk had been updated with a concrete sidewalk sometime prior to 1925, not long before this view was taken.

LARRY PLEASANT COLLECTION

CHARLES MILLER ENGEL COLLECTION – COURTESY RACHEL A. HARDWICK

OVERLEAF–UPPER VIEW: A town official in Rico came up with the idea one day that a horse race would attract a lot of needed money into the community—and the area just might become famous at the same time. This "grand event" took place on Labor Day of 1928. Glasgow Avenue was roped off, and the race was about to begin when this picture was taken. The business buildings in the scene were facing toward the east, and they were (from left to right): the stone Burley Block, next to two vacant lots where automobiles were parked, an original 1880 building, formerly housing the Rico Post Office (ca. 1890), but occupied by a restaurant at this time, the National Café, and the oldest brick building in Rico. Across Mantz Avenue, you can see: the old First National Bank building, which first appears on the Sanborn Map of 1886, a clothing store, and next to it is what is the Enterprise Cafe & Lounge as of 1996, while beyond this stone building was a one-story wooden false-front structure, which housed a saloon before Prohibition.

CHARLES MILLER ENGEL COLLECTION – COURTESY RACHEL A. HARDWICK

OVERLEAF–LOWER LEFT: Charles Engel, Sr., may have been in the second-floor bay window over the sidewalk to record the horse race coming to the finish line with his panoramic camera. This first-frame view shows a fourth-place horse galloping past on Glasgow Avenue. The Rico firehouse and bell tower were at the left, while the Burley Block was at the right.

OVERLEAF–LOWER RIGHT: The first three horses to cross the finish line are shown in this scene—with the judges standing in the middle of the intersection at Mantz Avenue. This snapshot may have been tripped by a cable release from a second camera at the same time the previous view was taken.

CHARLES MILLER ENGEL COLLECTION – COURTESY RACHEL A. HARDWICK

CAFE

WILLIAM A. GRAVES PHOTO

THIS AUTUMN SCENE of Rico was photographed during the late 1960's, shortly after Colorado Highway 145 had been paved through the town. The third public school in Rico, now the Rico Visitors Center, had a pitched roof at that time (near the center of this view, to the right of the Rico Hotel). The same building was photographed a few years later, as shown on Page 410, with the roof in its present state. Brightly colored roofs of "Navajo town" appear at the lower right. Compare this picture with the one on Pages 412–413, taken from Nigger Baby Hill.

CHARLES MILLER ENGEL COLLECTION –
COURTESY RACHEL A. HARDWICK

CHARLES ENGEL stood in the middle of Glasgow Avenue—almost in front of his general store—when he pointed his camera north to capture this scene in Rico around 1930, or a bit later. At the right, some youngsters were checking out the local movie billings posted in the window of the auto-parts store, while two older gentlemen posed on the sidewalk for the photographer. A 1929 Model "A" Ford, sporting a trunk on its nickel-plated luggage rack, was parked in front of the store, while farther down the main street a more expensive automobile stood at the curb.

YES, THERE ARE BEARS in the woods of Colorado. This cub was captured and chained to an aspen tree above Rico by boys from the town. He was later put on display in front of Engel's store as a curiosity. No record exists of what became of the poor fellow.

CHARLES J. LEE PHOTO – NORMA L. BUTLER COLLECTION

CHARLES MILLER ENGEL COLLECTION –
COURTESY RACHEL A. HARDWICK

THESE BEAR CUBS appear to be enjoying life in Rico during 1930, dressed up in children's clothing. It would be a scary ordeal to meet mama bear on the trail in the woods. (Incidentally, it is illegal to capture wild animals and treat them this way today.)

OVERLEAF: The autumn of 1995 was extremely late in arriving in Colorado's high country. This photograph was taken on October 7, when during a normal season, the aspen leaves would have all turned to gold. The beautiful snowcapped La Plata Mountains appear on the skyline. The towns of Hesperus and Mancos lay beyond these peaks, and the RGS went through both of these settlements en route to Durango.

DELL A. McCOY PHOTO

ESTAURANT

CHAS. ENGEL MERCANTILE CO. GENERAL MERCHANDISE

BOTH PHOTOS: CHARLES MILLER ENGEL COLLECTION – COURTESY RACHEL A. HARDWICK

THIS PHOTOGRAPH takes in Glasgow Avenue, as the camera was pointed toward the north, showing the right-hand side of Block 14 in the Sanborn map. An automobile-garage sign was hung on one of the old wooden buildings down the street, on the right. Several other old false-front business buildings do not show signs, except for the one on the end building, nearest Mantz Avenue. This sign reads "Lorig's Men's Wear," but this building previously had been used as an assay office, and it still stands today. The Rico Café was still in business, and the old Rico State Bank building now had a hotel on the second floor, and people were standing in front of the entrance. The St. Louis Smelting & Refining Company's office occupied space in the Burley Block building, at the left, and a restaurant was operating in the false-front building where an old Model "T" Ford was standing at the curb side.

DURING THE LATE 1920's, people had discovered they could travel about in the mountain country of southwestern Colorado in the dryer summer months—if they could stand the rough, rocky and often dusty gravel-surfaced roads. Older cars, like the two Ford Model "A" sedans parked at the curb, had bodies that were high enough off the ground to miss the rocks in the poor "highways" of that period. (The car parked between the two Fords may have been a Chevrolet.) Apparently, an automobile-supply dealer had set up shop next door to Engel's store. Mining boss Bob Pellet may have been the tall gentleman talking to another man in front of the store.

1929 The good times ended in Rico during 1929. The beginning of the infamous Great Depression caused a drop in metal prices, and most of the mines closed down or greatly cut back their production. The mainline of the RGS was blockaded by the Ames mud slide for most of the year. (See the story about the Ames mud slide in Volume III of this series.) Loss of income resulting from the blockade, as well as from reduced ore shipments forced the RGS into its second bankruptcy.

1930 The Great Depression got worse, and there was little demand for metals, as the Nation's factories shut down. During a two-year period, the annual value of ore produced in the Pioneer Mining District

(Continued on Page 484)

CHARLES J. LEE PHOTO – DAVID C. BUTLER COLLECTION

THIS SMALL AUTOMOBILE must have been built prior to the 1920's. The car had carbide-gas headlights and oil cowl lamps, and it was not equipped with a self-starter—as the hand crank looks well used. The axle length was unusually short, and the wheels would not have fit into the ruts of the dirt roads of this era! Down the hill (at the left) is a glimpse of the school and farther down is the Atlantic Cable tunnel.

THIS PHOTOGRAPH was taken from a higher elevation on Expectation Mountain and illustrates how Rico was built on the sedimentary area below the nearby mountains (on a terrace above the river). Blackhawk Mountain, at the left, is beyond Dolores Mountain, at the right. From this perspective, the task prospectors had in searching for mineral wealth becomes apparent. Mining dumps are visible on Dolores Mountain. Rico had this appearance ca. 1927, and by this time, the Pro Patria mill was the town's largest structure, with numerous additions having been completed.

CHARLES MILLER ENGEL COLLECTION
– COURTESY RACHEL A. HARDWICK

CHARLES MILLER ENGEL COLLECTION – COURTESY RACHEL A. HARDWICK

THE KOSTER MARKET was destroyed as the result of a fire that swept through the building in 1931. Fortunately, the firemen were able to subdue the flames and saved the building next door (which has the bay window that overhangs the sidewalk). Smoke and fire had engulfed the upper floor and roof when this picture was taken, sometime after the fire had started at the back of the building.

AS FIREMEN brought the blaze under control, smoke and steam poured out of the ruins. The roof had collapsed, but the metal stamped-brick siding helped to prevent the fire from spreading to the building next door. Probably because the fire happened during the day, the blaze was spotted early enough to get it under control. After dark, a fire can burn for a longer period, making it harder to control—and the flames can leap to the next buildings, sometimes leveling an entire block.

CHARLES MILLER ENGEL COLLECTION – COURTESY RACHEL A. HARDWICK

dropped from about $1,500,000 to a mere $200,000. Bob Pellet kept a small force of about 40 miners working, to avoid laying them off, even though he had to make up losses with his own money to pay them. Many residents left town, but others could not even afford to leave.

1931 Bob Pellet's losses continued to mount, and he finally was forced to close down his Rico operations, including the Pro Patria mill. Only $2,000 worth of ore was produced in the district during 1931. The RGS began a cost-reduction program by building and placing in operation the "Galloping Goose" railbuses, Nos. 1, 2 and 3, which replaced more costly steam-powered passenger trains.

1932 Rico's economy hit bottom in 1932. For the first time in over 50 years, no ore was produced during the year. At the same time, the town's population fell from about 2,500 to about 300 in 1932.

A severe snowstorm, which began on February 8, 1932, isolated Rico from

CHARLES MILLER ENGEL COLLECTION – COURTESY RACHEL A. HARDWICK

ON THE NEXT DAY, after the disastrous fire at the Koster Market, the windows were boarded up to protect the remaining salvageable contents. The building was a total loss, and it was subsequently torn down.

both north and south. A relief train spent five days clearing the line from Durango to Mancos, and another four days opening the track to Dolores and Rico. The snow-clearing train from the north had even less success. After fighting the snow on the south side of Lizard Head Pass for days, the crew had to give up. The plow-flanger was left at Burns (2.6 miles north of Rico), two engines were abandoned at Coke Ovens, and a Galloping Goose was required to stay at Lizard Head. The snow blockade lasted for over three months, and the line north of Rico was not reopened until May 16.

1933 Rico had been the only incorporated town in Dolores County since its founding in 1879. However, a small, but growing, settlement named Dove Creek—located near the west end of the county—began to challenge Rico in the 1930's. Unlike Rico, Dove Creek had a more stable economy, based on agriculture. In 1930, Dove Creek's population was only 120 people, and the town was not incorporated until 1939; however, in 1933, Dove Creek succeeded in "stealing" the county newspaper from Rico. For the first time in over 50 years, Rico was without its own newspaper.

ORVAL L. JAHNKE COLLECTION

THE RICO FIRE DEPARTMENT – **Hose Team No. 1 – proudly posed for action with their hook-and-ladder wagon and hose cart. This view probably was taken during a practice fire drill, which took place during the early 1930's. Next door, a signboard was hanging from the wall of the stone Burley Block structure, which read: "Office – St. Louis Smelting & Refining Co." The Burley Block still stands today (1996).**

1934 Mining was still very slow in the Rico area, but there was some renewed interest in gold mining, as a result of an increase in gold prices. Since 1850, the price of gold had been controlled by the Federal government at $20.67 per ounce. To help the Nation recover from its disastrous economic depression, the gov-

ernment allowed the price of gold to increase to $35.00 per ounce over a three-year period, beginning in 1933. A few of the mines in the district reopened in 1934, and they were worked mostly for gold and silver. However, the St. Louis Smelting & Refining Company shut down operations at the St. Louis Tunnel.

1935 Silver prices rose; however, ore production in the Rico area increased very little.

1936 The Obendorfer General Store closed down, after almost 50 years of family operation.

CHARLES MILLER ENGEL COLLECTION – COURTESY RACHEL A. HARDWICK

THE FIRE DEPARTMENT'S building at Rico (Hose Team No. 1) was photographed shortly after the building was repainted during the early 1930's. Some of the department's equipment was housed in the lean-to added to the left side of the building. And a fire siren had been placed atop the roof, replacing the old bell that had been suspended above the sidewalk in a wooden tower. Since the spindly trestlework at the Pro Patria mill can be seen behind these buildings, this pegs the date for the photograph at about 1926.

THE WINTER of 1931–1932 dumped about four feet of snow on Rico during one storm. Two 1920's automobiles (in the foreground) had been buried up to their hoods in snow. Familiar old buildings lined Glasgow Avenue, in the distance, including the fire department, near the center of this view.

MATT TONIOLLI PHOTO – COURTESY CECILIA BRIDGEWATER

CHARLES MILLER ENGEL COLLECTION – COURTESY RACHEL A. HARDWICK

THE CENTER SECTION of this photograph was enlarged from the view printed on Pages 490–491, which shows Rico's Glasgow Avenue business district, north and south of Mantz Avenue, ca. 1932. Some of these structures appear on the Sanborn Map of 1899. The aerial tramway, with its wire-rope cables and support towers, had been removed from the middle of Mantz Avenue by the time this picture was taken. The Rico Hotel—with its distinctive roof dormers—is at the upper right in this scene. Originally, this structure was called the "Ricato Hotel," perhaps an Italian name. At the opposite end of the main street was the Atlantic Cable shafthouse and other mine buildings, in the foreground. This mine's shafthouse and hoist house were on opposite sides of Silver Creek.

CHARLES MILLER ENGEL COLLECTION –
COURTESY RACHEL A. HARDWICK

RICO WAS PHOTOGRAPHED after the good times had ended, and the terrible Great Depression had begun. A strong indicator of this period is the fact that the RGS railroad yard was nearly empty of boxcars to transport ore concentrates, and there was no motive power visible at the little engine terminal. The streets of Rico were virtually deserted and devoid of automobiles in this 1932 scene. You can see the dormers of what is now the Rico Hotel at the end of Glasgow Avenue.

A wide swath in the Silver Creek drainage, to the right of the schoolhouse, is visible, produced during the flood of June 28, 1927. Notice the ox bow in the Dolores River, as it curved around the RGS enginehouse. This sharp bend in the Dolores River—the "Big Bend" referred to so often in the early accounts of Rico—caused the RGS flooding problems more than once. Mill waste from the Pro Patria was piped to their collection pond visible in the distance, below the road to Dolores.

CHARLES MILLER ENGEL COLLECTION –
COURTESY RACHEL A. HARDWICK

THE 1932 PHOTOGRAPH reproduced on the previous pages has been enlarged to illustrate an angle of Rico not found elsewhere in this volume. Most interesting in the left half of the view is the north side of the Pro Patria mill and the connecting trestle that spanned the depot access road at River Street. Built during the 1925 mining boom, the large building in the distant background was on the west side of Glasgow Avenue (part of the gravel-surfaced road to Dolores). It is believed to be a boarding house for mine-and-mill workers, which later became a dancehall.

AT THE LOWER END of the Pro Patria (Mantz Avenue) mill a boxcar had been spotted beside a coal shed used by a commercial coal dealer. The track where the car was sitting is a run-through siding, originally built to provide a track to spot boxcars for loading ore concentrates at the Pro Patria mill. A stub-end track was located beyond the coal shed, where there was a Conoco bulk-oil facility. (This track also served the Pro Patria mill.) The bulk-oil plant was built to serve the increased demands for gasoline and heating oil during the 1925 boom period. That two B&B work cars had been spotted on this spur indicate that it may have been closed by the Great Depression. At the RGS water tank you can see three pieces

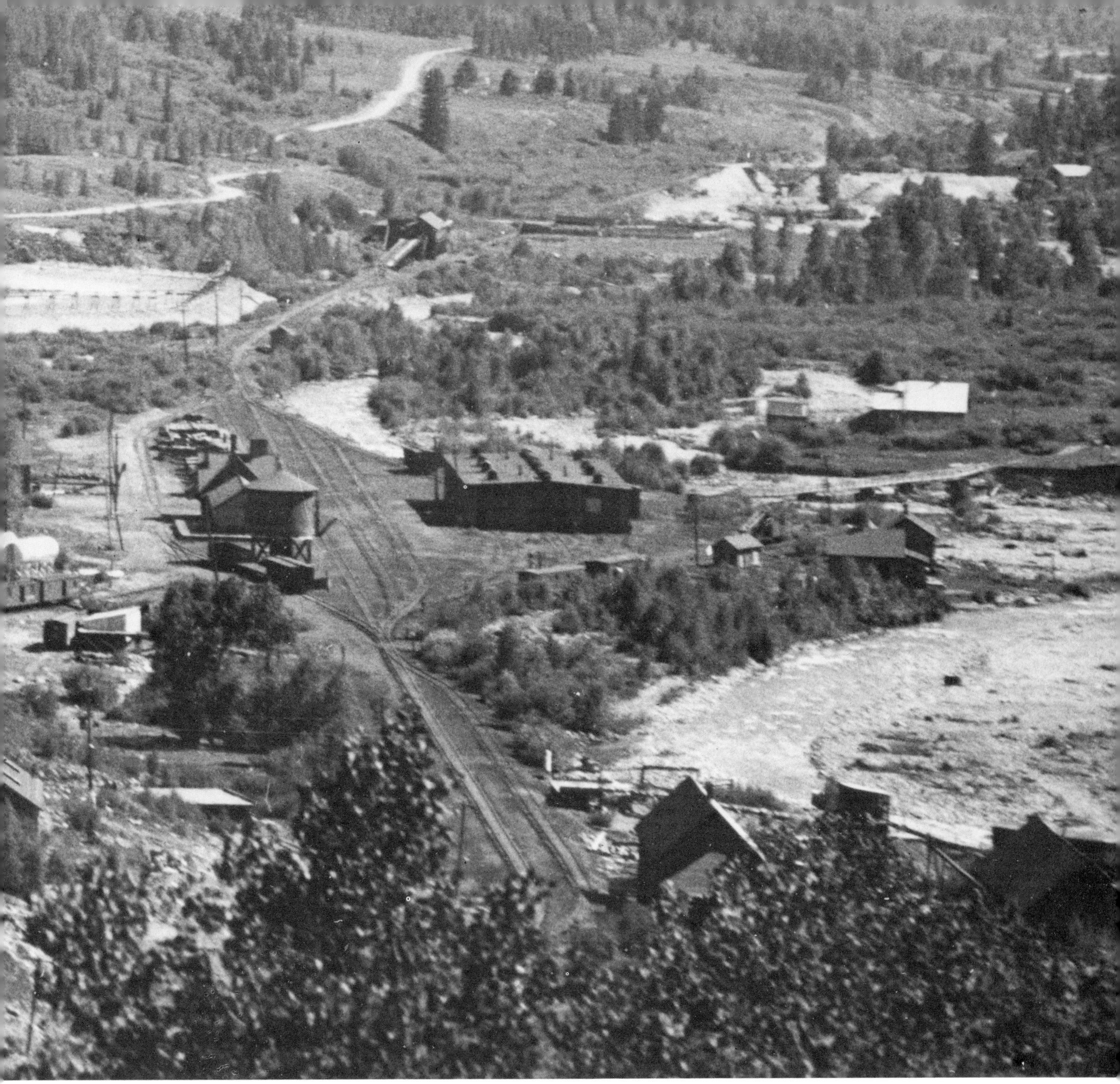

of rolling stock, a gondola, flatcar and boxcar.

THE RGS DEPOT is in view, partially hidden by the water tank, and to its left stood the idle St. Louis Smelting & Refining Company's electric power plant, abandoned after the high-voltage transmission line was built over Lizard Head Pass to Rico. Bridge timbers had been stacked between the depot's house track and the mainline. The sidings were empty, except for a lone boxcar at the south end of the enginehouse. Notice the Burnett & Clifton automatic coal chute, farther south of the RGS yard. The abandoned tailings pond, created by a retaining dam alongside the RGS mainline, is just beyond the curve at the south end of the yard. You can see the inlet pipe on raised supports along the embankment. This pipe came from the Pro Patria mill, which had been operated by the International Smelting Company. The dirt road to Dolores is visible beyond these objects. In the northern end of the yard, Milepost 66 is plainly in view, near the loading dock (along the right side of the run-through siding), in the lower right portion of the scene. A long coal shed had been located on this siding in the past, just north of the white milepost. Rio Grande Southern's employees' timetables show that the Rico yard could hold 144 cars.

In Conclusion...

Volume VI of this series will relate the story about the rejuvenation of mining in Rico during World War II. This will include the amazing account of Elizabeth Pellet's dramatic action that saved the Rio Grande Southern during the war years.

REFERENCES

BOOKS

An Empire of Silver, Robert L. Brown, Denver, Colorado, 1984

Beyond the Third Rail, Doris B. Osterwald, Lakewood, Colorado, 1994

Colorado Ghost Towns and Mining Camps, Sandra Dallas, Norman, Oklahoma, 1985

Colorado Year Book, State Planning Commission, Denver, Colorado, various years

Crofutt's Grip-Sack Guide of Colorado, George Crofutt, Omaha, Nebraska, 1885

Guide to the Colorado Ghost Towns, Perry Eberhart, Denver, Colorado, 1959

Minerals Yearbook, U.S. Bureau of Mines, Denver, Colorado, various years

Mines and Minerals of the Great American Rift, Richard Walker Holmes and Marrianna B. Kennedy, New York, New York, 1983

Silver San Juan, Mallory Hope Ferrell, Boulder, Colorado, 1973

That Pellet Woman!, Betty Pellet (with Alexander Klein), New York, New York, 1965

The R • G • S Story, Volumes I through III, Russ Collman, Dell A. McCoy and William A. Graves, Denver, Colorado, 1990 through 1993

The Rio Grande Southern Railroad, Josie Moore Crum, Durango, Colorado, 1961

ARTICLES, BULLETINS and PROFESSIONAL PAPERS

A History of Rico, Colorado: 1879–1971, James Anderson Reid, University of Texas, Austin, Texas, 1972

Engineering and Mining Journal, April, 1948

Establishment of Rico, A Mining Town in Southwestern Colorado, Doris Hopping, Texas Technological College, Lubbock, Texas, 1942

Investigation of the Low-Grade and Complex Ores in Colorado, Colorado Bureau of Mines, Denver, Colorado, 1920

Mining in Colorado, Charles W. Henderson, U.S. Geological Survey Professional Paper 138, 1926

Rico, A Brief History, Janice Fleming and M. Inez Towne, unknown location, 1987

Rico, Colorado, the Pride of the San Juan, 1879–1882, Mary Wilson, Texas Technological College, Lubbock, Texas, 1950

The Enterprise Mine, Transactions of the American Institute of Mining Engineers, Denver, Colorado, 1896

NEWSPAPERS and PERIODICALS

"Rico, Colorado—Once a Roaring Camp," Carl F. Mathews, ***Colorado Magazine*** (Colorado Historical Society quarterly), Denver, Colorado, 1951

"The Best of Hart's Stuff From Rico," Hartley Lee, From ***The Dolores Star,*** Dolores, Colorado, 1968

Articles in ***The Denver Post,*** Denver, Colorado, various years.

Articles in ***The Galloping Goose,*** The Rio Grande Southern Historical and Technical Society, La Mesa, California, 1989

An Article in the ***Narrow Gauge and Short Line Gazette,*** Los Altos, California, July/August, 1992

Articles in ***The Rico Democrat,*** Rico, Colorado, 1891

Articles in ***The Telluride Times–Journal,*** Telluride, Colorado, April 7, 1994

~ INDEX ~

**** Indicates photograph. All place names are at Rico and/or in the Pioneer Mining District unless otherwise stated.***